Cosmological Pattern of Microphysics in the Inflationary Universe

Fundamental Theories of Physics

An International Book Series on The Fundamental Theories of Physics:
Their Clarification, Development and Application

Volume 144

Cosmological Pattern of Microphysics in the Inflationary Universe

by

Maxim Yu. Khlopov
Center for Cosmoparticle Physics "Cosmion" of Keldysh Institute of Applied Mathematics,
Moscow, Russia
Moscow Engineering Physics Institute,
Moscow, Russia
Rome University "La Sapienza",
Rome, Italy

and

Sergei G. Rubin
Moscow Engineering Physics Institute,
Moscow, Russia

KLUWER ACADEMIC PUBLISHERS
DORDRECHT / BOSTON / LONDON

A C.I.P. Catalogue record for this book is available from the Library of Congress.

ISBN 1-4020-2649-8 (HB)
ISBN 1-4020-2650-1 (e-book)

Published by Kluwer Academic Publishers,
P.O. Box 17, 3300 AA Dordrecht, The Netherlands.

Sold and distributed in North, Central and South America
by Kluwer Academic Publishers,
101 Philip Drive, Norwell, MA 02061, U.S.A.

In all other countries, sold and distributed
by Kluwer Academic Publishers,
P.O. Box 322, 3300 AH Dordrecht, The Netherlands.

Printed on acid-free paper

Printed in the Netherlands.

Contents

We use the following notations throughout the book (if not directly stated otherwise):

$$\hbar = c = 1.$$

The signature of metric tensor in the Minkowski space is chosen in the form

$$(1, -1, -1, -1).$$

The connection between Newton constant G_N and Planck mass M_P is

$$G_N = M_P^{-2}.$$

The interval in the synchronous reference frame

$$ds^2 = dt^2 - \gamma_{\alpha\beta} dx^\alpha dx^\beta.$$

What nature delivers to us is never stale. Because what nature creates has eternity in it.

Isaac Bashevis Singer

Preface

Modern cosmology is a quickly developing field of research. New technical devices and tools supply the community with new experimental data measured with high accuracy. The self-consistent explanation of these data needs theoretical models that are based on hypothetical predictions of particle theory. In their turn, such predictions imply cosmology for their probe. Specific studies of the cosmological consequences of particle theory, linking them to their observable signatures, are actual. This boiling kettle of theoretical research and experimental efforts produces ideas that will be preserved for following generations.

The aim of this book is to acquaint the reader with some of these ideas, offering nontrivial ways to probe the physical basis of modern cosmology. An extensive review of the newest ideas in modern cosmology, e.g., related with the development of the M-brane theory, lies beyond the scope of our book, which is aimed at providing a firmly established system of probes for these ideas, linking their predictions to their possible experimental test. We use the framework of inflationary paradigm to reveal the phenomena that can shed light on the physical origin of the observed Universe, of its matter content and large-scale structure. The crucial role of quantum fluctuations in creation of our Universe and in possible features, reflecting cosmological impact of microphysics, is discussed. These features are shown to be accessible to experimental test in the near future.

The regular course of lectures "Introduction to Cosmoparticle physics" in the Moscow Engineering Physics Institute and the courses "Physics of Primordial Universe" in the 1 and 3 Rome Universities, given by one of the authors (M.Yu.Kh.), helped us to make the book content eligible for a wide audience of students and PhD students, specializing in physics and astronomy.

We hope that this book will help the reader to be more easily involved in the fascinating problems of the mutual relationship between micro- and macro-worlds.

Authors

Chapter 1

PRINCIPLES OF COSMOPARTICLE PHYSICS

Cosmoparticle physics originates from the well-established relationship between microscopic and macroscopic descriptions in theoretical physics. Historically, it reminds us of the links between statistical physics and thermodynamics, or between electrodynamics and theory of electron. To the end of the XX Century a new level of this relationship was realized, linking the science of the Universe as a whole and physics of elementary particles.

Now it has become evident that large-scale properties of the observed Universe cannot be understood without the proper model of elementary particles. The modern Standard Model of the Big Bang Universe is based on inflationary cosmology with baryosynthesis, dark matter and possibly dark energy. Such a model can find its physical grounds only beyond the world of known elementary particles in the hypothetical predictions of particle theory. Particle theory that will provide such grounds should use in its turn cosmological tests as the important and in many cases unique way to probe its predictions.

The convergence of the frontiers of our knowledge in micro- and macro-worlds leads to the wrong circle of problems, illustrated by the mystical Uhroboros (self-eating snake). The Uhroboros puzzle may be formulated as follows: *The theory of the Universe is based on the predictions of particle theory, that need cosmology for their test*. Cosmoparticle physics [1, 2, 3] offers the way out of this wrong circle. It studies the fundamental basis and mutual relationship between micro- and macro-worlds in the proper combination of physical, astrophysical and cosmological signatures.

1. Particle physics in the Big Bang Universe

Let us specify in more detail the links between the properties of fundamental particles and their cosmological effects.

1

1.1   Particles – gauge symmetry of the Standard Model

To the end of the XX Century the set of experimentally proven ideas about elementary particles and their transformations was fixed in the Standard Model, which is based on the extension of the principle of gauge symmetry of quantum electrodynamics in the case of weak and strong interactions (see [4] for review).

This set of ideas implies the fundamental notion of particle symmetry, ascribing the observed difference in particle properties and interactions to the difference in symmetry properties, as well as to the mechanism of symmetry breaking.

The fundamental role of particle symmetry follows naturally from the basic principles of quantum field theory (QFT). The way QFT describes particles in terms of creation and annihilation operators made a revolutionary change in the notion of the "elementary particle". From antiquity to the XX century the basic idea of the elementary particle was to consider it being eternal. Elementary pieces of a conserved quality could not be destructed or created, and the conservation of quality was the natural consequence of eternity of its elementary bricks. So, the conservation of electric charge simply followed from the eternal nature of elementary charged particles – electrons.

The formalism of QFT allows electrons to be created and annihilated. It revolves the picture and attributes the priority to charge conservation. Therefore, electric charge should be conserved in all the processes of creation and annihilation of charged particles. The theoretical reasoning for charge conservation comes from the gauge invariance of quantum electrodynamics (QED) (see [5] for review). Strict U(1) gauge symmetry, assumed for free charged particles, made it possible to introduce the electromagnetism as gauge interaction, mediated by the gauge boson – quantum of the electromagnetic field.

This picture could be naturally extended to the processes of particle transformation, in which a particle of one type converts itself into a particle of another type, provided that both types of particles are related by symmetry. Say, the process of weak interaction, in which incoming neutrino converts into outgoing electron, can be described in a way similar to QED, if we assume a symmetry between the neutrino and the electron and treat them as different states of one (lepton) field. Such gauge SU(2) symmetry for free leptons implies the existence of an intermediate gauge boson of weak interaction, created or annihilated in the elementary act of weak interaction, when the initial neutrino is converted into the final electron.

Weak interaction of elementary particles is short-ranged, because, unlike the massless photon, W-boson of weak interaction is massive. The mass of W-boson reflects the scale of the symmetry breaking. It is just this scale that determines the masses of quarks and leptons. Another difference is the group of symmetry: U(1) is an Abelian group, and SU(2) is a non-Abelian group.

Gauge theory, based on the non-Abelian group of symmetry, possesses confinement. The scale of confinement determines the fundamental energy scale, even if the non-Abelian symmetry is unbroken. This property is important in the modern theory of strong interaction – quantum chromodynamics (QCD). The non-Abelian nature of $SU(3)_c$ gauge group of QCD implies QCD confinement at the energy scale $\Lambda_{QCD} \sim 300$ MeV. QCD confinement explains the absence of free quarks – owing to confinement of their colors they can be found only in the bound states. The energy scale of confinement determines the effective mass of quarks in the bound state. It is just this constituent quark mass that determines the mass of the proton.

The Standard Model assumes that the known elementary particles of matter (the three families of leptons and quarks) possess $SU(3)_c \otimes SU(2) \otimes U(1)$ local gauge symmetry, implying the existence of gauge bosons of strong ($SU(3)_c$ – gluons), and electroweak ($SU(2) \otimes U(1)$ – W-, Z-bosons and photon) interaction. The $SU(2) \otimes U(1)$ symmetry of electroweak interaction is broken by the Higgs mechanism of spontaneous symmetry breaking. It implies the existence of elementary scalar particles, called Higgs bosons.

The majority of the experimental data about elementary particles can be reproduced by the Standard Model. However, it seems to be evidently incomplete, and this opens the door to new physics.

1.2   Particles – Beyond the Standard Model

The new physics arises from the necessity to extend the Standard Model. The white spots in the representations of symmetry groups, considered in the extensions of the Standard Model, correspond to new unknown particles. The extension of the symmetry of gauge group puts into consideration new gauge fields, mediating new interactions. Breaking of the global symmetry results in the existence of massless Goldstone boson fields.

For a long time the necessity of extending the Standard Model had purely theoretical reasons. Aesthetical arguments favored embedding of the symmetry group of the Standard Model within a larger group of unifying symmetry in order to reach full unification of fundamental interactions.

Practically, introduction of new particles and fields was used to save the Standard Model from internal theoretical inconsistencies. So, introduction of axion provides solution for the problem of strong CP violation in QCD [6, 7, 8] (see [3, 9] for review), and supersymmetric partners compensate divergences in the quantum corrections to the mass of Higgs boson (see review and references in [10]).

Theoretical description of neutrino mass also implies new physics (see review in [11, 3]). In particular, new particles – heavy right-handed neutrinos – are introduced to explain the smallness of neutrino mass in the seesaw mechanism of neutrino mass generation. A new physical property – Majorana mass, violating conservation of lepton numbers arises in this mechanism. The neut-

So, the very weakly interacting particles with the annihilation cross-section $\sigma < 1/(T_r M_P)$, as well as very heavy particles with the mass $m \gg T_r$ can not be in thermal equilibrium, and the detailed mechanism of their production should be considered to calculate their primordial abundance.

Decaying particles with the lifetime τ, exceeding the age of the Universe, t_U, $\tau > t_U$, can be treated as stable. By definition, primordial stable particles survive to the modern epoch and should be present in the modern Universe. The net effect of their existence is given by their contribution to the total cosmological density. They can dominate in the total density being the dominant form of cosmological dark matter, or they can represent its subdominant fraction. In the former case the particles determine dynamics of galaxy and LSS formation. In the latter case more detailed analysis of their distribution in space, of their condensation in galaxies, of their capture by stars, Sun and Earth, as well as of the effects of their interaction with matter and of their annihilation provides more sensitive probes for their existence.

In particular, hypothetical stable neutrinos of the 4th generation with mass of about 50 GeV are predicted to form the subdominant form of the modern dark matter, contributing less than 0.1% to the total density. However, direct experimental search for cosmic fluxes of weakly interacting massive particles (WIMPs) may be sensitive to the existence of such component [62], [16], and maybe even favors it [16]. It was shown in [25, 63, 64, 65] that annihilation of 4th neutrinos and their antineutrinos in the Galaxy can influence and even explain the observed spectrum of galactic gamma-background and of cosmic ray positrons. 4th neutrino annihilation inside the Earth should lead to the flux of underground monochromatic neutrinos of known types, which can be traced in the analysis of the already existing and future data of underground neutrino detectors [64].

New particles with electric charge and/or strong interaction can form anomalous atoms and be contained in the ordinary matter as anomalous isotopes. For example, if the lightest quark of 4th generation is stable, it can form stable charged hadrons, serving as nuclei of anomalous atoms of, e.g., crazy helium [66].

Primordial unstable particles with a lifetime less than the age of the Universe, $\tau < t_U$, cannot survive to the present time. But, if their lifetime is sufficiently large to satisfy the condition $\tau \gg (M_P/m) \cdot (1/m)$, their existence in the early Universe can lead to direct or indirect traces. Cosmological flux of decay products contributing to the cosmic and gamma ray backgrounds represents the direct trace of unstable particles. If the decay products do not survive to the present time, their interaction with matter and radiation can cause indirect traces in the light element abundance or in the fluctuations of thermal radiation. If the particle lifetime is much less than 1 second the multi-step indirect traces are possible, provided that particles dominate in the Universe before their decay. On the dust-like stage of their dominance, black hole formation takes place, and the spectrum of such primordial black holes (PBH) [67, 68, 69]

Chapter 2

BASIS OF INFLATION

Inflationary phenomena were discovered in seventies years of the previous century but the true triumph of inflation came in eighties. Its success in the explanation of the observable Universe was so impressive that the majority of scientists had no doubts on its correctness, at least as the basic principle. Unfortunately (or maybe, on the contrary, fortunately) the mechanism of inflation may be put into practice by a variety of ways. It led to overproduction of the inflationary models, a number of which increase constantly. Below we discuss the properties of inflationary scenario on the basis of chaotic inflation. Some other models are considered in Chapter 10 to give imagination on the beauty and rich possibilities of the inflationary paradigm.

1. Equations for uniform media

First of all, we have to discuss the relativistic framework of modern cosmology and to consider cosmological evolution of media with different properties. The only requirement is their homogeneity. An inclusion of gravity leads to nontrivial dynamical processes in such media. Thus we have to start with Einstein's equations that couple gravity with matter and play a determinative role in cosmology. Metric tensor $g_{\mu\nu}$ is the dynamical variable that represents gravitational field. By definition, it determines an interval between two space-time points marked by 4-vectors x^μ and x'^μ in curved space–time

$$ds^2 = g_{\mu\nu}dx^\mu dx^\nu, \qquad (2.1)$$

where $dx^\mu = x^\mu - x'^\mu$. Einstein's equations have the form

$$G_\mu^\nu \equiv R_\mu^\nu - \frac{1}{2}\delta_\mu^\nu R = 8\pi G T_\mu^\nu + \Lambda \delta_\mu^\nu,$$

which can be found in any textbook devoted to gravity. Here R_μ^ν is the Ricci tensor, R is scalar curvature, both of which depend on the metric tensor and

T^ν_μ is the energy–momentum tensor determined by properties of a medium. Recent observations indicate possible existence of Λ-term and we include it in the equations from the beginning. The equivalent form of Einstein's equations is

$$R_{\mu\nu} - \frac{1}{2}g_{\mu\nu}R = 8\pi G T_{\mu\nu} + \Lambda g_{\mu\nu} \qquad (2.2)$$

The stress tensor of ideal liquid has the form [30]

$$T^\mu_\nu = (p + \rho)u^\mu u_\nu - p\delta^\mu_\nu, \qquad (2.3)$$

where p is pressure of the medium, ρ is its energy density and u is a 4-velocity of the medium. In a local comoving coordinate system (i.e. those measured by a local observer who experiences the action of gravitational forces only) 4-velocity is $u = (1, 0, 0, 0)$ by definition and the stress tensor possesses only diagonal components

$$T^0_0 = \rho; \; T^r_r = T^\theta_\theta = T^\varphi_\varphi = -p. \qquad (2.4)$$

Consider the case of a uniform medium which is important for cosmology as the realistic zero-order approximation in the description of the Universe. The fact is that the Universe is homogenous only in average. For example, the modern Universe contains large-scale inhomogeneities, such as galaxies and galaxy clusters and may be considered as uniform only after an averaging by a scale of hundreds megaparsec. In this case metric tensor $g_{\mu\nu}$ could be simplified. Here we only outline a proof. Using coordinate transformations one can reduce most general forms of an interval to a form [31], [32]

$$ds^2 = dt^2 - a(t)^2 \left[f^2(r)dr^2 + r^2 d\Omega^2 \right].$$

The Ricci tensor depends on two functions – $a(t)$, $f(r)$ that could be found from Eq. (2.2). The scale factor $a(t)$ plays a significant role in the cosmology. Uniform liquid is locally at rest so that $u = (1, 0, 0, 0)$ and hence nonzero components of the energy–momentum tensor are only those represented in expression (2.4). To determine the form of the function $f(r)$ one only needs equalities $R_{rr} = R_{\theta\theta} = R_{\varphi\varphi}$ [32]. The final result is $f(r) = 1/(1 - kr^2)$ with constant $k = 0, \pm1$, so that interval is usually written as

$$ds^2 = dt^2 - a(t)^2 \left[\frac{dr^2}{1 - kr^2} + r^2(d\theta^2 + \sin^2\theta \; d\varphi^2) \right]. \qquad (2.5)$$

This metric is fixed by the parameter k and function $a(t)$ and is known as the Friedmann–Robertson–Walker (FRW) metric. The form of the metric tensor is obtained by comparison of expressions (2.1) and (2.5)

$$\begin{aligned} g_{00} &= 1, \quad g_{rr} = \frac{-a(t)^2}{1 - kr^2}, \\ g_{\theta\theta} &= -a(t)^2 r^2, \quad g_{\varphi\varphi} = -a(t)^2 r^2 \sin^2\theta. \end{aligned} \qquad (2.6)$$

Tensor Ricci can be easily expressed now in the terms of the scale factor

$$R_0^0 = 3\ddot{a}/a; \tag{2.7}$$
$$R_r^r = R_\theta^\theta = R_\varphi^\varphi = \ddot{a}/a + 2\dot{a}^2/a^2 + 2k/a^2.$$

There are only three essentially different values of the parameter k, $k = 0, \pm 1$. Time evolution of the scale factor $a(t)$ is determined from Einstein's equations and depends on properties of media described by the stress tensor $T_{\mu\nu}$. Space points are marked by variables r, θ, φ. Their physical meaning is clarified if one calculates a square of a surface at specific moment t: $S = 4\pi [a(t)r]^2$. For $r = Const$ physical radius of the sphere grows as $a(t)$. Physical frame is used when measuring in terms of

$$R(t) \equiv a(t)r. \tag{2.8}$$

If two observers with local velocities equal to zero are disposed at distance r they measure the physical distance growing with time according to (2.8). The value $R(t)$ is known as the 'physical distance' and it does not coincide with instantaneous 'physical' radius of the sphere

$$R_{inst} = a(t) \int_0^r \frac{dr}{\sqrt{1 - kr^2}}. \tag{2.9}$$

This distance is measured by a set of observers who are placed along the radius. Its analytical form depends on the value of the parameter k

$$R_{inst} = \begin{cases} a(t)\arcsin(r), & k = 1, \\ a(t)r, & k = 0, \\ a(t)arcsh(r), & k = -1. \end{cases}$$

A third kind of the distance is measured by a traveller who moves from a center of the surface. It can be easily found for the traveller who moves with a speed of light along the radius ($d\theta = d\varphi = 0$). Due to the fact that for the light $ds = 0$ and from Eq. (2.5) it is obvious that

$$\int_0^t \frac{dt}{a(t)} = \int_0^r \frac{dr}{\sqrt{1 - kr^2}}$$

that immediately gives physical distance travelled by light during time interval $(0, t)$

$$R(t)_{hor} \equiv a(t)r = \begin{cases} a(t)\sin\left(\int_0^t dt/a(t)\right), & k = 1, \\ a(t)\int_0^t dt/a(t), & k = 0, \\ a(t)sh\left(\int_0^t dt/a(t)\right), & k = -1. \end{cases} \tag{2.10}$$

Here is the appropriate place to determine a notion of horizon. We have to warn about two meanings of this word, both of which are widely used. The first one is "horizon is a distance travelled by light during a specific time". For example, the horizon of our Universe is about 10^{28}cm. This value is increasing constantly. Another meaning of horizon is "geometrical set of points which could be reached by light during an infinitely large time interval". Schwarzschild radius of a black hole is the typical illustration for the second definition. The value $R(t)_{hor}$ determines horizon size of the Universe in its first meaning.

After some algebra with Einstein equations one can obtain using Eq. (2.6) first-order differential equations that are valid for a homogeneous Universe

$$\frac{\dot{a}^2}{a^2} + \frac{k}{a^2} = \frac{8\pi G}{3}\rho + \frac{\Lambda}{3}, \tag{2.11}$$

$$\frac{d\rho}{da} + \frac{3(\rho + p)}{a} = 0. \tag{2.12}$$

De Sitter [248] has considered hypothetic space with zero stress energy–momentum tensor and $\Lambda \neq 0$. Being as simple as possible, this space possesses interesting and important features which are widely used in modern theories. The first of Eq. (2.11) is simplified significantly

$$\dot{a}^2 - H_0^2 a^2 = -k, \ H_0^2 \equiv \frac{\Lambda}{3} = Const. \tag{2.13}$$

There are three types of solutions corresponding to three possible values of constant k: 1) $k = 0$ – Flat Universe –

$$a(t) = Const \cdot \exp(H_0 t), \tag{2.14}$$

2) $k = -1$ – Open Universe –

$$a(t) = H_0^{-1} \sinh(H_0 t + Const'), \tag{2.15}$$

3) $k = 1$ – Closed Universe –

$$a(t) = H_0^{-1} \cosh(H_0 t + Const''). \tag{2.16}$$

The constants can be found from initial conditions. The values of $Const'$ and $Const''$ are usually chosen to be zero and $Const = H_0^{-1}$. In this case all the three possible forms of the scale factor tend to the unique one at large times

$$a(t) = H_0^{-1} \exp(H_0 t), \quad t >> H_0^{-1}. \tag{2.17}$$

The physical distance between two points increases exponentially as is evident from the expression (2.8) and from the form of the scale factor in de Sitter space. On the other hand, these points are characterized by comoving distances r_1 and r_2 that do not vary with time. This remark indicates once more the difference between physical and comoving distance. It is very instructive to consider a distance travelled by light both in the comoving coordinate and the physical frame. Keeping in mind flat Universe ($k = 0$) with the scale factor $a(t) = H_0^{-1}\exp(H_0 t)$, we obtain

$$L_{comov}(t, t') = \int_t^{t'} d\tau/a(\tau) = e^{-H_0 t} - e^{-H_0 t'} \tag{2.18}$$

for the light motion from the initial moment t to the final moment t' in the comoving frame. Horizon size

$$L_{hor} \equiv L_{comov}(t, t' \to \infty) = e^{-H_0 t} \tag{2.19}$$

is finite and moreover, the later the light was emitted, the smaller the path it will cover. It may be said that the horizon is decreasing with time in the comoving frame. The result for the physical frame is as follows

$$L_{phys}(t, t') = a(t')L_{comov}(t, t') = H_0^{-1}\left[\exp(H_0(t' - t)) - 1\right].$$

The horizon tends to infinity according to

$$L_{phys,hor}(t, t' \to \infty) = H_0^{-1}\exp(H_0 t'). \tag{2.20}$$

and depends on the difference $t' - t$ only. The horizon size is a very important value that strongly influences all physical phenomena. The horizon size equals infinity when physical processes run in the Minkowski space. The situation differs drastically for the FRW and for inflationary stages. As we will see below an interplay between the horizon size and the spatial scale of physical processes must be taken into account at these stages. One has to keep in mind from the beginning what frame is chosen for given consideration.

2. Inflation

Decades have passed since it was understood that our Universe has a very productive and interesting history before the Big Bang. It was realized that the expansion of the Universe is not able to explain many intrinsic problems of its evolution. Rather a complete list of them is represented in [57], [3] as well as in many other textbooks devoted to the very early Universe. The main conclusion is that a period of very quick expansion of the Universe must have taken place before the Big Bang. It could be easily achieved by postulating de Sitter space instead of the Minkowski. On the other side, the Universe expands rather slowly in the modern epoch and the geometry of the space is almost flat. The De Sitter space possesses higher symmetry of space–time than the Minkowski space. Such symmetry may be realized if the energy density of physical vacuum is nonzero.

The first ideas to combine the above-mentioned desired features of the cosmological evolution were based on the postulate that the initial state of the

Universe was maximally symmetric and on the possible interpolation from this state to the FRW regime of expansion [48, 49] (see [110] for review).

A.I. Bugrii and A.A. Trushevsky [96] found the possibility to realize the regime of quick expansion in the course of high temperature phase transition in the hadronic era. Though based on the unrealistic extrapolation of pre-QCD physics of strong interactions, it may have been the first attempt to realize inflation as the effect of high temperature phase transition, predicted by a particle physics model.

It was found in [50, 51] that De Sitter vaccum-dominated stage may be realized as the R^2 effect of quantum corrections to the gravitational field, where R is the scalar curvature.

However, it was the work of A. Guth in 1981 [52], aimed to solve the problem of magnetic monopoles (see further Section 5 of this Chapter), that revealed the set of the general internal problems of the Big Bang cosmology and proclaimed the essential features of the inflationary scenario. It initiated the transition from the principal possibility to the vital necessity of inflation as the element of the Standard Big Bang Cosmology.

The transparent and appealing idea of inflation, its principal ability to provide the self-consistent framework for the Big Bang Cosmology is so impressive that the scientific majority is now convinced of its correctness at least in general issues in spite of some remaining difficulties of its realization [58].

Scalar field coupled with gravity provides us with the simplest way to describe the physics of inflation. The standard Lagrangian density of scalar field coupled with gravity is

$$L = \sqrt{-g} \left\{ \frac{R}{16\pi G} + \frac{1}{2} g^{\mu\nu} \partial_\mu \varphi \partial_\nu \varphi - V(\varphi) \right\}, \qquad (2.21)$$

where G is Newton constant. Equation of motion of scalar field can be written in the form

$$\partial_\mu \sqrt{-g} g^{\mu\nu} \partial_\nu \varphi + \sqrt{-g} V'(\varphi) = 0. \qquad (2.22)$$

In the FRW Universe nonzero values of the metric tensor $g^{\mu\nu}$ are (see (2.6))

$$g^{00} = \frac{1}{g_{00}} = 1, \quad g^{11} = \frac{1}{g_{11}} = -\frac{1 - kr^2}{a(t)^2};$$

$$g^{22} = \frac{1}{g_{22}} = -a(t)^{-2} r^{-2}, \qquad (2.23)$$

$$g^{33} = \frac{1}{g_{33}} = -a(t)^{-2} r^{-2} \sin^{-2} \theta;$$

$$\sqrt{-g} = \sqrt{-g_{00} g_{rr} g_{\theta\theta} g_{\varphi\varphi}} = -\frac{a(t)^3}{\sqrt{1 - kr^2}} r^2 \sin \theta.$$

In the important case of uniform distribution of the field φ, i.e. $\varphi = \varphi(t)$, the equation (2.22) is simplified significantly

$$\ddot{\varphi} + 3H\dot{\varphi} + V'(\varphi) = 0, \quad H \equiv \frac{\dot{a}}{a}. \tag{2.24}$$

The Hubble parameter H is one of the most important parameters which influenced the evolution of our Universe. The energy density of the scalar field is equal to $\rho = \frac{1}{2}\dot{\varphi}^2 + V(\varphi)$ and Eq. (2.11) in the form

$$H^2 = \frac{8\pi G}{3}\left(\frac{1}{2}\dot{\varphi}^2 + V(\varphi)\right) \tag{2.25}$$

may be considered as the second equation of the system to find dynamical variables $\varphi(t), a(t)$. The term $\Lambda/3$ was included in the potential V.

It was noticed that in the case of slow motion of field φ the behavior of the system (2.24), (2.25) is very similar to that for de Sitter space even if $\Lambda = 0$. Indeed, slow motion takes place if "friction term" $3H\dot{\varphi}$ is large enough, i.e.

$$3H\dot{\varphi} >> \ddot{\varphi} \tag{2.26}$$

that leads to the inequality

$$\dot{\varphi}^2 << V(\varphi). \tag{2.27}$$

The latter results in $V \simeq Const$ and, keeping in mind Eq. (2.25), in approximately constant Hubble parameter

$$H \equiv \frac{\dot{a}}{a} \simeq \sqrt{\frac{8\pi G}{3}V(\varphi)}. \tag{2.28}$$

We obtain exponential growth of scale factor $a(t) \propto \exp(Ht)$ as in the exact de Sitter case. Scalar field dynamics is much more rich than the simple de Sitter case and reveals new, interesting features. In particular, the field φ governed by the equation of motion, slowly moves to the potential minimum and hence inequality (2.27) becomes wrong inevitably at some small values of the potential. Thus, exponential growth is not eternal and inflation is finished at some values of the field close to the minimum of the potential. The field evolution during inflation is governed by equation

$$3H\dot{\varphi} + V'(\varphi) = 0 \tag{2.29}$$

obtained from Eq. (2.24), where the secondary derivative is omitted due to slow motion of the field φ.

Slow-roll conditions can be derived as follows. From Eq. (2.29) represented in the form

$$\dot{\varphi} = -\frac{V'(\varphi)}{3H(\varphi)}$$

one can easily obtain expression for the second derivative

$$\ddot{\varphi} = \frac{V'(\varphi)}{3H^2(\varphi)}H'(\varphi)\dot{\varphi} - \frac{V''(\varphi)}{3H(\varphi)}\dot{\varphi}.$$

To supply the slow-roll condition both terms in the right-hand side must be small as compared with the term $3H\dot{\varphi}$. Bearing in mind expression (2.28) for the Hubble parameter, slow-roll condition could be represented in the standard form

$$\varepsilon \equiv \frac{M_P^2}{16\pi}\frac{V'(\varphi)^2}{V(\varphi)^2} << 1; \ \eta \equiv \frac{M_P^2}{8\pi}\left|\frac{V''(\varphi)}{V(\varphi)}\right| << 1. \tag{2.30}$$

It is worth estimating the value of the field at which $\epsilon = 1, \eta = 1$. When the field reaches this value the inflation is finished. For the potential in the form $V(\varphi) = \lambda\varphi^n$ slow-rolling is succeeded by a quick classical motion of the inflaton when

$$\varphi_{end} \approx \frac{n}{4\sqrt{\pi}}M_P.$$

Evidently, the inflation takes place at those field values which yield sufficiently large value of potential.

The scale factor evolves according to Eq. (2.28)

$$a(t) = H(\varphi_{in})^{-1}\exp\left[\int_{t_{in}}^{t} H(\varphi)dt\right]. \tag{2.31}$$

When the inflation is finished, say, at time t_e, initial space domain of a size $H(\varphi_{in})^{-1}$ has been expanded up to the size $R(t_e) = a(t_e)H(\varphi_{in})^{-1}$. The commonly used value of the Hubble parameter at the end of inflation is $H_e \sim 10^{13} GeV$. For most estimations it is enough to suppose approximately $H(\varphi_{in}) \approx H_e = Const$ during the last stage of inflation. Another useful value is e-folding, N that is determined as

$$N \equiv \ln\left[\frac{a(t_f)}{a(t_{in})}\right]. \tag{2.32}$$

This value indicates a factor of the Universe expansion during time interval (t_{in}, t_f) in logarithmic scale. It can be expressed in terms of inflaton φ in the following manner

$$N \equiv \ln\left[\frac{a(t_f)}{a(t_{in})}\right] = \int_{t_{in}}^{t_f} Hdt = \int_{\varphi_{in}}^{\varphi_f} H\frac{d\varphi}{\dot{\varphi}} =$$

$$-\int_{\varphi_{in}}^{\varphi_f} \frac{3H(\varphi)^2 d\varphi}{V'(\varphi)} = -\frac{8\pi}{M_P^2}\int_{\varphi_{in}}^{\varphi_f} \frac{V(\varphi)}{V'(\varphi)}d\varphi. \tag{2.33}$$

Equations (2.28) and (2.29) may be solved for different forms of potentials. As an example, if the potential has the form

$$V(\varphi) = \frac{1}{2}m^2\varphi^2 \tag{2.34}$$

the solution is

$$\varphi(t) = \varphi_{in} - \frac{mM_P}{\sqrt{12\pi}}t. \tag{2.35}$$

Be reminded that this solution is valid until inequality (2.26) is fulfilled. If the last is true, the stress tensor has diagonal form approximately

$$T \cong \begin{pmatrix} V(\varphi) & 0 & 0 & 0 \\ 0 & -V(\varphi) & 0 & 0 \\ 0 & 0 & -V(\varphi) & 0 \\ 0 & 0 & 0 & -V(\varphi) \end{pmatrix}.$$

We have to notice that any media with stress tensor $T_{\mu v} \propto g_{\mu v}$ implies inflation.

3. Scale factor

Our Universe is supposed to be uniform and isotropic in the majority of models. In this case physical distance R between any two points is governed by scale factor a. Simple formula

$$R(t) = a(t)r \tag{2.36}$$

expresses physical distance in terms of comoving distance r. At the modern epoch this expression is valid at the scale much larger than the galaxy scale. Below we consider a time dependence of the scale factor at the main stages of cosmological evolution.

The main equations for the discussion below follow from Einstein equations (2.11), (2.12). There are three unknown functions $a(t), \rho(t)$ and $p(t)$ and a third equation is needed to solve the problem. Widespread choice is the connection between energy density and pressure, representing the equation of state of the Universe. This connection, which we write in the form

$$p + \rho = \gamma\rho, \tag{2.37}$$

is valid in many (but not in all) cases. Numerical value of the parameter γ depends on properties of a medium in question. We will see below that the scale factor grows rapidly with time so that the curvature term (the second term in the left-hand side of the Eq. (2.11)) can be freely omitted. As a result, the main system of equations acquires the final form

$$\dot{a}^2 = \frac{8\pi G}{3}\rho a^2, \tag{2.38}$$

$$\frac{d\rho}{da} = -3\gamma\frac{\rho}{a} \tag{2.39}$$

with the obvious solution

$$\rho = \frac{C}{a^{3\gamma}}.$$ (2.40)

Unknown constant C may be defined using initial condition: $C = \rho_{in} a_{in}^{3\gamma}$, where index $'in'$ determines initial value for the considered stage that is supposed to be equal to the final value of a previous stage. The transitions between neighboring stages is not sharp, of course, which is often neglected. Substituting expression (2.40) into the first equation of system (2.38) one finds time dependence of scale factor

$$a(t) = \left[a(t_{in})^{3\gamma/2} + \frac{2}{3\gamma} \sqrt{\frac{8\pi G}{3} C} (t - t_{in}) \right]^{\frac{2}{3\gamma}},$$

where time t_{in} marks the beginning of a considered stage. The final formula may be readily written as

$$a(t) = a_{in} \left[1 + \frac{2}{3\gamma} \sqrt{\frac{8\pi G \rho_{in}}{3}} (t - t_{in}) \right]^{\frac{2}{3\gamma}}.$$ (2.41)

The majority of processes that happened in the early Universe indicate quick growth of the scale factor so that a more simple formula appears to be more practical

$$a(t) \simeq a_{in} \left[\frac{2}{3\gamma} H_{in} \cdot (t - t_{in}) \right]^{2/3\gamma}.$$ (2.42)

Here $H_{in} \equiv H(t_{in}) = \sqrt{8\pi G \rho_{in}/3}$ is the Hubble parameter at the beginning of a specific stage. The value of the parameter γ is strongly dependent on the stage of the Universe. There are four main stages of evolution that should be present with necessity in the inflationary Universe: inflation, reheating (or preheating), stage of radiation dominance and matter-dominance stage.

Inflationary stage was discussed above in detail. The scale factor of this stage is described by formula (2.31).

3.1   Reheating

The period of inflation is finished when the friction term appears to be small. After that the inflaton field starts oscillating coherently about its potential minima. This process is accompanied by high energy particle emission and leads to heating the Universe. The stage of reheating (or preheating, if the Universe has previously never been hot) is finished when the energy density of particles is comparable with the energy density of the field oscillations. Before this we may roughly suppose dominance of the energy density of field oscillations. It means that the equation

$$\ddot{\varphi} + 3H(t) + V'(\varphi) = 0$$ (2.43)

is still valid. Using formulae for pressure $p = \frac{1}{2}\dot{\varphi}^2 - V(\varphi)$ and energy density $\rho = \frac{1}{2}\dot{\varphi}^2 + V(\varphi)$, this equation is easily transformed into the already known form (2.12)

$$\frac{d\rho}{dt} = -3H(t)(p + \rho). \tag{2.44}$$

Auxiliary condition $\rho + p = \gamma\rho$ takes place only approximately if one averages by a period of oscillations and finds model-dependent parameter γ. Indeed,

$$\gamma \simeq \frac{\int\limits_0^T (\rho + p)dt}{\int\limits_0^T \rho \, dt} = \frac{\int\limits_0^T \dot{\varphi}^2 \, dt}{\int\limits_0^T \left[\frac{1}{2}\dot{\varphi}^2 + V(\varphi)\right] dt} = \frac{\int\limits_0^{\varphi_{max}} \dot{\varphi}^2 \, d\varphi}{\int\limits_0^{\varphi_{max}} d\varphi \left[\frac{1}{2}\dot{\varphi}^2 + V(\varphi)\right]/\dot{\varphi}} \tag{2.45}$$

If the oscillations are very quick compared with expansion of the Universe we may use the law of energy conservation in the form

$$\dot{\varphi} = \sqrt{2\left[V(\varphi_{max}) - V(\varphi)\right]}$$

. The potential may be approximated by a polynomial $V(\varphi) \cong \lambda\varphi^v$ in a vicinity of its minima and we come to the simple formula

$$\gamma \cong \frac{2v}{v + 2}. \tag{2.46}$$

Let us determine the scale factor as a function of time. At the beginning of the reheating, scale factor was the same as at the end of inflation, $a(t_{in}) = a_I(t_e) = H(\varphi_U)^{-1}e^{N_U}$. We suppose that the visible Universe emerges N_U e-folds before the end of inflation with the size $H(\varphi_U)^{-1}$ and at the field value φ_U. Energy density equals to the potential energy at the end of inflation, $\rho(t_{in}) = \rho(t_e) = V(\varphi_e)$. Finally we obtain scale factor $a_{reh}(t)$ at the stage of reheating in the form (2.41) with $a_{in} = H(\varphi_U)^{-1}e^N$ and $\rho_{in} = V(\varphi_e)$. When this stage is finished (at the time t_{reh}), initial space domain of a size H_e^{-1} has been expanded up to the size $R(t_e) = a_{reh}(t_{reh})H_e^{-1}$. A more practical expression for the formula (2.42) may be written in the form

$$a_{reh}(t) = \left(\frac{2}{3\gamma}H_e\right)^{2/3\gamma} H(\varphi_U)^{-1}e^N t^{2/3\gamma}.$$

The reheating stage gives most uncertain estimation of the scale factor as compared with other stages. Indeed, factor γ, being effective value, reflects such features as the form of the potential and the decay rate of the inflaton into light particles, what is strongly model-dependent. The products of decay heat the medium which gives additional uncertainties.

3.2 Radiation – dominated stage

Up to now we were able to express energy density in terms of field theory to solve Eq. (2.43) and to find behavior of scale factor. The decay of field oscillations transforms their energy density into the energy density of high energy particles, dominating in the Universe after the field oscillations faded out. At this stage it seems reasonable to consider the hot Universe introducing temperature and utilizing standard results of statistical physics and thermodynamics. The problem is that we can use the concept of temperature only in equilibrium which is not strictly true in our case due to expansion of the Universe. Fortunately, the characteristic processes are so quick that the state is very close to equilibrium. Indeed, consider the collision time of electron and photon $t_{\gamma e}$ and cosmological time $t_{cosm} \sim M_P/T^2$ at the moment with temperature T. The collision time is evaluated as $t_{\gamma e} \sim n\sigma v$, where $n \sim T^3$ is electron density, $\sigma \sim \alpha^2/T^2$ is the Compton cross-section and $v \simeq 1$ is the electron velocity. One could deal with 'temperature' only if $t_{\gamma e} \ll t_{cosm}$. It takes place if $T \ll \alpha^2 M_P \sim 10^{17} GeV$. Meanwhile the temperature at the FRW stage hardly exceeded $10^9 GeV$ which means that the concept of temperature could be used freely. To find scale factor $a_{RD}(t)$ one can apply formula (2.41) with $\gamma = 4/3$ for the radiation-dominated Universe. This value of factor γ follows from the equation of state $p = \rho/3$ for the gas of highly relativistic particles and from the formula (2.37). The initial conditions for this stage are: $t_{in} = t_{roh}$, $\rho_{in} = \rho(t_{roh})$ and $a_{in} = a(t_{roh})$, where t_{roh} corresponds to the end of the reheating stage. The final expression for the scale factor could be written in the form (2.55). Energy density of relativistic plasma is connected with the temperature T in the standard way

$$\rho = \frac{\pi^2}{30} g_* T^4, \tag{2.47}$$

where g_* is the number of relativistic species with the account of their statistic weights. Thus, if quick oscillations of the inflaton field heat the medium to the temperature T_{reh}, formula (2.47) determines energy density $\rho(t_{reh})$ which is the final one for the reheating stage and the initial one for the radiation-dominated stage. Let us show also that the entropy is constant after reheating. Indeed, in our case $\rho \sim a^{-4}$ ($\gamma = 4/3$), on the other hand $\rho \sim T^4$ and hence

$$aT = Const. \tag{2.48}$$

Keeping in mind connection of entropy density and temperature

$$s = \frac{2\pi^2}{45} g_* T^3, \tag{2.49}$$

one immediately obtains

$$S \sim sa(t)^3 = Const. \tag{2.50}$$

This statement confirms the conclusion made above that the processes of particle interaction which could change the entropy of the system are much slower compared with the expansion of the Universe. Another useful formula is also evident from Eq. (2.48)

$$T(t) = \frac{a(t_{reh})}{a(t)} T_{reh}. \tag{2.51}$$

This connection between the scale factor and the temperature is very important for cosmological estimations. The temperature dependence on time is ruled by the formula

$$T = \left(\frac{45}{32\pi^2 g_*}\right)^{1/4} \sqrt{\frac{M_P}{t}}. \tag{2.52}$$

This formula is obtained from the combination of formulae (2.47), (2.38) and (2.55).

3.3   Matter-dominated stage

The matter-dominated period of the Universe can be treated as a dust-like period with good accuracy. It means that pressure p equals zero and the parameter $\gamma = 1$ in Eq. (2.37) (see [3] for more details). In full analogy with the radiation-dominated stage one can find scale factor $a_{MD}(t)$ by applying formula (2.41) with $\gamma = 1$ for the pressureless Universe. The observational data do not exclude and various cosmological scenarios do admit the existence of early dust-like stages (see review in [3]). But in any case the relationship between the energy densities of matter, ρ_m, and radiation, ρ_γ, in the modern Universe, $\rho_m \gg \rho_\gamma$, indicates that radiation dominance has been inevitably finished, when the matter dominance stage began. The initial conditions for this stage are

$$t_{in} = t_{RD}, \quad \rho_{in} = \rho(t_{RD}) = \frac{\pi^2}{30} g_* T_{RD}^4, \quad a_{in} = a_{RD}(t_{RD}),$$

where t_{RD} corresponds to the end of the radiation-dominated period. Formula (2.48) is also valid provided T is the temperature of relic photons. Let us bring together the main formulae for the scale factor based on more vivid expression (2.42), valid at $t \gg t_{in}$.

Inflationary stage

$$a(t) \equiv a_I(t) = H_U^{-1} \exp\left(\int H dt\right) \approx H_e^{-1} e^{N_U}. \tag{2.53}$$

Here $N_U \approx 60$, $H_e \approx 10^{13} GeV$ for ordinary inflationary models.

Reheating stage

$$a(t) \equiv a_{reh}(t) = a_I(t_e) \left(\frac{2}{3\gamma} H_e\right)^{2/3\gamma} (t - t_e)^{2/3\gamma}. \tag{2.54}$$

Radiation-dominated stage

$$a(t) \equiv a_{RD}(t) = a_{reh}(t_{reh}) \left[\frac{1}{2} H(t_{reh}) \right]^{1/2} (t - t_{reh})^{1/2}. \qquad (2.55)$$

Matter-dominated stage

$$a(t) \equiv a_{MD}(t) = a_{RD}(t_{RD}) \left[\frac{2}{3} H(t_{RD}) \right]^{2/3} (t - t_{RD})^{2/3} \qquad (2.56)$$

It is worth mentioning that recent observational data are widely interpreted in favor of the existence of a "dark energy", a medium with negative pressure, such as a cosmological term Λ. Moreover, its contribution to the total energy density of the Universe is estimated at about 70%. The simplest supposition is that the energy density of this medium does not vary with time, i.e. that it is just the cosmological term (see however the discussion in [94, 95] and references in the review [93]). If it is like this, the immediate conclusion is that we are coming into a new de Sitter stage. Indeed, the matter and radiation contribution into the total energy density decreases with time, while the energy density associated with Λ term remains constant. It means that the Hubble parameter tends to constant as well

$$H \rightarrow \sqrt{\frac{8\pi}{3} \frac{\rho_\Lambda}{M_P^2}}. \qquad (2.57)$$

The inverse Hubble parameter characterizes the size of causally connected volume. Simple estimation of this value gives

$$H^{-1} \approx 10^{28} cm.$$

Note that this value coincides approximately with the size of the visible part of our Universe and hence we never obtain any information that is contained in a larger volume. The distance R_0 between two pointlike objects governed only by gravitation in the modern epoch t_0 is

$$R_0 = a(t_0)r \qquad (2.58)$$

(see (2.8)). It is the value R_0 that is measured these days. What could one say about values $a(t_0)$ and r? The formulae for the scale factor written above are useful if only one specific stage is analyzed. To determine the value $a(t_0)$ one needs to know all values in formulae (2.53), (2.54), (2.55) and (2.56). Meanwhile, even the value of $a_{RD}(t_{RD})$ that seems to be fixed by the observations of CMB and large-scale structure can actually vary owing to the possible existence of unstable dark matter. This value also depends on the choice of the model for the dark energy. The parameters of inflationary and post-inflationary stages are much more model-dependent, and the main uncertainty in these parameters is related with the reheating stage. Another question is connected with the numerical value of coordinate r. These problems could be avoided if we consider expression (2.58) as a substitution of the variables – instead of mathematical coordinate r we will use observational value R_0. In this case physical distance between the two points at arbitrary time is

$$R(t) = \frac{a(t)}{a(t_0)} R_0.$$

For example, distances at the matter-dominated period looks like

$$R(t) = \left(\frac{t - t_{RD}}{t_0 - t_{RD}} \right)^{2/3} R_0 \simeq \left(\frac{t}{t_0} \right)^{2/3} R_0.$$

Very often dimensionless value

$$\mathrm{a}(t) \equiv \frac{a(t)}{a(t_0)}$$

is denoted as the scale factor. Now let us answer the question: "What is the connection between the scale factor and temperature?" We already have the answer for the radiation-dominated period – see formula (2.48). The only which remains is to determine this dependence at the matter-dominated period. To proceed, we have to notice that in this period recombination took place, when protons and electrons were coupled into hydrogen atoms and thus decoupled from radiation. So, photons do not interact with matter starting from the period of recombination. Up to this period they were in the thermal equilibrium with plasma. Their distribution was nothing but a Planck one and to the moment of recombination, corresponding to the temperature T_{rec}, it took a form

$$dN(t_{rec}) = V_{rec} \frac{E_{rec}^2}{\pi^2} \frac{dE_{rec}}{\exp(E_{rec}/T_{rec}) - 1}. \tag{2.59}$$

Here dN is a number of the photons with energy between E_{rec} and $E_{rec} + dE_{rec}$ within a volume V_{rec}. After the recombination their interaction with surrounding neutral atoms is negligible. The energy of a photon decreases with time according to the following

$$E(t) = p(t) = \frac{2\pi}{\lambda(t)} = \frac{2\pi}{\frac{a(t)}{a(t_{rec})} \lambda(t_{rec})} = \frac{a(t_{rec})}{a(t)} E_{rec}. \tag{2.60}$$

The number of free particles is a conserved value, so that

$$dN(t) = dN(t_{rec}).$$

The volume V grows with time as

$$V(t) = \left(\frac{a(t)}{a(t_{rec})} \right)^3 V_{rec}.$$

Combining all the formulae written above one obtains the photon distribution at the instant t

$$dN(t) = V(t)\frac{E(t)^2}{\pi^2}\frac{dE}{\exp\left[E(t)\frac{a(t)}{a(t_{rec})}/T_{rec}\right] - 1}. \qquad (2.61)$$

It is evident now that Planck distribution of relic photons still takes place with the temperature

$$T(t) = \frac{a(t_{rec})}{a(t)}T_{rec}. \qquad (2.62)$$

Thus the law (2.48) takes place both in radiation-dominated and matter-dominated stages of the evolution of our Universe. The constant could be determined from the normalization to the modern epoch $Const = a_0 T_0$.

4. Does expansion during the de Sitter stage really exist?

De Sitter space may be imagined as the space with gravity being produced by constant potential density $V = \Lambda = Const$. One of the possible forms of the interval is

$$ds^2 = dt^2 - e^{2Ht}\left[dr^2 + r^2(d\theta^2 + \sin^2\theta\, d\varphi^2)\right], \qquad (2.63)$$

where $H = \sqrt{\Lambda/3}$ (see (2.13)). Note that this widespread form of the interval for flat Universe ($k = 0$) implies that scale factor is dimensionless value $a(t) = e^{Ht}$ while coordinate radius is a dimensional one. It does not lead to misunderstanding usually. It can be easily shown that particle trajectory $x^i(t) = const$ in comoving coordinates (t, x^i) is the solution of geodesic equation of motion

$$\frac{d^2 x^i}{ds^2} + \Gamma^i_{\mu\nu}\frac{dx^\mu}{ds}\frac{dx^\nu}{ds} = 0,$$

because $\Gamma^i_{tt} = 0$ for metric (2.63). It means that a distance between two particles at rest does not vary with time in comoving coordinates. On the other side, physical frame implies measuring in terms of physical coordinates $x^i_{phys} = H^{-1}e^{Ht}x^i$ that indicates unambiguously an increase of the distance between the two particles. This contradiction could be strengthened if one realizes that the form (2.63) for the interval is only one among many. For example, one of the possible forms is

$$ds^2 = \left(1 - H^2 r^2\right)dt^2 - \left(1 - H^2 r^2\right)^{-1}dr^2 - r^2(d\theta^2 + \sin^2\theta\, d\varphi^2),$$

which is obviously static. There exist other coordinate systems where distances decrease with time. How can we understand whether our Universe really expands if distances depend on a coordinate system? To solve this problem we have to find another indicator of space expansion besides the distance. This indicator should not depend on a frame. For this, consider a system of test particles uniformly distributed in the Universe with some constant density and small but nonzero interaction between them. If we find out a decline of interaction with time, it may be interpreted as the uniform expansion of the Universe what leads to increasing of the interparticle distances. To this end, consider a behavior of a horizon with time in the comoving coordinates. The most simple way to determine the size of the horizon is to find a distance where light travels from the moment t. By definition, the propagation of light corresponds to the interval $ds = 0$. Hence, Eq. (2.63) gives the connection $dt = e^{Ht} dr$ and

$$\int_0^{R_H} dr = \int_t^\infty dt' e^{-Ht'}.$$

Horizon size

$$R_H = H^{-1} e^{-Ht} \tag{2.64}$$

decreases with time. But average distance between the test particles in the comoving space is shown to be constant. Consequently, when the horizon appears to be smaller than this distance, the particles turn out to be placed in a causally disconnected area and the interaction between them must be absent. Hence, initial nonzero interaction between particles tends to zero with time. It indicates (or determines if one wishes) that the Universe is really expanding during the de Sitter stage. Evidently, this statement should not depend on a coordinate system – the conclusion is the same for physical and comoving frames.

5.   Why do we need inflation?

5.1   Flatness problem

The old Big Bang cosmology suffered several problems that cannot be recovered in its framework. The inflationary paradigm resolves these problems in a natural manner. Here we consider only a couple of them – the so-called "flatness" problem, i.e. the problem of why the density is so close to the critical one, and the monopole problem to show how it works. Let us consider the equation (2.11) with $\Lambda = 0$. Critical energy density is determined as that which leads to flat Universe, i.e. $k = 0$. It means that

$$\rho_{crit} \equiv \frac{3}{8\pi} M_P^2 H^2 \tag{2.65}$$

(Let us be reminded that $G = 1/M_P^2$ and $\dot{a}/a = H$). Eq. (2.11) leads immediately to the equality

$$\left| \frac{\rho_{crit} - \rho(t)}{\rho_{crit}} \right| = \dot{a}(t)^{-2} \tag{2.66}$$

for $k = \pm 1$. Modern observational value of energy density is close to the critical density [74], so that the inequality

$$\left| \frac{\rho_{crit} - \rho(t_{now})}{\rho_{crit}} \right| = \dot{a}(t_{now})^{-2} < 1 \qquad (2.67)$$

is established rather firmly. Combining the two equations (2.66), (2.67) one can easily obtain

$$\left| \frac{\rho_{crit} - \rho(t)}{\rho_{crit}} \right| < \frac{1}{\dot{a}(t)^2} = \frac{\dot{a}(t_{now})^2}{\dot{a}(t)^2}.$$

To simplify the situation, suppose that there is only a radiation-dominated stage so that $a(t) \propto \sqrt{t}$. The temperature at the beginning of this stage was not less than 10^6 GeV, while modern temperature is about 10^{-13} GeV. Hence, the last inequality leads to a very unnatural situation at the beginning of the RD stage:

$$\left| \frac{\rho_{crit} - \rho(t_{RD})}{\rho_{crit}} \right| < \frac{\dot{a}(t_{now})^2}{\dot{a}(t_{RD})^2} = \frac{t_{RD}}{t_{now}} = \left(\frac{T_{now}}{T_{RD}} \right)^2 \sim 10^{-19}. \qquad (2.68)$$

The question is why real energy density of our Universe was so close to the critical one in the past?

There are two possible answers to this question. The first one is concerned with the more general problem of fine-tuning of parameters of the Universe, discussed in Chapter 9. Another way is to reveal some mechanism that gives rise to such a small value. Fortunately, the inflationary paradigm supplies us with a mighty tool to resolve problems such as these. Let our Universe have the inflation stage with scale factor (2.31) $a(t) = H^{-1} \cosh(Ht)$, with the Hubble parameter $H = Const$ that is approximately valid in most inflationary models. In this case the behavior of ratio (2.66) with time is

$$\left| \frac{\rho_{crit} - \rho(t)}{\rho_{crit}} \right| = \sinh^{-2}(Ht).$$

This function tends to zero exponentially while the inflation lasts and we realize that the inflationary paradigm is able to explain small values in estimation (2.68).

Other problems, such as the problem of primordial density fluctuations in the Universe, a horizon problem and a monopole problem are also solved by supposition of the inflationary stage.

5.2   Monopole problem

As we have already mentioned above, the early approaches to inflation have at most demonstrated the principal possibility of quick cosmological expansion. They were motivated either by aesthetical reasoning to relate the beginning of the Universe with the maximally symmetric state, or offered some possible mechanisms for quick expansion. The common understanding that inflation should be the necessary element of the Standard Big Bang scenario came after the problem of relic magnetic monopole overproduction [42, 43, 44] (see [3] for review) which was revealed as the dramatic disaster for the old Big Bang cosmology and after inflation was offered [52] as the resolution for this trouble. The possible existence of magnetic monopoles – isolated poles of magnet – was discussed at each step of the development of the theory of electromagnetism. So, Coulomb has offered the inverse distance-squared law for magnetostatic force between "magnetic charges", being similar to the electrostatic force between electric charges. In the quantum theory of electromagnetism Dirac [38, 39] has found that quantization of electric charge inevitably leads to the existence of the Dirac monopole – isolated pole of magnet with the magnetic charge

$$g = \pm \frac{\hbar c}{2e},$$
(2.69)

where e is the electric charge of electron. The mass of the Dirac monopole was a free parameter. One could ascribe the absence of Dirac monopoles among the particles, created at accelerator experiments, to such a large value of monopole mass that it corresponds to the energy threshold of their creation, exceeding the energy, accessible at the given accelerator. It made the magnetic monopole search the challenge for each new accelerator, at which higher energy range was reached. Since magnetic charge conservation should be as strict as the conservation of electric charge, the lightest particle, possessing the magnetic charge, should be absolutely stable. A monopole should be created in pairs with its antiparticle (antimonopole) that bears magnetic charge of the opposite sign. Monopole and antimonopole should annihilate into particles, having no magnetic charge. According to the old Big Bang scenario, in the early Universe, when the temperature exceeded the energy threshold of monopole–antimonopole pair production, i.e. at $T \gg m$, where m is the monopole mass, these pairs should have been in the equilibrium with the relativistic plasma. When in the course of expansion the temperature fell down below m, monopole abundance should have been frozen out. It should have happened when the rate of cosmological expansion exceeded the rate of monopole–antimonopole annihilation. Owing to their absolute stability primordial monopoles (and antimonopoles) should have been retained in the Universe. They should have been present in the modern Universe. Their absence in the terrestrial and lunar matter, in cosmic rays, as well as the very existence of galactic magnetic fields puts severe constraint on the modern abundance of relic magnetic monopoles (see review and references in [3]). t'Hooft [40] and Polyakov [41] have found that the Dirac monopole should inevitably appear as a topologically nontrivial solution of the field equation for the Higgs field that breaks spontaneously a non-Abelian group of symmetry, unifying electromagnetism with other gauge interactions. The necessary condition is that the unifying group of symmetry is compact. The corresponding "hedgehog" solution [41] should have the mass

$$m \sim \frac{\Lambda}{e},\qquad(2.70)$$

where Λ is the energy scale of the symmetry breaking and e is the unit electric charge. In that case, magnetic monopoles cannot exist in the early Universe at the high temperatures, exceeding the critical temperature of symmetry breaking phase transition, i.e. at $T > T_c \sim \Lambda$. In the course of the phase transitions monopoles (and antimonopoles) should be formed as topological defects [37]. All these possible ideas on the monopole productions in the early Universe were taken into account in the calculation [42] of the frozen out concentration of relic monopoles in the Universe. It was shown that the relative frozen out concentration of magnetic monopoles $r = n_m/n_\gamma$, where n_γ is the concentration of relic photons, is given by (see the details in [3])

$$r = \frac{m}{g^5 M_P} \approx 10^{-9}\frac{m}{10^{16}GeV}.\qquad(2.71)$$

Here g is the monopole magnetic charge, given for the the the the the the the the the the the Dirac monopoles by the Eq. (2.69). The result (2.71) was shown to be independent of the mechanism of monopole production, provided that the initial monopole concentration (originated from the equilibrium with plasma or from creation of topological defects in high temperature phase transition) exceeded this value. The U(1) symmetry group of electromagnetism is not embedded within a compact group in the Standard Model, but it is with necessity embedded in such group in GUT models. It makes the existence of magnetic monopoles with the mass $m \sim 10^{16}GeV$ the general consequence of such models. The old Big Bang scenario assumed that the temperature in the early Universe could be as high as the critical temperature for the phase transition with GUT symmetry breaking. Then magnetic monopoles should have been created and their frozen out concentration should have been given by (2.71). Substituting the value of GUT monopole mass $m \sim 10^{16}GeV$ into the Eq. (2.71) one could easily find [43, 44] that the modern concentration of magnetic monopoles should be as high as the concentration of baryons, while their mass is by 16 orders of magnitude higher than the mass of proton! It was just this contradiction that created the problem of magnetic monopole overproduction in the old Big Bang scenario. The inflationary solution for this problem, offered by Guth [52], assumed that the GUT phase transition is strongly first-order. It resulted in inflation, driven by the potential of the GUT Higgs field. Due to inflation the initial concentration of topological defects (monopoles and antimonopoles) is then exponentially small. Due to the supercooling, caused by inflation, the temperature after the phase transition did not approach the GUT critical temperature, which also suppressed the initial monopole concentration much below the value (2.71). These principal features of Guth solution for magnetic monopole problem – inflationary regime in the period of GUT phase transition and low temperature after preheating – are retained in all the models of inflationary cosmology. It makes inflationary cosmology free from the problem of magnetic monopole overproduction. But there are still left the questions on the mechanisms of production and actual abundance of magnetic monopoles in the inflationary Universe (see [3] for review).

5.3 Main properties of the Universe with inflation at the beginning

Let us consider two space points separated from the beginning by a physical distance l_0 in a causally connected region. The last is characterized by the Hubble parameter H, so that

$$l(t = 0) = l_0 < H^{-1},$$

Here and below we suppose the Hubble parameter being a constant for simplicity. The coordinate distance r does not depend on time and the time dependence of these quantities is ruled only by the scale factor $a(t)$

$$l(t) = a(t)r; \quad H(t) \equiv \frac{\dot{a}(t)}{a(t)}. \tag{2.72}$$

Using these formulae one can easily find the ratio $l(t)/H(t)^{-1}$ of the physical distance between the particles and horizon

$$\frac{l(t)}{H^{-1}} = \dot{a}(t)r. \tag{2.73}$$

Scale factor during the inflationary stage is

$$a(t) \cong H^{-1} \exp(Ht)$$

and ratio (2.73) has the form

$$\left(\frac{l(t)}{H(t)^{-1}} \right)_{infl} \cong re^{H_0 t}$$

and we reveal that the size between the two points grows exponentially comparing with the size of horizon H^{-1}. Evidently, these two space points appear to be causally disconnected, at some moment t_1, i.e. $l(t_1) > H(t_1)$ even if they were produced in a causally connected region from the beginning. It is said that the distance between the two points crosses the horizon.

When the inflation is finished, after reheating the scale factor behaves like

$$a(t) \propto t^\beta,$$

where the parameter is in the range $0 < \beta < 1$ for any stage which takes place after the inflation (see formulae (2.55), (2.56)). According to formula (2.73) we have

$$\left(\frac{l(t)}{H(t)^{-1}} \right)_{FRW} \propto t^{\beta-1}, \quad 0 < \beta < 1$$

and hence the size of the horizon $H(t)^{-1}$ grows with time quicker than the distance between the two points in the FRW Universe. We come to a very important feature of the Universe with inflation at a first stage – any extensive phenomenon produced at the inflationary stage stretches its size far from the horizon. After the inflation, the size of the horizon is increased quicker compared with the space scale of the phenomenon. Thus, there is some time, say t_2, when the horizon crosses the scale l for the second time and the points are included again within a causally connected area.

Chapter 3

QUANTUM FLUCTUATIONS
DURING INFLATION

Quantum field theory teaches us that a classical motion of a system is disturbed by quantum fluctuations. In the Minkowski space their role is rather weak because quantum corrections are proportional to Planck constant $\hbar$. In addition, according to Heisenberg's uncertainty principle, the larger the fluctuation, the smaller time it exists. A much more interesting picture was discovered in the inflationary stage. As it was shown in Chapter 2, this stage may be approximated by de Sitter space. The most important property of inflation is that any inhomogeneity grows in space, going far beyond the horizon size. The fluctuations are also the specific sort of inhomogeneities. It seems reasonable that their destiny differs from the destiny of the fluctuations in Minkowski space. In de Sitter space, quantum fluctuations do not die out. On the contrary, their size in space increases exponentially as compared with the size of horizon and they contribute to classical constituent of the field. This process reminds us of a pair creation in strong fields. The energy is conserved due to work produced by the field. In our case this field is evidently a gravitational one. In this chapter we consider shortly important results on quantum fluctuations during the inflationary stage that are supported by modern observations.

1. Birth of quantum fluctuations

The simplest and widespread considered case is scalar field φ which is governed by equation

$$\ddot{\varphi} + 3H\dot{\varphi} - e^{-2Ht}\Delta\varphi + V'(\varphi) = 0. \tag{3.1}$$

The Eq. (3.1) follows from equations (2.22), (2.22) and (2.14) after simple calculations. Slow variation of Hubble parameter $H = \dot{a}/a$ during the inflationary stage is neglected. To proceed, let us decompose the field into a "classical" – Φ – and a "quantum" – q – part

$$\varphi(\mathbf{x}, t) = \Phi(\mathbf{x}, t) + q(\mathbf{x}, t). \tag{3.2}$$

This decomposition is quite conventional. Classical part Φ is associated with smooth, slow motion of a field. The most natural way to extract the classical, coarse-grained part is to associate it with small momenta $\mathbf{k}$. Such a problem can be solved for example by Fourier transform

$$\varphi(\mathbf{x}, t) = \int \frac{d^3 k}{(2\pi)^{3/2}} \left[a_{\mathbf{k}}(t) e^{-i(\mathbf{kx})} + a_{\mathbf{k}}^{\dagger}(t) e^{i(\mathbf{kx})} \right]; \qquad (3.3)$$

$$\Phi(\mathbf{x}, t) = \int_{|\mathbf{k}| < k^*} \frac{d^3 k}{(2\pi)^{3/2}} \left[a_{\mathbf{k}}(t) e^{-i(\mathbf{kx})} + a_{\mathbf{k}}^{\dagger}(t) e^{i(\mathbf{kx})} \right],$$

$$q(\mathbf{x}, t) = \int_{|\mathbf{k}| > k^*} \frac{d^3 k}{(2\pi)^{3/2}} \left[a_{\mathbf{k}}(t) e^{-i(\mathbf{kx})} + a_{\mathbf{k}}^{\dagger}(t) e^{i(\mathbf{kx})} \right],$$

with a suitably chosen boundary momentum k^*. The last is specific for each problem. Functions $a_{\mathbf{k}}(t) \sim e^{ik_0 t}$ in Minkowski space, but simple plane wave basis is not appropriate for de Sitter space. Thus our nearest aim is to choose the proper basis. The most natural one is a set of solutions of Laplace equation

$$\Box g(\mathbf{x}, t) = 0$$

in de Sitter space. This equation coincides with Eq. (3.1) in its comprehensive form for zero potential V. After Fourier transformation

$$\tilde{g}_{\mathbf{p}}(t) = \int d^3 x e^{i(\mathbf{px})} g(\mathbf{x}, t)$$

the equation acquires the form

$$\frac{\partial^2 \tilde{g}_{\mathbf{p}}(t)}{\partial t^2} + 3H \frac{\partial \tilde{g}_{\mathbf{p}}(t)}{\partial t} + (H\mathbf{p})^2 e^{-2Ht} \tilde{g}_{\mathbf{p}}(t) = 0, \qquad (3.4)$$

where H is the Hubble parameter. Note that momentum $\mathbf{p}$ is dimensionless, as well as the comoving coordinates $\mathbf{x}$. Very often it is substituted by a value with proper dimension $\mathbf{P} \equiv H\mathbf{p}$. The set of solutions to Eq. (3.4) can be expressed in terms of Hankel functions [57]

$$H_{3/2}^{(2)}(y) = \left[H_{3/2}^{(1)}(y) \right]^* = -\sqrt{\frac{2}{\pi y}} e^{-iy} \left(1 + \frac{1}{iy} \right).$$

One of the most suitable solutions has the form

$$\tilde{g}_{\mathbf{p}}(t) = \frac{\sqrt{\pi}}{2} H \eta^{3/2} \left[c_1(p) H_{3/2}^{(1)}(\eta P) + c_2(p) H_{3/2}^{(2)}(\eta P) \right].$$

Here 'conformal' time

$$\eta = -H^{-1} e^{-Ht}, \qquad (3.5)$$

was introduced. This variable often simplifies equations and is used widely in those analytical calculations where de Sitter space plays a significant role.

The solution contains two unknown constants which would be determined by auxiliary conditions. They could be chosen in such a way to coincide with the Minkowski case at $p \to \infty$, i.e. at small distances: $c_1 = 0, c_2 = -1$. Finally, the set of orthonormal functions in the comoving frame is [57]

$$\tilde{g}_{\mathbf{p}}(t) = \frac{iH}{P^{3/2}\sqrt{2}} \left(1 + \frac{P}{iH}e^{-Ht}\right) \exp\left(\frac{iP}{H}e^{-Ht}\right). \tag{3.6}$$

If a fluctuation has characteristic (comoving) size l, the functions of set (3.6) with momenta $p \sim 1/l$ dominate in the contribution to the Fourier transformation of the field configuration that represents the fluctuation. So it is worth discussing the time and momentum dependence of this function. The time behavior of the function $\tilde{g}_{\mathbf{p}}(t)$ depends on a momentum interval. One can easily see that a threshold value of the momentum is

$$P^*(t) \equiv He^{Ht}. \tag{3.7}$$

More definitely, oscillations die out at small momenta $P << P^*$ and the function tends to constant

$$\tilde{g}_{\mathbf{p}}(t) \simeq \frac{iH}{P^{3/2}\sqrt{2}}, \quad P << P^*, \tag{3.8}$$

while for large momenta $P >> P^*$ oscillations still take place. To clarify the physical meaning of the picture described above, let us express all values in terms of physical coordinates, $\mathbf{R}_{phys} = a(t)\mathbf{r}$. Evidently, physical momentum $\mathbf{P}_{phys}$ is connected with comoving one, $\mathbf{p}$ as follows, $\mathbf{P}_{phys} = \mathbf{p}/a(t) = \mathbf{P}/(a(t)H)$. In this chapter we are working in pure de Sitter space and the scale factor $a(t)$ was chosen in the form $a(t) = H^{-1}e^{Ht}$. The threshold value of physical momentum P^*_{phys} expressed in terms of physical values is

$$P^*_{phys} \equiv \frac{P^*}{a(t)H} = H.$$

This very simple and at the same time important result indicates that a value of any fluctuation does not vary if their characteristic size $L_{phys} \sim P^{-1}_{phys}$ is greater than the horizon size H^{-1}. Time dependence of any quantum fluctuation could be determined qualitatively using expression (3.6). It can be easily seen that an amplitude of the quantum fluctuations with arbitrary momenta tends to constant (3.8) with time. On the other hand, the physical size L_{phys} of fluctuation grows exponentially

$$L_{phys} \sim P^{-1}_{phys} = \frac{a(t)H}{P} = \frac{1}{P}e^{Ht}. \tag{3.9}$$

This behavior is rather different from that in the Minkowski space where the lifetime of the fluctuations is about $1/\Delta E$ (ΔE is an energy of the fluctuation).

In de Sitter space, quantum fluctuation exponentially stretches its size according to expression (3.9). At the same time, its amplitude is determined by expression (3.8). Now we can come back to the problem of the partition of the scalar field φ into classical and quantum parts. Following [56], its quantum part – see (3.3) – can be taken in the form

$$q(\mathbf{x},t) \equiv \int \frac{d^3\mathbf{p}}{(2\pi)^{3/2}} W(P,t) \left[\hat{a}_\mathbf{p} \tilde{g}_\mathbf{p}(t) e^{-i(\mathbf{px})} + \hat{a}_\mathbf{p}^\dagger \tilde{g}_\mathbf{p}^*(t) e^{i(\mathbf{px})} \right]. \qquad (3.10)$$

Here we inserted the creation and annihilation operators $\hat{a}_\mathbf{p}^\dagger, \hat{a}_\mathbf{p}$ as in the standard method of quantization. Instead of cutting off momenta by condition $P > P^*$ we use the function $W(P,t)$ with properties $W(P \to 0,t) \to 0; W(P \to \infty,t) \to 1$. A suitable form is

$$W(P,t) = \theta\left(P - \varepsilon H e^{Ht}\right); \quad \varepsilon \ll 1. \qquad (3.11)$$

As we will see later, final physical results do not depend on small but arbitrary value of the parameter ε. Substituting expressions (3.2) and (3.10) into Eq. (3.1) we obtain

$$\frac{\partial \Phi}{\partial t} - \frac{1}{3H} \left[e^{-2Ht} \Delta\Phi - \frac{\partial V(\Phi)}{\partial \Phi} \right] = y(\mathbf{x},t); \qquad (3.12)$$

$$y(\mathbf{x},t) \equiv \left(\frac{1}{3H} \frac{\partial^2}{\partial t^2} - \frac{\partial}{\partial t} + \frac{1}{3H} e^{-2Ht} \Delta \right) q(\mathbf{x},t).$$

This equation was simplified: we have omitted second time derivative due to slow roll approximation and neglected higher powers of function $y(\mathbf{x},t)$. Eq. (3.12) describes classical motion of the field Φ under permanent influence of random 'force' y. The last is supposed to be small so that we may find a solution to the equation in a form [249]

$$\Phi = \Phi_{\text{det}} + \phi. \qquad (3.13)$$

The deterministic part of the classical field Φ_{det} is governed by the equation

$$\frac{\partial \Phi_{\text{det}}}{\partial t} - \frac{1}{3H} \left[e^{-2Ht} \Delta\Phi_{\text{det}} - \frac{\partial V(\Phi_{\text{det}})}{\partial \Phi_{\text{det}}} \right] = 0, \qquad (3.14)$$

while its random part ϕ depends strictly on the quantum fluctuations according to linear equation

$$\frac{\partial \phi}{\partial t} - \frac{1}{3H} \left[e^{-2Ht} \Delta\phi - V''(\Phi_{\text{det}})\phi \right] = y(\mathbf{x},t) \qquad (3.15)$$

(here we consider a limit $\Phi_{\text{det}} \gg \phi$ that is valid if random 'force' $y(\mathbf{x},t)$ is small). Performing calculations of the random 'force' caused by quantum fluctuations as follows (see [249] for details)

$$
\begin{aligned}
y(\mathbf{x},t) &\equiv \left(\frac{1}{3H}\frac{\partial^2}{\partial t^2} - \frac{\partial}{\partial t} + \frac{1}{3H}e^{-2Ht}\Delta\right)q(\mathbf{x},t) = \\
&= \left(\frac{1}{3H}\frac{\partial^2}{\partial t^2} - \frac{\partial}{\partial t} + \frac{1}{3H}e^{-2Ht}\Delta\right)\int\frac{d^3\mathbf{p}}{(2\pi)^{3/2}}\theta\left(P - \varepsilon He^{Ht}\right)\cdot \\
&\quad \left[\hat{a}_{\mathbf{p}}\tilde{g}_{\mathbf{p}}(t)e^{-i(\mathbf{px})} + \hat{a}_{\mathbf{p}}^\dagger\tilde{g}_{\mathbf{p}}^*(t)e^{i(\mathbf{px})}\right] \simeq \\
&\simeq i\frac{H^3\varepsilon}{\sqrt{2}}e^{Ht}\int\frac{d^3p}{(2\pi p)^{3/2}}\delta\left(P - \varepsilon He^{Ht}\right)\left[\hat{a}_{\mathbf{p}}e^{-i(\mathbf{px})} - \hat{a}_{\mathbf{p}}^\dagger e^{i(\mathbf{px})}\right].
\end{aligned} \tag{3.16}
$$

Here we used Eq. (3.4) what simplified the expression significantly. Second time derivative is small being proportional to ε^2. The value $P = \sqrt{\mathbf{P}^2} = \sqrt{(\mathbf{p}H)^2}$. Approximation (3.8) was taken into account in the last line. Validity of this approximation is justified by smallness of momenta due to the argument of δ-function. Another important value is the correlator

$$
D(\mathbf{x},t,\mathbf{x}',t') \equiv \left\langle 0\left|y(\mathbf{x},t),y(\mathbf{x}',t')\right|0\right\rangle.
$$

Using expression (3.16) obtained above and properties of the creation and annihilation operators $a, a^\dagger$ one can easily obtain analytical expression for this value

$$
D(\mathbf{x},t,\mathbf{x}',t') = \frac{H^3}{4\pi^2}\delta(t-t')\frac{\sin\varepsilon|\mathbf{x}-\mathbf{x}'|e^{Ht}}{\varepsilon|\mathbf{x}-\mathbf{x}'|e^{Ht}}, \quad \varepsilon \ll 1. \tag{3.17}
$$

Uniform case

According to expression (3.17), the correlator $D(\mathbf{x},t,\mathbf{x}',t')$ appears to be a very sharp function of distance $|\mathbf{x}-\mathbf{x}'|$. The same reason permits us to neglect spatial derivatives in Eq. (3.15) and we come to much more simple equations without spatial dependence. Thus uniform distribution, $\Phi = \Phi(t)$ has physical meaning to be considered. It is governed by more simple equation

$$
\frac{\partial\Phi}{\partial t} + \frac{1}{3H}\frac{\partial V(\Phi)}{\partial\Phi} = 0, \tag{3.18}
$$

$$
\frac{\partial\phi}{\partial t} + \frac{m^2}{3H}\phi = y(t). \tag{3.19}
$$

Here we have denoted

$$
m^2 \equiv V''(\Phi_{\text{det}}).
$$

This value is strictly constant for the simplest form of the potential

$$
V(\phi) = V_0 + \frac{1}{2}m^2\phi^2.
$$

and it is almost constant during inflation in more general cases. The correlator (3.17) of random function $y(t)$ may be approximated as follows

$$\langle y(t_1)y(t_2) \rangle = D(\mathbf{x}, t, \mathbf{x}, t') = \frac{H}{4\pi^2}\delta(t_1 - t_2)$$

in the limit $\varepsilon \ll 1$. Delta function in the right-hand side of this expression indicates that random function $y(t)$ is distributed according to Gauss's law with the density

$$W(y) = Const\, exp\left[-\frac{1}{2\sigma^2}\int y(t)^2 dt\right], \qquad \sigma = \frac{H^{3/2}}{2\pi}.$$

Probability distribution of function ϕ is proportional to that of function $y(t)$ due to their linear relationship given by (3.19). It means that the probability to find function $\phi(t)$ inside some small interval is equal to, see [250]

$$dP(\phi) = Const \cdot \mathrm{D}\phi\, exp\left[-\frac{1}{2\sigma^2}\int\left[\frac{\partial\phi}{\partial t} + \frac{m^2}{3H}\phi\right]^2 dt\right].$$

The measure $\mathrm{D}\phi \equiv \prod_{i=1}^{N} d\phi(t_i)$, $N \to \infty$. Now we are ready to obtain the probability to find field value ϕ_2 at an instant t_2 provided a value ϕ_1 at an instant t_1 is known. Evidently, we have to integrate over all values of the field inside the time interval (t_1, t_2) except the values $\phi_1 \equiv \phi(t_1)$, $\phi_2 \equiv \phi(t_2)$ and come to the expression

$$dP(\phi_2, t_2; \phi_1, t_1) = \tag{3.20}$$

$$Const \cdot d\phi_2 \int_{\phi_1}^{\phi_2} \mathrm{D}\phi\, exp\left[-\frac{1}{2\sigma^2}\int_{t_1}^{t_2}\left[\frac{\partial\phi}{\partial t} + \frac{m^2}{3H}\phi\right]^2 dt\right].$$

The constant factor in this equation is determined by normalization condition

$$\int_{-\infty}^{\infty} d\phi_2 P(\phi_2, t_2; \phi_1, t_1) = 1.$$

Functional integral (3.20) can be calculated exactly in a standard manner [250] by finding extremal trajectory of the integral in the exponent

$$\ddot{\phi} - \mu^2\phi = 0; \qquad \mu \equiv \frac{m^2}{3H}$$

with boundary conditions

$$\phi(t_1) = \phi_1; \phi(t_2) = \phi_2.$$

Solution to this equation is

$$\phi(t) = Ae^{\mu t} + Be^{-\mu t}$$

$$A = \frac{\phi_2 - \phi_1 e^{-\mu T}}{2sh(\mu T)}, \qquad B = \frac{-\phi_2 + \phi_1 e^{\mu T}}{2sh(\mu T)}; \qquad T = (t_2 - t_1)$$

Substituting it into the integral in the exponent of expression (3.20) one obtains desired probability

$$dP(\phi_2, t_1 + T; \phi_1, t_1) = d\phi_2 \cdot \sqrt{\frac{r}{\pi}} \exp\left[-r\left(\phi_2 - \phi_1 e^{-\mu T}\right)^2\right],$$

$$r \equiv \frac{\mu}{\sigma^2} \frac{1}{1 - e^{-2\mu T}};$$

$$\mu = \frac{m^2}{3H} \simeq Const; \sigma = \frac{H^{3/2}}{2\pi} \simeq Const. \tag{3.21}$$

In the limit of massless field we obtain more simple formula

$$dP(\phi_2, t_1 + T; \phi_1, t_1) = d\phi_2 \sqrt{\frac{2\pi}{H^3 T}} \exp\left[-\frac{2\pi^2}{H^3 T}\left(\phi_2 - \phi_1\right)^2\right]. \tag{3.22}$$

This formula is widely used in the inflation scenarios where a motion of fields must be slow and hence the second derivative of potential is negligible.

The picture of field evolution looks as follows. The field consists of two parts, see expression (3.13). Deterministic part Φ_{det} moves according to classical equation of motion (3.18) and is permanently disturbed by random 'force' y. As it is shown above, its influence is described by random part ϕ that is distributed with probability density (3.21). One can calculate average value $<\phi^2>$ to estimate a deviation from Φ_{det} with time. Explicit form of the probability (3.21) permits doing it rather easily with the answer

$$<\phi(t)^2> = \int_{-\infty}^{\infty} \phi^2 dP(\phi, t_1 + t; \phi_1, t_1) = \tag{3.23}$$

$$= \frac{1}{2r} = \frac{\sigma^2}{2\mu}\left(1 - e^{2\mu t}\right)$$

This formula can be significantly simplified in the case of massless field. It is often fulfilled approximately during inflation because the latter takes place only if $m << H$. Expanding the exponent in Eq. (3.23) we come to the result

$$\sqrt{<\phi(t)^2>} = \sigma = \frac{H}{2\pi}\sqrt{Ht}. \tag{3.24}$$

In terms of *e*-folds, $N \equiv Ht$ we obtain formula that will be used widely in the following

$$\sqrt{<\phi(t)^2>} = \sigma = \frac{H}{2\pi}\sqrt{N}. \tag{3.25}$$

As a particular result, one can conclude that a fluctuation with an amplitude $\sim H/2\pi$ is formed in time interval $t \sim H^{-1}$, $(N = 1)$. The expression (3.7) gives us the moment when the fluctuation ceases its variation, so that its space size does not vary after the same time interval $t \sim H^{-1}$.

The space size of this fluctuation could also be estimated. For this, one has to notice that correlator (3.17) is not small at $t \sim H^{-1}$ if the comoving distance

$|\mathbf{x} - \mathbf{x}'| \lesssim 1$. It means that the fluctuations with comoving size of the order of unity are essential. Their physical size grows with time in an ordinary manner

$$L_{fluct} = a(t)|\mathbf{x} - \mathbf{x}'| \approx H^{-1}e^{Ht} \tag{3.26}$$

and it is equal H^{-1} at time $t \sim H^{-1}$. This point is very important for the further applications. Note that results (3.25) and (3.26) could be approximately reproduced rather easily even in the Minkowski space. Indeed, starting from Lagrangian (2.21) for massless field, the estimation of action

$$S = \int d^4x\sqrt{-g}\frac{1}{2}g^{\mu\nu}\partial_\mu\varphi\partial_\nu\varphi$$

looks as follows.

$$S \sim \frac{(\Delta\varphi)^2}{H^{-2}}H^{-4}.$$

Here we denote as $\Delta\varphi$ the fluctuation that is formed during time interval $t \sim H^{-1}$. It was taken into account also that a size of the fluctuation is of order H^{-1} for massless field distributed with speed of light. The probability of such a fluctuation is not small if the action $S \sim 1$ and we come to the estimation

$$\Delta\varphi \sim H$$

which coincides with more accurate result (3.25) in order of magnitude. There is another way of performing the calculations which may be often met in literature. If one notes that equation (3.19) is represented by nothing but Langevin equation it immediately follows that probability distribution must satisfy Fokker–Planck equation

$$\frac{\partial P}{\partial t} = \frac{H^3}{8\pi^2}\frac{\partial^2 P}{\partial\phi^2} + \frac{m^2}{3H^2}\frac{\partial^2(\phi P)}{\partial\phi^2}. \tag{3.27}$$

Our expression (3.21) for probability is the solution of this equation ($dP = Pd\phi$ in our notations).

2.   Classical evolution of quantum fluctuations

One of the conclusions of previous sections is the following. A density fluctuation, being produced by quantum fluctuation, sharply increases its size. During inflation, an amplitude of the fluctuation evolves independently after its size prevails on the horizon. It takes place up to an instant of second crossing of the horizon which happens after the end of inflation. Let us briefly discuss evolution of fluctuations between the two crossings of a horizon. As a result of profound discussion, a relation was established that is used widely in modern literature

$$\frac{\delta\rho}{p+\rho} \simeq Const \tag{3.28}$$

This relation is correct during the period of evolution of the fluctuation between two horizon crossings. Here we show a simple approximate way to obtain this formula. To this end let us start with equation (2.12) in the form

$$\frac{d\rho}{p + \rho} = -\frac{3da}{a}.$$

Suppose that one-to-one correspondence $\rho \leftrightarrow p$ exists (may be different for different intervals). Hence, the pressure is some function $p(\rho)$ and we can integrate this equation

$$\int_{\rho(t_0)}^{\rho(t)} \frac{d\rho}{p(\rho) + \rho} = -3 \int_{a(t_0)}^{a(t)} \frac{da}{a}, \tag{3.29}$$

where t_0 is some moment during inflation stage and t is an arbitrary moment such that $t - t_0 \gg 1/H_e$, H_e is the Hubble parameter at the end of inflation. We would like to warn the reader that the last expression is not an exact one. The fact is that there is no one-to-one correspondence $\rho \leftrightarrow p$ during short periods of first-order phase transitions that may have taken place in the past as possible cosmological effects of particle symmetry breaking. If quantum fluctuation produced energy density perturbation $\delta\rho(t_0)$ in the manner discussed in the previous section, the last equation must be rewritten for that space domain which was occupied by the fluctuation

$$\int_{\rho(t_0)+\delta\rho(t_0)}^{\rho(t)+\delta\rho(t)} \frac{d\rho}{p(\rho) + \rho} = -3 \int_{a(t_0)+\delta a(t_0)}^{a(t)+\delta a(t)} \frac{da}{a}. \tag{3.30}$$

Now let us attribute to t_0 and t the meaning of the first and the second crossing of the horizon by this fluctuation. One could expand both sides of Eq. (3.30) into a sum of three integrals to obtain

$$\int_{\rho(t_0)+\delta\rho(t_0)}^{\rho(t_0)} \frac{d\rho}{p(\rho) + \rho} + \int_{\rho(t)}^{\rho(t)+\delta\rho(t)} \frac{d\rho}{p(\rho) + \rho} =$$
$$-3 \int_{a(t_0)+\delta a(t_0)}^{a(t_0)} \frac{da}{a} - 3 \int_{a(t)}^{a(t)+\delta a(t)} \frac{da}{a}.$$

Here Eq. (3.29), valid for the volume of larger size, was taken into account to cancel third integrals in both sides of the equations. Taking in mind smallness of the fluctuations, these integrals can be easily estimated and we come to equation

$$\left(\frac{\delta\rho}{p + \rho} + 3\frac{da}{a} \right)_{t_0} \simeq \left(\frac{\delta\rho}{p + \rho} + 3\frac{da}{a} \right)_{t}.$$

Scale factor a grows very quickly so that we could neglect second terms in both sides of this equation thus coming to the desired equation (3.28). Formula

(3.28) written in the form

$$\left(\frac{\delta\rho}{p+\rho}\right)_{t_f} \simeq \left(\frac{\delta\rho}{p+\rho}\right)_{t_{in}} \tag{3.31}$$

could be useful to obtain characteristics of modern large scale fluctuation. Here t_{in} is time of the fluctuation formation at inflationary stage and its first crossing of the horizon, t_f is the time of the second horizon crossing at Friedmann–Robertson–Walker stage. The left-hand side of this equality can be expressed in terms of inflationary parameters and variables while the right-hand side is simplified at the matter-dominated stage due to the absence of pressure, $p = 0$. Expressions for $\delta\rho$ and $p + \rho$ may be easily written in the inflationary stage in terms of the scalar field (inflaton) $\varphi = \varphi(t_{in})$

$$\delta\rho = V'(\varphi)\delta\varphi,$$
$$p + \rho = \dot{\varphi}^2.$$

The quantity $\delta\rho/\rho|_{t_f}$ which is important to evaluate is now expressed in terms of inflaton field

$$\left(\frac{\delta\rho}{\rho}\right)_{t_f} \simeq \left(\frac{V'(\varphi)\delta\varphi}{\dot{\varphi}^2}\right)_{t_{in}}$$

These estimations are not exact and it is enough to limit ourselves with the massless case for estimation of fluctuations – $\delta\varphi \approx H(\varphi)/2\pi$. Using equations of motion for the inflaton field we obtain

$$\left(\frac{\delta\rho}{\rho}\right)_{t_f} = \left(\frac{9H(\varphi)^3}{2\pi V'(\varphi)}\right)_{t_{in}}. \tag{3.32}$$

On the inflationary stage the Hubble parameter H depends on the potential $V(\varphi)$ and we have the formula for calculation of the amplitude of energy density fluctuation.

$$\left(\frac{\delta\rho}{\rho}\right)_{t_f} = \frac{9}{5}\left(\frac{8}{3}\right)^{3/2}\frac{\sqrt{\lambda}}{v}\varphi^{\frac{v+2}{2}} \tag{3.33}$$

The fluctuation is characterized by the value of the field φ at the moment of first crossing of horizon. In this formula we have included the factor $2/5$ to take into account the fact that the Universe was in matter-dominated stage during second horizon crossing [57] though exact value of the numerical factor is not important.

Another important parameter of the fluctuations is their size, l. The last is connected with the moments t_{in} and t_f of the first and the second horizon crossings correspondingly. The evaluation of the size of fluctuation for time t such that $t_{in} < t < t_f$ could be done by normalizing to the size of our Universe, L_U. Namely,

$$l = L_U \exp(N - N_U), \tag{3.34}$$

where N is number of e-folds for the scale l during inflation, index "U" denotes the scale of our Universe, $L_U \approx 10^4 Mpc \approx 10^{28} cm$.

On the other hand, the number of e-folds is expressed in terms of the inflaton field φ according to Eq. (2.32)

$$N \approx \frac{4\pi}{M_P^2 v} \varphi^2,$$

where potential is chosen in the form $V(\varphi) = \lambda \varphi^v$ and the inequality $\varphi \equiv \varphi_{in} >> \varphi_f$ is supposed. Substituting this formula and formula (3.34) into the expression (3.32) we come to the connection of the amplitude of the fluctuation $\left(\frac{\delta\rho}{\rho}\right)_{t_f}$ and its size l [251],

$$\left(\frac{\delta\rho}{\rho}\right)_{t_f} = \left(\frac{\delta\rho}{\rho}\right)_{t_U} \left(1 + \frac{1}{N_U} \ln\left(l/L_U\right)\right)^{\frac{v+2}{4}}.$$

The second term in brackets is small compared with unity if one takes into account that $N_U \approx 60$ and $l/L_U \geq 0.01$ (only large scales are considered). We came to an important conclusion that the amplitude of large-scale fluctuations depends on their size very weakly. It is said that the spectrum of the fluctuations is almost flat.

Chapter 4

STRONG PRIMORDIAL INHOMOGENEITIES AND GALAXY FORMATION

The modern theory of the cosmological large-scale structure is based on the assumption that this structure is formed as the result of development of gravitational instability from small initial perturbations of density or gravitational potential. As a rule, these perturbations are Gaussian, but some versions of non-Gaussian perturbations are also discussed.

In this chapter we first analyze the problem, inherent to practically all the cosmological cold dark matter models of invisible axion, that concerns primordial inhomogeneity in the distribution of the energy of coherent oscillations of the axion field. This problem, referred to as the problem of *archioles*, invokes a non-Gaussian component in the initial perturbations for axionic cold dark matter.

Archioles are the formation that represents a replica of percolation Brownian vacuum structure of axionic walls bounded by strings, which is fixed in the strongly inhomogeneous primeval distribution of cold dark matter. They reflect the unstable structure of topological defects, arising in the succession of phase transitions in which symmetry of vacuum state changes. Such phase transitions, resulting in formation and de-formation of topological defects, do not necessarily mean the existence of high temperature stage, on which the symmetry is restored. So, the structure of archioles can appear in the result of non-thermal symmetry breaking effects on the post-inflationary preheating stage of inflaton field oscillations.

Non-thermal phase transitions on inflationary stage can lead to spikes in the amplitude of density fluctuations. Primordial black holes of arbitrary large mass can originate from such spikes. Moreover, even in the absence of spikes in the spectrum of density fluctuations, symmetry breaking on inflationary stage can give rise to interesting alternative scenarios of structure formation that relate the mechanism responsible for galaxy formation to unstable large-scale structures of topological defects.

Such new mechanism describing the formation of protogalaxies is considered in this chapter. It is based on the second order phase transition at the inflationary stage and on the mechanism of a domain wall formation upon the end of inflation. It leads to the formation of massive black hole clusters that can serve as nuclei for the future galaxies. The number of black holes with the mass $M \sim 10^3$ Solar masses and more could be comparable with the number of Galaxies in the visible part of the Universe.

The discussed mechanisms shed new light on the problem of primordial black hole formation. Widespread opinion is that this process should take place within the cosmological horizon. Since the mass within it is small in the very early Universe, it seems to imply the smallness of black hole masses formed at this stage of the cosmological evolution. However, if the appropriate conditions for black hole formation are originated during inflationary stage (spikes in the spectrum of density fluctuations or closed domain walls), they could be extended to much larger scales than the scale of cosmological horizon. It makes possible to form primordial black holes of arbitrary large mass (with appropriate values of the model parameters) during FRW stages before galaxy formation.

We have to underline that the following discussion leads to existence of rather massive black holes before star formation. It will be shown that this mechanism of black hole formation could be realized in most modern models of inflation. Primordial fractal structure of galaxies is predicted in the framework of developed models. This approach gives basis for a new scenario of the galaxy formation in the Big Bang Universe.

The discussion of physical basis for these scenarios begins our systematic treatment of strong primordial inhomogeneities which can appear in the inflationary Universe as the reflection of particle symmetry breaking pattern.

1.　　Primordial archiole structure

1.1　　Formation of archioles at high temperature

In the standard invisible axion scenario [9] the breaking of the Peccei–Quinn symmetry is induced by the complex $SU(3) \otimes SU(2) \otimes U(1)$ – singlet Higgs field ϕ with a "Mexican hat" potential

$$V(\phi) = \frac{\lambda}{2} \left(\phi^+ \phi - F_a^2 \right)^2. \qquad (4.1)$$

Such field can be represented as $\phi = F_a \exp(i\vartheta)$, where $\vartheta = a/F_a$ and a is the angular Goldstone mode – axion. QCD instanton effects remove the vacuum degeneracy and induce effective potential for ϑ

$$V(\vartheta) = \Lambda_1^4 (1 - \cos(\vartheta N)). \qquad (4.2)$$

Below, following [78, 217], we will simply assume for the standard axion that $N = 1$ and $\Lambda_1 = \Lambda_{QCD}$. In the context of the Big Bang scenario it is usu-

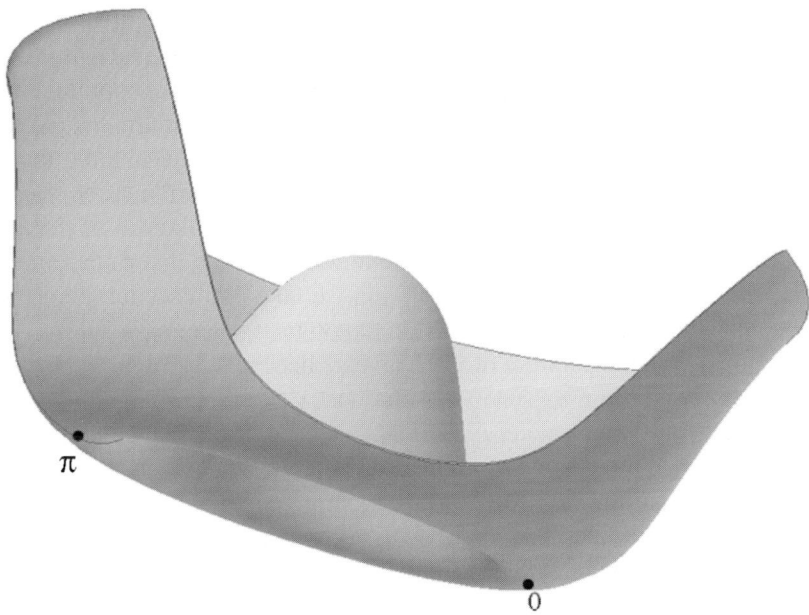

π

0

Figure 4.1. Mexican hat potential, the view from a side. The tilt provides maximum of the potential when moving along the circle valley.

ally assumed that the phase transition with $U(1)$ – symmetry breaking occurs when the Universe cools below the temperature $T \cong F_a$. Thus, in this case the crucial assumption is that from the moment of the PQ phase transition and all the way down to the temperatures $T \cong \Lambda_{QCD}$, the bottom of the potential (4.1) is exactly flat and there is no preferred value of a during this period (the term given by (4.2) vanishes). Consequently, at the moment of the QCD phase transition, when the instanton effects remove vacuum degeneracy, a rolls to the minimum and starts coherent oscillations (CO) about it with energy density [9, 217]

$$\rho_a(T, \vartheta) = 19.57 \left(\frac{T_1^2 m_a}{M_P} \right) \left(\frac{T}{T_1} \right)^3 T^2 F_a^2. \qquad (4.3)$$

The coherent axion field oscillations turn on at the moment $\tilde{t} \approx 8.8 \cdot 10^{-7}$s.

Note, that the existence of the term (4.2) at $T \gg \Lambda_{QCD}$ could remove vacuum degeneracy and switch on axion field oscillations much earlier. The condition $H \sim m_a$, where m_a is the axion mass, is fulfilled in this case at $T \sim \sqrt{M_P m_a}$, and the axion field would have started to move to its true vacuum states, giving rise to its CO.

It is generally assumed, that PQ transition takes place after inflation and the axion field starts oscillations with a different phase in each region, causally connected at $T \cong F_a$, so one has the average over all the values to obtain the modern axion density. Thus, in the standard cosmology of an invisible axion, it is usually assumed that the energy density of coherent oscillations is distributed uniformly and that it corresponds to the averaged phase value of $\bar{\vartheta} = 1$ ($\bar{\rho}_a = \rho(\bar{\vartheta})$). However, the local value of the energy density of coherent oscillations depends on the local phase ϑ that determines the local amplitude of these coherent oscillations. It was first found in [77], that the initial large-scale (LS) inhomogeneity of the distribution of ϑ must be reflected in the distribution of the energy density of coherent oscillations of the axion field. Such LS modulation of the distribution of the phase ϑ and consequently of the energy density of CO appears when we take into account the vacuum structures leading to the system of axion topological defects.

As soon as the temperature of the Universe becomes less than F_a, the field ϕ acquires the vacuum expectation value (VEV) $\langle\phi\rangle = F_a \exp(i\vartheta)$, where ϑ varies smoothly at the scale F_a^{-1}. The existence of noncontractable closed loops that change the phase by $2\pi n$ leads to emergence of axion strings. These strings can be infinite or closed. The numerical simulation of global string formation [243] revealed that about 80% of the length of strings corresponds to infinite Brownian lines. The remaining 20% of this length is contributed by closed loops. Infinite strings form a random Brownian network with the step $L(t) \approx t$. After string formation, when the temperature becomes as low as $T \approx \Lambda_{QCD}$, the term (4.2) makes a significant contribution to the total potential so that the minimum of energy corresponds to a vacuum with $\vartheta = 2\pi k$, where k is an integer – for example, $k = 0$. However, the vacuum value of the phase ϑ cannot be zero everywhere, since the phase must change by $\Delta\vartheta = 2\pi$ upon a loop around a string. Hence, we come from the vacuum with $\vartheta = 0$ to the vacuum with $\vartheta = 2\pi$ as the result of such circumvention. The vacuum value of ϑ is fixed at all points with the exception of the point $\vartheta = \pi$. At this point, a transition from one vacuum to another occurs, and the vacuum axion wall is formed simultaneously with CO turning on.

The width of such wall, bounded by strings, is $\delta \cong m_a^{-1}$. Thus, the initial value of ϑ must be close to π near the wall, and the amplitude of CO in Eq. (4.3) is determined by the difference of the initial local phase $\vartheta(x)$ and the vacuum value, which is different from the one of the true vacuum only in a narrow region within the wall of thickness $\delta \cong m_a^{-1}$. Therefore in this region we can write [77] $\vartheta(x) = \pi - \varepsilon(x)$, where $\varepsilon(x) = 2\tan^{-1}(\exp(m_a x))$ and $x \cong m_a^{-1}$. Thereby the energy density of CO in such regions is given by

$$\rho^A \approx \pi^2 \bar{\rho}_a. \tag{4.4}$$

So we obtain, following [77, 78, 217], that the distribution of CO of the axion field is modulated by nonlinear inhomogeneities in which relative density contrasts are $\delta\rho/\rho > 1$. Such inhomogeneities were called *archioles* [77].

In other words *archioles* are a formation that represents a replica of the percolational Brownian vacuum structure of axionic walls bounded by strings and which is fixed in the strongly inhomogeneous initial distribution of axionic CDM. The scale of this modulation of density distribution exceeds the cosmological horizon because of the presence of 80% infinite component in the structure of axionic walls bounded by strings. The superweakness of the axion field self-interaction results in the separation of archioles from the vacuum structure of axionic walls–bounded–by–strings. So these two structures evolve independently.

The structure of walls, bounded by strings, disappears rapidly due to disintegration into separate fragments and successive axion emission.

The structure of archioles remains frozen at the RD stage. On the large scales, the structure of archioles is an initially nonlinear formation. It is a Brownian network of quasi-one-dimensional filaments of dust-like matter with the step

$$L^A(t) = \lambda \tilde{t}, \tag{4.5}$$

where $\lambda \cong 1$. At the moment of creation $\tilde{t}$, the linear density of this quasilinear filamentary formation is given by

$$\mu_A = \pi^2 \bar{\rho}_a \tilde{t} \delta. \tag{4.6}$$

In accordance with this, the cosmological evolution of archioles in the expanding Universe is reduced to the extension of lines along only one direction.

The spectrum of inhomogeneities that the density develops in response to the large-scale Brownian modulation of the distribution of CO of axion field was studied in [78]. Density perturbations, associated with Brownian network of archioles, may be described in the terms of a two-point autocorrelation function [78]. To obtain such autocorrelation function, it is necessary to perform averaging of energy density of infinite Brownian lines over all lines and over the Winner measure, which corresponds to the position along a Brownian line (see [78, 217]).

The two-point autocorrelation function in the Fourier representation has the form [78, 217]

$$\langle \frac{\delta \rho}{\rho_0}(\vec{k}) \frac{\delta \rho}{\rho_0}(\vec{k}') \rangle = 12 \rho_A \mu_A k^{-2} \delta(\vec{k} + \vec{k}') \tilde{t}^{-1} f^{-2} t^4 G^2, \tag{4.7}$$

where ρ_0 is background density, $f_{MD} = 3/(32\pi)$ for dust-like stage, $f_{RD} = (6\pi)^{-1}$ for RD stage, G is the gravitational constant, ρ_A is the total energy density of the Brownian lines. The mean-square fluctuation of the mass is given by [78, 217]

$$\left(\frac{\delta M}{M} \right)^2 (k, t) = 12 \rho_A \mu_A \tilde{t}^{-1} f^{-2} G^2 k t^4. \tag{4.8}$$

1.2   Cosmological impact of archioles

Let us consider, following [217] a region characterized at instant t by a size l and a density fluctuation Δ. For anisotropy of relic radiation we then obtain

[217]

$$\frac{\delta T}{T} \cong -\Delta \left(\frac{l}{t}\right)^2.$$
(4.9)

If $l = t$, we have $|\delta T/T| \cong |\Delta|$; that is, the anisotropy of relic radiation is equal to the density contrast calculated at the instant when the size of the region is equal to the size of the horizon (Sachs–Wolf effect). To estimate the quadrupole anisotropy that is induced in relic radiation by the structure of archioles, we must find the amplitude of perturbations on the scale of the modern horizon

$$\left(\frac{\delta M}{M}\right)^2 = 2.1 \cdot 10^{-25} \left(\frac{F_a}{10^{10}GeV}\right)^4 \left(\frac{t_{RD}}{1s}\right)^{2/3} \left(\frac{t_{pres}}{1s}\right)^{1/3} (k_{hor}t_{pres}).$$
(4.10)

Thus Sachs–Wolf quadrupole anisotropy of relic radiation induced by archioles will be [217]

$$\frac{\delta T}{T} \cong 2.3 \cdot 10^{-6} \left(\frac{F_a}{10^{10}GeV}\right)^2.$$
(4.11)

According to Relic-1 and COBE data (see for example [253, 254]), the measured quadrupole anisotropy of relic radiation is at the level of

$$\frac{\delta T}{T} \approx 5 \cdot 10^{-6}.$$
(4.12)

If we take into account the uncertainties of the consideration [78, 217] such as the uncertainties in correlation length scale of Brownian network ($\lambda \approx 1 \div 13$) and in temperature dependence of axion mass, we can obtain a constraint [78, 217] on the scale of symmetry breaking in the model of invisible axion

$$F_a \leq 1.5 \cdot 10^{10}GeV \div 4 \cdot 10^9 GeV; \qquad m_a \geq 410\mu eV \div 1500\mu eV. \quad (4.13)$$

This upper limit for F_a is close to the strongest upper limits in [244, 245, 246], obtained by comparing the density of axions from decays of axionic strings with the critical density, but it has an essentially different character.

The point is that the density of axions formed in decays of axionic strings depends critically on the assumption about the spectrum of such axions (see [244, 245]) and on the model of axion radiation from the strings (see [246]). For example, Davis [244] assumed that radiated axions have a maximum wavelength of $\omega(t) \cong t^{-1}$, while Harari and Sikivie [245] have argued that the motion of global strings was overdamped, leading to an axion spectrum emitted from infinite strings or loops with a flat frequency spectrum $\propto k^{-1}$. This leads to an uncertainty factor of $\simeq 100$ in the estimation of the density of axions from strings and to the corresponding uncertainty in the estimated upper limit on F_a

$$F_a \leq 2 \cdot 10^{10}\varsigma GeV; \qquad m_a \geq 300/\varsigma \mu eV. \quad (4.14)$$

Here, $\varsigma = 1$ for the spectrum from Davis [244], and $\varsigma \approx 70$ for the spectrum from Harari and Sikivie [245].

In their treatment of axion radiation from global strings, Battye and Shellard [246] found that the dominant source of axion radiation are string loops rather than long strings, contrary to what was assumed by Davis [244]. This leads to the estimations

$$F_a \leq 6 \cdot 10^{10} GeV \div 1.9 \cdot 10^{11} GeV; \qquad m_a \geq 31\mu eV \div 100\mu eV. \quad (4.15)$$

Arguments of [78, 217] that lead to the constraint Eq. 4.13 are free from these uncertainties, since they have a global string decay model – independent character.

At the smallest scales, corresponding to the horizon in the period $\tilde{t}$, evolution of archioles just in the beginning of axionic CDM dominance in the Universe (at redshifts $z_{MD} \cong 4 \cdot 10^4$) should lead to formation of the smallest gravitationally bound axionic objects with the minimal mass $M \simeq \rho_a \tilde{t}^3 \simeq 10^{-6} M_\odot$ and of typical minimal size $\tilde{t}(1 + z_A)/(1 + z_{MD}) \cong 10^{13} cm$. One can expect the mass distribution of axionic objects at small scale to peak around the minimal mass, so that the existence of halo objects with the mass $(10^{-6} M_\odot \div 10^{-1} M_\odot)$ and size $10^{13} \div 10^{15} cm$ is rather probable, what may have interesting application to the theoretical interpretation of MACHOs microlensing events.

Another interesting aspect of archioles is related with their possible impact on the formation of antimatter domains in the baryon asymmetrical Universe. As it was revealed in [242], the phase $\vartheta(x)$ that determines further the amplitude of axionic CO plays the role of spatial dependent CP-violating phase in the period starting from Peccei–Quinn symmetry breaking phase transition until the axion mass is switched on at $T \approx 1$ GeV. The net phase changes continuously and, if baryosynthesis takes place in the considered period, axion-induced baryosynthesis implies continuous spatial variation of the baryon excess given by [85]:

$$b(x) = A + b \sin \vartheta(x). \quad (4.16)$$

Here A is the baryon excess induced by the constant CP-violating phase, which provides the global baryon asymmetry of the Universe and b is the measure of axion-induced asymmetry. If $b > A$, antibaryon excess is generated along the direction $\vartheta = 3\pi/2$. The stronger the inequality $b > A$, the larger interval of ϑ around the layer $\vartheta = 3\pi/2$ provides generation of antibaryon excess [85]. In the case $b - A = \delta \ll A$ the antibaryon excess is proportional to δ^2 and the relative volume occupied by it is proportional to δ.

The axion-induced antibaryon excess forms the Brownian structure looking like an infinite ribbon along the infinite axion string (see [77, 78]). The minimal width of the ribbon is of the order of horizon in the period of baryosynthesis and is equal to M_P/T_{BS}^2 at $T \approx T_{BS}$. At $T < T_{BS}$ this size experiences red shift and is equal to

$$l_h(T) \approx \frac{M_P}{T_{BS} T}. \quad (4.17)$$

This structure is smoothed by the annihilation at the border of matter and anti-matter domains. When the antibaryon diffusion scale exceeds $l_h(T)$ the infinite structure decays on separated domains. The distribution on domain sizes turns to be strongly model dependent and was calculated in [85]. The possible effect of such domains in the modern Universe will be discussed in Chapter 6.

1.3   Nonthermal effects of symmetry breaking

The inclusion of obtained restriction into the full cosmoparticle analysis can provide detailed quantitative definition of the cosmological scenario, based on the respective particle physics model. Consider, for example, a simple variant of gauge theory of broken family symmetry (TBFS) [238], which is based on the Standard Model of electroweak interactions and QCD, supplemented by spontaneously broken local $SU(3)_H$ symmetry for quark–lepton families. This theory provided natural inclusion of Peccei–Quinn symmetry $U(1)_H \equiv U(1)_{PQ}$, being associated with heavy "horizontal" Higgs fields and it gave natural solution for QCD CP violation problem. The global $U(1)_H$ symmetry breaking results in the existence of axion-like Goldstone boson – archion, a.

TBFS turned out to be a simplest version of the unified theoretical physical quantitative description of all main types of dark matter (HDM–massive neutrinos, axionic CDM and UDM in the form of unstable neutrinos [238, 239, 28]) and the dominant form of the dark matter was basically determined by the scale of the "horizontal" symmetry breaking V_H, being the new fundamental energy scale of the particle theory. For given value of V_H the model defined the relative contribution of hot, cold and unstable dark matter into the total density. Since in the TBFS the scale of horizontal symmetry breaking V_H was associated with F_a, from Eq. (4.13) followed the same upper limit on V_H.

However, this limit assumed that the considered inflationary model permits topological defects and hence archioles formation due to the sufficiently high reheating temperature $T_{RH} \geq V_H$. In the inflationary model, which occurs in TBFS, we can achieve $T_{RH} \sim 10^{10} GeV$.

The "horizontal" phase transitions on inflationary stage lead to the appearance of characteristic spikes in the spectrum of initial density perturbations. These spike-like perturbations, on scales that cross the horizon 60 e-folds before the end of inflation re-enter the horizon during the radiation or matter-dominant stage and could in principle collapse to form primordial black holes.

The minimal interaction of "horizontal" scalars of TBFS $\xi^{(0)}, \xi^{(1)}, \xi^{(2)}$ with inflaton allows us to include them in the effective inflationary potential [255]:

$$V(\phi, \xi^{(0)}, \xi^{(1)}, \xi^{(2)}) = -\frac{m_\phi^2}{2}\phi^2 + \frac{\lambda_\phi}{4}\phi^4 - \sum_{i=0}^{2} \frac{m_i^2}{2}\left(\xi^{(i)}\right)^2 \quad (4.18)$$

$$+ \sum_{i=0}^{2} \frac{\lambda_\xi^{(i)}}{4}\left(\xi^{(i)}\right)^4 + \sum_{i=0}^{2} \frac{v_\xi^2}{2}\phi^2\left(\xi^{(i)}\right)^2.$$

The last term in the potential (4.18) implies the effect of symmetry restoration at large amplitudes of inflaton field, similar to the effect of symmetry restoration at high temperatures, induced by thermal corrections in QFT. At high amplitudes of inflaton field ϕ, VEVs of $\xi^{(i)}$ are zero, but when the inflaton amplitude rolls down below the critical value $\phi_{ci} < m_i/v_\xi$ the transition to the phase with broken symmetry takes place on the inflationary stage. Following [256] such phase transition results in the spike in the spectrum of density fluctuations.

The analysis of processes of primordial black holes formation from density fluctuations, which can be generated by "horizontal" phase transitions at the inflationary stage gave rise to an upper limit on the scale of horizontal symmetry breaking [255]

$$V_H \leq 1.4 \cdot 10^{13} GeV. \tag{4.19}$$

Therefore, the range between the two upper limits Eq. (4.13) and Eq. (4.19) turned out to be not closed, and the following values seem to be possible

$$10^{11} GeV \leq V_H \leq 10^{13} GeV. \tag{4.20}$$

The indicated range corresponds to the case when all the horizontal phase transitions take place on the post-inflationary stage of the inflaton field oscillations and $\phi_{c_2} \ll M_P$. In this case the inflaton field ϕ oscillates with initial amplitude $\sim M_P$. According to [240, 255] it means that any time the amplitude of the field becomes smaller then $\phi_{c_2} \ll M_P$, the last (axion $\xi^{(2)}$) phase transition with symmetry breaking occurs, and topological defects are produced. Then the amplitude of the oscillating field ϕ becomes greater than ϕ_{c_2}, and the symmetry is restored again. However, this regime does not continue too long. Within a few oscillations, quantum fluctuations of the field $\xi^{(2)}$ will be generated with the dispersion $\langle \left(\xi^{(2)} \right)^2 \rangle \simeq v_\xi^{-1} \lambda_\phi^{1/2} M_P^2 \ln^{-2} 1/v_\xi^2$ [240]. For

$$m_2^2 \leq v_\xi^{-1} \lambda_\phi^{1/2} \lambda_\xi M_P^2 \ln^{-2} 1/v_\xi^2 \tag{4.21}$$

these fluctuations will keep the symmetry restored. The symmetry breaking will be finally completed when $\langle \left(\xi^{(2)} \right)^2 \rangle$ will become small enough. Thus such phase transition leads to formation of topological defects and archioles without any need for high-temperature effects.

Substituting the typical values for potential (4.18) such as $m_2^2 \approx 10^{-3} V_H^2$, $\lambda_\xi \simeq 10^{-3}$, $v_\xi \simeq 10^{-10}$, $\lambda_\phi \simeq 10^{-12}$ (see [255]) we will obtain that the condition (4.21) means that for the scales

$$V_H \leq 2M_P \tag{4.22}$$

the phenomenon of non-thermal symmetry restoration takes place in the simplest inflationary scenario based on TBFS. Owing to this phenomenon oscillations of the field $\xi^{(2)}$ do not suppress the topological defects and archioles production for the range (4.20). So the range (4.20) turned out to be closed by comparison of BBBR quadrupole anisotropy, induced by archioles, with the COBE data. As a result, the upper limit on the scale of horizontal symmetry breaking is given by (4.13).

The existence of neutrino oscillations, as indicated by [12, 13, 14, 15], may rule out the simplest version of TBFS, considered here. However, the phenomenon of archioles is general for a wide class of models with broken U(1) symmetry, leading to large-scale primordial inhomogeneity of dark matter in the form of scalar field CO.

Note that formally the limit (4.22) admits the scales as high as the scales of GUT symmetry breaking. However, non-thermal restoration of GUT symmetry at post-inflationary stage would inevitably lead to magnetic monopole over-production (see Chapter 2), which puts constraints on the realization of GUT physics in the framework of inflationary models.

There is another possible form of strong primordial inhomogeneities, arising from the pattern of U(1) symmetry breaking in the inflationary Universe. It is the structure of primordial massive black hole clouds, to which we turn in the successive sections of this chapter.

2.   Massive primordial black holes

Now there is no doubt that the centers of almost all galaxies contain massive black holes [257]. An original explanation of the formation of such supermassive black holes assumes the collapse of a large number of stars in the galaxy centers. However, the mechanism of the galactic nuclei formation is still unclear. According to Veilleux [258], there are serious grounds to believe that the formation of stars and galaxies proceeded simultaneously. Stiavelli [259] considered a model of the galaxy formation around a massive black hole and presented arguments in favor of his model. Each of the two approaches has certain advantages, while being not free of drawbacks.

Below we will consider a new model of very early formation of galactic nuclei from primordial black holes (PBH), which serve as the nucleation centers in the subsequent formation of galaxies. This mechanism may prove to be free from disadvantages inherent in the models based on the concept of a single PBH being a nucleus of the future galaxy. Its foundation is the new mechanism [79] of the PBH formation that opens the possibility of the massive black hole formation in the early Universe. In the framework of this mechanism black holes are formed as a result of a collapse of closed walls arising from the succession of second order phase transitions after the end of inflation. The masses of such black holes may vary within broad limits, up to a level of the order of $\sim 10^6$ Solar masses.

Let us assume that a potential possesses at least two different vacuum states. Then there are two possible distributions of these states in the early Universe. The first possibility is that the Universe contains approximately equal numbers of both states, which is typical for phase transitions at high temperatures or for non-thermal post-inflationary phase transitions, discussed in the previous section. The alternative possibility corresponds to the case when the two vacuum states are formed with different probabilities. In this case, islands of less probable vacuum state surrounded by the sea of another, more probable, vacuum state appear. As it was shown in [79], an important condition for this distribution is the existence of valleys in the scalar field potential during inflation. Then the background de Sitter fluctuations lead to the formation of islands representing one vacuum in the sea of another vacuum. The phase transition takes place after the end of inflation in the FRW Universe.

We will show below that other potentials exhibit the ability to produce closed walls. The conclusion is that this effect is almost inevitable for potentials possessing at least two minima. If this is so, the two vacuum states are separated by a wall after the phase transition. The size of this wall may be significantly greater as compared to the cosmological horizon at that period of time. After the whole wall "enters" the horizon, it begins to contract because of the surface tension. As a result, provided that friction is absent and the wall does not radiate a considerable part of its energy in the form of scalar waves, almost all energy of this closed wall may be concentrated within a small volume inside the gravitational radius. This is the sufficient condition for the black hole formation.

The mass spectrum of black holes formed by this mechanism depends on parameters of the scalar field potential determining the direction and size of the potential valley during inflation and the post-inflationary phase transition. The presence of massive PBHs is a new factor in the development of gravitational instability in the surrounding matter and may serve as a base for new scenarios of the formation and evolution of galaxies. Although we deal here with the so-called pseudo-Nambu–Goldstone field, the proposed mechanism possesses a sufficiently general character.

2.1   Closed wall formation

Now we will describe a mechanism accounting for the appearance of massive walls with the size markedly greater than the horizon at the end of inflation. Let us consider a complex scalar field φ with the Lagrangian

$$L_\varphi = \frac{1}{2}|\partial\varphi|^2 - V(|\varphi|), \tag{4.23}$$

where $\varphi = re^{i\vartheta}$. The potential is chosen in the form

$$V(|\varphi|) = \lambda(|\varphi|^2 - f^2/2)^2 + \delta V(\vartheta), \tag{4.24}$$
$$\delta V(\vartheta) = \Lambda^4\left(1 - \cos\vartheta\right).$$

Here λ and f are parameters of the Lagrangian. Instanton effects are responsible for the additional term proportional to additional parameter Λ^4 (see details and refs in [84]). As is evident from Figure 4.1 this potential possesses the saddle point at the phase $\vartheta = \pi$.

We assume the mass of the radial field component r, i.e. the value

$$m_r^2 \equiv d^2 V/dr^2 |_{r=f/\sqrt{2}},$$

to be sufficiently large, so that the complex magnitude of field acquires the value somewhere in the circle valley $|\varphi| = f/\sqrt{2}$ before the end of inflation. A wide range of models, predicting the additional term $\delta V(\vartheta)$, is discussed in [84].

Since the minimum of potential (4.24) is almost degenerate, the field has the form

$$\varphi \simeq f/\sqrt{2} \cdot \exp(i\vartheta(x)). \tag{4.25}$$

For the following considerations, it should be noted that, using expression (4.24) in the inflation period, we ignored the term reflecting the contribution of instanton effects to the Lagrangian renormalization (see also [84]). Substitution of the expression (4.25) into Lagrangian (4.24) gives effective Lagrangian

$$L_\vartheta = \frac{1}{2}(\partial\chi)^2 - \Lambda^4 (1 - \cos(\chi/f)) \tag{4.26}$$

for the dynamical variable $\chi = \vartheta f$ acquiring the meaning of almost massless field χ.

Since the parameter Λ appears as a result of the instanton effects and renormalization, its value cannot be large and we assume that $\Lambda \ll H, f$. The term (4.26) begins to play a significant role in the post-inflationary stage, when the Hubble parameter decreases with time (e.g., $H = 1/2t$ during the radiation-dominated stage, $H = 2/3t$ during the matter-dominated stage).

Let us assume that the whole part of the Universe observed within the contemporary horizon was formed N_U e-folds before the end of inflation. As was demonstrated in [260], the quantum field fluctuations during inflation were rapidly transformed into a classical field component. Values of the massless field χ in the neighboring causally-disconnected space points differ on the average by $\delta\chi = H/2\pi$ after a single e-fold. In the next time step $\Delta t = H^{-1}$ (i.e. during the next e-fold) each causally-connected domain is divided into $\sim e^3$ causally-disconnected subdomains; the phase in each of the new domains differs by $\sim \delta\vartheta = \delta\chi/f = H/2\pi f$ from that at the preceding step. Thus, more and more domains appear with time in which the phase differs significantly from the initial value.

A principally important point is the appearance of domains with the phases $\vartheta > \pi$. Appearing only after a certain period of time during which the Universe exhibited exponential expansion, these domains turn out to be surrounded by a space with the phase $\vartheta < \pi$. As we show below, the existence of these domains leads in the following to the formation of large-scale structures. Note that the phase fluctuations during the first e-folds may transform eventually into fluctuations of the cosmic microwave radiation, thus leading to restrictions to the scale f. This difficulty can be avoided by taking into account interaction of the field φ with the inflaton field (i.e. by making parameter f a variable – see Chapter 5, Section 3).

Initially, the potential (4.24) possessed a $U(1)$ symmetry and the phase ϑ corresponded to a massless scalar field. Owing to the potential term in (4.26), the symmetry is broken after the end of the inflation period: the potential of the χ field acquires discrete minima at the points $\vartheta_{min} = 0, \pm 2\pi \pm 4\pi...$, and the field acquires the mass $m_\chi = 2f/\Lambda^2$. According to the classical equation of motion, the phase performs decaying oscillations about the potential minimum, the initial values being different in various space domains. Moreover, domains with the initial phase $\pi < \vartheta < 2\pi$ perform oscillations about the potential minimum at $\vartheta_{min} = 2\pi$, whereas the phase in the surrounding space tends to a minimum at the point $\vartheta_{min} = 0$. Upon ceasing of the decaying phase oscillations, the system contains domains characterized by the phase $\vartheta_{min} = 2\pi$ surrounded by the space with $\vartheta_{min} = 0$. Apparently, on moving in any direction from inside to outside of the domain, we will unavoidably pass through a point with $\vartheta = \pi$ because the phase varies continuously. This implies that a closed surface characterized by the phase $\vartheta_{wall} = \pi$ must exist. The size of this surface depends on the moment of domain formation in the inflation period, while the shape of the surface may be arbitrary. The principal point for the subsequent considerations is that this surface is closed.

After heating of the Universe, the evolution of domains, formed with the phase $\vartheta > \pi$ and sharply increased in volume during the inflation period, proceeds on the background of the Friedmann expansion and is described by the relativistic equation of state. First, an equilibrium state with the "vacuum" phase $\vartheta = 2\pi$ inside the domain and the $\vartheta = 0$ phase outside is established at $T \sim \sqrt{M_P m_\chi}$, as was mentioned in the previous section and will be shown below – see formula (4.38). A closed wall corresponding to the phase is formed in the transition region with a width of $\sim 1/m_\chi \sim f/\Lambda^2$, which separates the domain from the surrounding space. The surface energy density of the wall $\sigma \sim f\Lambda^2$.

The field configuration of the plane walls is well-known [261]. We can apply it to describe the distribution of phase across approximately plane border of domain

$$\vartheta(z,t) = -4 \arctan\left\{ \exp\left[\frac{\Lambda^2}{f}(z - z_0) \right] \right\}. \tag{4.27}$$

Energy of this field configuration is concentrated in plane of width $d = 2f/\Lambda^2$ (some kind of "wall") and center z_0. Fortunately the wall width is determined by microscopical parameters and appears to be much smaller than the characteristic size of the wall, because the latter depends on the classical history of the wall formation. Thus, considering local process, such as interaction of the wall with particles, we can consider the wall being plane. Meantime, large-

Another restriction arises if one notices that the large wall starts to determine the local expansion when its energy (4.28) within the cosmological horizon exceeds the energy of relativistic plasma in this scale ($R = t$) $E_V = \rho \frac{4\pi}{3} R^3$. It must happen because the energy density of plasma decreases with time after the reheating stage. Equating both expressions for energy E_V and E_w we obtain connection of the energy density and the scale of the wall

$$R = \frac{3\sigma}{\rho}. \qquad (4.34)$$

On the other hand it seems reasonable to suppose that up to this instant the wall provided small energy excess and hence did not influence the expansion rate of surrounding media. In this case a value of the wall radius R is proportional to a size of the Universe and its dependency on temperature T is described by formula (4.32). Combining formulas (4.34), (4.28) written above and connection between energy density ρ of the Universe at that period and its temperature T (2.47) one obtains the maximal size of the wall

$$R_2 \simeq \left(\frac{\pi^2}{90\sigma} g_*\right)^{1/3} (L_U T_0)^{4/3} e^{\frac{4}{3}(N - N_U)}. \qquad (4.35)$$

In our units ($\hbar = c = 1$) $L_U T_0 \approx 10^{29}$.

Finally, the maximal size of the wall is determined according to the condition (4.29). This value depends on the number N of e-fold when the domain surrounded by the wall was nucleated. Evidently $R_1 > R_2$ for large N. It means that the size of large walls are described by the expression (4.35) whereas the expression (4.33) is important for determination of the size of small walls.

The above considerations do not take into account the effect of a gravity field on the wall dynamics. Therefore, the obtained relationships are valid provided that the initial wall size is much greater than the gravitational radius. Formally, the gravitational radius could exceed the wall size for sufficiently large domains. For a wide range of reasonable values of f the size of such domains exceeds the cosmological horizon in the period of wall formation, and special investigation is needed to clarify the question on their successive evolution. In this study, we suppose that intrinsic gravity does not affect the wall evolution.

Now we proceed to the study of PBH cluster formation in the early Universe.

The walls are formed in the very early Universe after the end of inflation and long before star formation. Namely, it happened when a friction term $3H\dot{\chi}$ in equation of motion for the field χ becomes comparable with 'force' term $m_\chi^2 \chi$, i.e. at the moment t^* when

$$m_\chi \simeq H(t^*) = \sqrt{\frac{8\pi}{3} \frac{\rho(t^*)}{M_P^2}}. \qquad (4.36)$$

Energy density ρ at the radiation-dominated stage is connected with temperature and time according to formulae

$$\rho = \frac{\pi^2}{30}g_*T^4; \quad T = \left(\frac{45}{32\pi^2 g_*}\right)^{1/4}\sqrt{\frac{M_P}{t}}. \tag{4.37}$$

Combining expressions (4.36) and (4.37) one obtains time t^* and temperature T^* of the walls formation

$$t^* = \frac{1}{m_\chi}\sqrt{\frac{\pi}{8}}; \quad T^* = \left(\frac{45}{4\pi^3 g_*}\right)^{1/4}\sqrt{M_P m_\chi}. \tag{4.38}$$

For example, field χ starts oscillating with subsequent walls formation at temperature $T^* \sim 10^5\,GeV$ if model-dependent parameter $m_\chi = 10eV$. Then the wall expands up to the scale R_1 or R_2 with subsequent contraction.

Let us be reminded that the above estimation assumes constant parameter Λ in Lagrangian (4.23). In some realistic models, such as the model of invisible axion, the instanton effects, generating Λ, depends on temperature, so that m_χ is "switched on" at some temperature T_s ($T_s \sim 800MeV$ for invisible axion), which is much smaller than T^*.

One could worry about a temperature at the moment, when walls acquire maximal size. As was discussed above, the size of large walls is described by expression (4.35). Such walls start shrinking when the size of horizon $l_{hor}(t)$ becomes comparable with the size R_2. The temperature at this moment can be obtained after simple algebra

$$T_{wall} \sim (L_U T_0)^{-2/3} M_P^{1/2} f^{1/6}\Lambda^{1/3}\exp\left[\frac{2}{3}(N_U - N)\right]. \tag{4.39}$$

Using this estimation, we could find the temperature when the largest size of walls are formed. As an example, let us choose the topical values of the parameters $f = 10^{14}GeV$, $\Lambda = 10^3 GeV$, $N = 40$, to obtain $T_{wall} \sim 0.2GeV$ (remember that $N_U = 60$ throughout the book). The size of the wall at this moment is $R_2 \sim 10^{20}GeV^{-1}$, while the Universe size is about $\sim 10^{29}GeV^{-1}$.

Note that the real matter of worry in the realistic models, underlying the considered mechanism, will be the proper mechanism, providing the sufficiently effective decay of field oscillations.

For the oscillations, starting at $T \sim T^*$, the contribution into the total density is in this period small $\sim (\Lambda/T^*)^4 \sim (f/M_P)^2$. However, the oscillation energy density decreases in the course of expansion as $\propto a^{-3}$, whereas the one of the relativistic matter is $\propto a^{-4}$, so that at $T \sim T^*(\Lambda/T^*)^4 \sim \Lambda(f/M_P)^{3/2}$ it starts to dominate in the Universe. In the above example with $f = 10^{14}GeV$, $\Lambda = 10^3 GeV$ the matter dominance of field oscillations should come at $T \sim 30keV$. If the oscillations do not decay, this dominance should have continued up to the present time, to result in their modern density, exceeding the critical one by more than 4 orders of the magnitude, which is evidently ruled out by observations. Moreover the observational data put severe constraints on even short periods of matter dominance, predicted at RD stage after $1s$ (i.e. at $T < 1MeV$). The lifetime estimation $\tau \sim 1/m_\chi$ gives promising value of order $10^{-19}s$. On the other hand, the χ-particles lifetime, naively estimated as $\tau = 64\pi f^2/m_\chi^3 \sim 10^{30}s$, exceeds the age of the Universe, appealing to special theoretical mechanisms for their sufficiently rapid decay.

2.3   Wall deceleration in plasma

The first- and second-order phase transitions lead to the formation of field walls separating one vacuum of this field from another. One of these mechanisms was described in the preceding section. In turn, the walls are moving at a subluminal velocity and interact with the surrounding plasma. Depending on the character of this interaction and the shape of the field potential, there are two possible situations. In the first case, the plasma particles pass through the wall, falling into a different vacuum and acquiring a certain mass. This situation corresponds to an electroweak phase interaction [262], whereby the corresponding Higgs field is responsible for a mechanism of the fermion mass generation. In the opposite case, the particle mass is not changed upon going from one vacuum to another (an example is offered by the case of interaction with an axion wall). In the former case, the interaction with the medium leads to a significant deceleration of the domain wall, while in the latter case, the walls are virtually transparent for the medium for the reasonable values of parameters

All considerations are conveniently conducted in the rest frame of the wall. The probability of a particle scattering from the plane wall at rest is described by standard formula of the quantum theory

$$dw = dn(k)2\pi\delta(\varepsilon - \varepsilon')|M|^2 \frac{d^3k'}{2\varepsilon V(2\pi)^3\varepsilon'}, \qquad (4.40)$$

where $dn(k)$ is the momentum distribution of incident particles and M is the matrix element for the particle transition from a state with the energy ε and momentum k to the state with the energy ε' and momentum k' upon interaction with the potential $U = U(z)$ describing the plane wall. The pressure produced by incident particles upon the wall is related to the rate of their momentum transfer to the wall,

$$p = \frac{1}{S}\int dw \cdot q_z, \quad q_z = k'_z - k_z, \qquad (4.41)$$

where S is the wall area.

Let us consider the Lagrangian of the particle–wall interaction. The wall in question represents a classical configuration of the phase ϑ of complex field φ. Thus we have to consider interaction of the field φ with surrounding fermions

ψ. For definiteness, let us choose Lagrangian in the form

$$L = \frac{1}{2}|\partial\varphi|^2 - V(|\varphi|) + 2\overline{\psi}_R\gamma^\mu\partial_\mu\psi_L + g\varphi\overline{\psi}_R\psi_L + h.c. \qquad (4.42)$$

The Lagrangian is invariant under global chiral transformations

$$\psi_L \rightarrow \psi_L\exp(\alpha/2); \quad \psi_R \rightarrow \psi_R\exp(-\alpha/2);$$
$$\varphi \rightarrow \varphi\exp(-\alpha).$$

This fact could be used to extract phase explicitly [84]. Namely, let us be reminded that the field φ is in the bottom of potential (4.24) i.e. has the form

$$\varphi = f/\sqrt{2}\exp(i\chi(x)/f).$$

Substituting this value along with replacement

$$\psi_L \rightarrow \psi_L\exp(i\chi(x)/2f); \psi_R \rightarrow \psi_R\exp(-i\chi(x)/2f)$$

into Lagrangian (4.42) one can easily come to an effective Lagrangian of the phase $\vartheta(x)$ (the field $\chi(x) = f\vartheta(x)$ being treated now as dynamical variable) and fermion fields ψ

$$L_{eff} = \frac{1}{2}(\partial\chi)^2 + \frac{i}{f}(\partial_\mu\chi)\overline{\psi}_R\gamma^\mu\psi_L + \frac{gf}{\sqrt{2}}\overline{\psi}_R\psi_L + h.c. \qquad (4.43)$$

In our case the field χ represents classical configuration (4.27) of the complex field phase interacting with fermions, such as

$$L_{int} = \kappa J_z\partial_z\chi(z); \quad J_\mu = \overline{\psi}\gamma_\mu\psi; \quad \kappa = i/f. \qquad (4.44)$$

Calculating a matrix element for the particle scattering on the wall with the transition from the initial momentum k to final momentum k',

$$M = \langle k'|\int L_{int}d^4x|k\rangle, \qquad (4.45)$$

we obtain

$$|M|^2 = 8(4\pi)^6\kappa^2 S\delta^{(2)}(\vec{q}_\parallel)k_z^2\frac{1}{ch^2(k_zd\pi)}. \qquad (4.46)$$

In deriving formula (4.46), we took into account that the laws of the energy–momentum conservation lead to the following relationships: $k'_z = \pm k_z$; $q_\parallel \equiv k'_\parallel - k_\parallel = 0$ according to which a nonzero contribution to the pressure is only due to the reflected particles with $k'_z = -k_z$. Therefore, the pressure of incident particles on the wall can be written as

$$p = \frac{4}{\pi^2}\kappa^2\int\frac{k_z^2}{ch^2(\pi k_zd)}(k_z - k'_z)\cdot \qquad (4.47)$$
$$\delta(\varepsilon - \varepsilon')\delta\left(k_\parallel - k'_\parallel\right)\frac{dn(k)}{V}\frac{d^3k'}{\varepsilon\varepsilon'}$$

Let us determine the distribution of the incident particles with respect to the transverse momentum $dn(k)$. In the rest frame of plasma

$$dn_0(k_0) = C \exp\{-E_0(\mathbf{k_0})/T\} \frac{d^3 k_0 V}{(2\pi)^3}. \qquad (4.48)$$

Here and below, the subscript '0' denotes quantities determined in the rest frame of plasma. Assuming the plasma temperature T to be significantly greater as compared to the fermion masses and normalizing it to the total particle density, $n_{tot} \approx N(g*)T^3, (N(g*) \approx 5)$, we obtain $C = 20\pi^2$. In addition, it is evident that

$$dn(k) = dn_0(k_0); \qquad (4.49)$$

where the incident particle momentum in the rest frames of wall and of plasma (in the latter frame, the wall moves at a velocity v) are related as

$$k_{0\parallel} = k_\parallel,$$
$$k_{0z} = \gamma(k_z + v\varepsilon),$$
$$E_0 = \gamma(vk_z + \varepsilon),$$

where $\gamma = [1 - v^2]^{-1/2}$.

Integrating the pressure (4.41) with respect to the momentum of the incident particle, we obtain

$$p = |k_z| \frac{32 C\kappa^2}{(2\pi)^5} \gamma \cdot$$
$$\left[\int \frac{d^3 k}{\varepsilon^2} (\varepsilon + vk_z) \frac{k_z^2}{\mathrm{ch}^2(\pi k_z d)} \exp\{-\gamma(\varepsilon + vk_z)/T\} - \right.$$
$$\left. \int \frac{d^3 k}{\varepsilon^2} (\varepsilon - vk_z) \frac{k_z^2}{\mathrm{ch}^2(\pi k_z d)} \exp\{-\gamma(\varepsilon - vk_z)/T\} \right]. \qquad (4.50)$$

This formula was derived with allowance for the Lorentz invariance of the phase volume $d^3 k'/\varepsilon'$. Numerical calculation of the integral in (4.50) presents no difficulties, but the analytical estimation of the pressure produced by the medium upon the wall are also topical. For this purpose, note that the walls have been formed at temperatures (4.39) and the wall thickness is $d = f/2\Lambda^2$. Therefore, for any reliable values of parameters there is a large parameter $Td \gg 1$, using which we may obtain a sufficiently firm estimation of the integral. According to (4.50), the most effective scattering takes place for an incident particle momentum of $k_z \sim 1/\pi d \ll T$. At the same time, it is evident that $k_\parallel \sim \varepsilon \sim \gamma T \gg k_z$.

Using these relationships, we may estimate the integrals in (4.50). Main contribution comes from the first term. The final expression for the pressure produced by the surrounding medium upon the relativistic domain wall is determined mostly by the first integral and it is given by

$$p \approx \frac{20\kappa^2}{\pi^7} \frac{\gamma}{d^4}. \tag{4.51}$$

This expression is valid at $\gamma \gg 1$. Evidently the pressure equals zero for a wall at rest. We can slightly modify the final formula

$$p \approx \frac{20\kappa^2}{\pi^7} \frac{\gamma - 1}{d^4}, \tag{4.52}$$

which is valid both for high and small velocities of the wall. This value has to be compared with internal pressure p_{int} of the wall with radius R, $p_{int} \approx \sigma/R$. Their ratio is given by

$$\frac{p}{p_{int}} \sim \frac{\kappa^2/d^4}{\sigma/R} \approx R\frac{\Lambda^6}{f^5}.$$

Characteristic scale when internal and external pressures become equal is of the order of

$$R_f \sim \frac{f^5}{\Lambda^6}.$$

It means that the friction is important for walls with the size $R_f \sim (10^{32} \pm 10^{90})GeV^{-1}$ if we limit ourselves with intervals $(10^{14} \div 10^{18})GeV$ for parameter f and $(1 \div 10^6)GeV$ for parameter Λ. The scale R_f is too big for any reliable models and we could neglect the friction as it was mentioned above.

2.4 Distribution of black holes

a. Mass distribution of black holes in the Universe

In the following the size distribution of the islands is found numerically. To this purpose it is necessary to study the inhomogeneities of phase induced by fluctuations during inflation stage. It has been well-established that for any given scale $l = k^{-1}$ large-scale component of the phase value ϑ is distributed in accordance with Gauss's law [263], [264], [265] (see also Chapter 3)

$$P(\vartheta, N) = \frac{1}{\sqrt{2\pi\sigma_N}} \exp\left\{-\frac{(\vartheta_U - \vartheta)^2}{2\sigma_N^2}\right\}. \tag{4.53}$$

Here N is the number of e-foldings before the end of inflation, ϑ_U is initial phase of the Universe. This value does not depend on parameters of the model. It varies randomly in the range $0 < \vartheta_U < 2\pi$ for distant parts of the Universe. The dispersion could be expressed in the following manner

$$\sigma_N^2 = \frac{H^2}{4\pi^2 f^2}(N_U - N), \tag{4.54}$$

In the inflationary Universe the cosmological scale l corresponds to the period $t_l = N_l H^{-1}$ of inflationary stage.

Suppose that at e-fold N before the end of inflation a space volume $V(\bar{\vartheta}, N)$ has been filled with the phase value $\bar{\vartheta}$. Then the condition that a new volume filled with average phase $\bar{\vartheta}$ will be produced at the e-fold $N - 1$ obeys the following iterative expression

$$\Delta V(\bar{\vartheta}, N - 1) = (V_U(N) - e^3 V(\bar{\vartheta}, N)) P(\bar{\vartheta}, N - 1) \langle \delta \vartheta \rangle. \tag{4.55}$$

Here $V_U(\vartheta, N) \approx e^{3N} H^{-3}$ is the volume of the Universe at N e-foldings with average phase ϑ and $\langle \delta \vartheta \rangle = \sigma_1 = H/(2\pi f)$. Keeping in mind that causally connected volume has the size $\sim H^{-3}$ one can easily find a number of domains with phase $\bar{\vartheta}$ which were produced at e-fold number N

$$\Delta K_{\bar{\vartheta}}(N) = H^3 \Delta V(\bar{\vartheta}, N - 1) \tag{4.56}$$

The total volume with the phase $\bar{\vartheta}$ can be readily written

$$V(\bar{\vartheta}, N - 1) = e^3 V(\bar{\vartheta}, N) + \Delta V(\bar{\vartheta}, N - 1). \tag{4.57}$$

One can easily calculate the size distribution of domains filled with appropriate value of phase in dependence of N using iterative procedure described by the expressions (4.55), (4.56), (4.57).

In case $\bar{\vartheta} = \pi$ we obtain distribution of those domains which are able to be surrounded by a wall and hence collapse into BHs. As was discussed above, mass of BH is the function of N.

b. Correlation in space distribution of secondary black holes. Nearest vicinity.

We have discussed above the mechanism [80] of massive PBH formation in the Universe. It can be demonstrated that this model with reasonable parameters readily provides for the formation of $\sim 10^{11}$ massive (10^{30}–10^{40}g each) black holes, which is equal to the number of galaxies in the visible Universe. In that analysis, we did not take into account correlations between the formation of a massive black hole and the appearance of smaller black holes surrounding it. This correlation, being inherent to this mechanism, is related primarily to certain features of the above-discussed process of the formation of domains with the phases $\vartheta > \pi$, see Figure 4.1. Apparently, the appearance of such domains creates prerequisites for the formation of new smaller domains inside them. Below, we determine the mass distribution of these daughter domains in the nearest vicinity of specific BH [81].

Consider a region with a size of the order of H^{-1} and a phase within $\pi < \vartheta_0 < \pi + \delta$ (where $\delta = H/2\pi f$ is the average phase jump during the H^{-1} time period) formed during the inflation period as a result of fluctuation in a certain (larger) region of space with the phase $\vartheta < \pi$. During the next e-fold, this space domain will be divided on e^3 subdomains and some of them will acquire the phase ϑ_1 in the interval $\pi - \delta < \vartheta_1 < \pi$. Upon the subsequent phase transition, these domains will be separated by walls from the external region. Similar transitions, with crossing the phase $\vartheta = \pi$ in the reverse direction will take place in each subdomain during the next e-fold. Thus, a structure of the fractal type appears which reproduces itself at each time step in the decreasing scale.

Let N denote the number of subdomains formed in each step, around which a wall may form later. Apparently, this value obeys the inequality $1 < N << e^3$. In the subsequent estimates, we will assume that $N \approx 2 \div 3$. Since each causally connected domain touches approximately six neighboring domains, we can hardly expect N to be greater because of their total number $\sim e^3 \approx 20$. The mass of the future black hole (if it forms) is determined by the area of a closed surface with the phase $\vartheta = \pi$. The ratio of areas of the initial (mother) and daughter domains is readily estimated: the initial area after a single e-fold is $S_0 \approx e^2 H^{-2}$ and the daughter subdomain area is $S_1 \approx H^{-2}$. Therefore, the ratio of masses of the black holes originated from the two sequential generations of subdomains is

$$M_j/M_{j+1} \approx S_j/S_{j+1} \approx e^2, \tag{4.58}$$

for their relative number assumed to be

$$N_{j+1}/N_j = N. \tag{4.59}$$

As it is readily seen, the number and mass of black holes appearing upon the j-th e-fold after the initial domain formation are determined by parameters of the largest black hole genetically related to the primary domain in which the phase exceeded π for the first time. It is evident that

$$N_j \approx N^j; \quad M_j \approx M_0/e^{2j}. \tag{4.60}$$

Excluding j from these relationships, we obtain the desired black hole mass distribution in a cluster:

$$N_{cl}(M) \approx (M_0/M)^{\frac{1}{2} \ln N}. \tag{4.61}$$

The total mass of the cluster can be expressed through the mass M_0 of the largest initial black hole:

$$M_{tot} = M_0 + NM_1 + N^2 M_2 + ... =$$
$$M_0 + Ne^{-2} M_0 + \left(Ne^{-2}\right)^2 M_0 + ... = M_0[1 - N/e^2]^{-1}$$

As is seen, the total mass of the black hole cluster is only one-and-a-half to two times greater than the largest initial black hole mass. The number of daughter black holes depends on the factors considered in the next section.

The inflationary mechanism described above leads to the occurrence of fractal structure of the closed walls. After the end of inflation, as soon as the size of horizon becomes larger than the characteristic size of closed walls, the walls begin to shrink. The energy of each wall, proportional to the area of their surface, concentrates in small spatial domains (in the following they are considered as pointlike objects) [81]. These high density clots of energy could serve in the following for star and/or galaxy formation [3]. Hence, according to the given models, the distribution of stars and galaxies should also carry fractal character. It is important to note that the total surface of walls in specific volume is proportional to the total energy of objects, while the number of walls is equal to the number of dense clots.

The arguments, used above to deduce formula (4.58), indicate that fractal structure of domains could be produced. Denote the number of secondary walls by N and their average size by ξR, $\xi > 1/e$ ($\xi \neq 1/e$, due to a possible merging of causally disconnected subdomains with one common wall). In each of these subdomains, N new smaller closed walls of size $\xi^2 R$ arise during the next time step. Denote by "b" the size of rather small walls compared with the initial one. The sizes of bigger walls will be measured in the units of "b". This means that we may terminate the process after a step n such that $a \equiv \xi^n R$. The total area of the closed walls with bigger sizes in the initial volume is the sum of areas with closed walls of size greater than b. The simple summation leads to the following result for total square of the walls

$$S \approx R^2 q(q^n - 1)/(q - 1), \quad q \equiv \xi^2 N . \tag{4.62}$$

This expression can be written in the form

$$S \approx (R/b)^D , \tag{4.63}$$

where D is the fractal dimension. Equating these two expressions, one obtains

$$D = 2 + \frac{\ln\left(q^{\frac{q^{\frac{\ln(b/R)}{\ln \xi} - 1}}{q-1}}\right)}{\ln(R/b)} . \tag{4.64}$$

This quantity is constant only when the ratio R/b is large, it is different for $q < 1$ and $q > 1$. It can be easily verified that $D \to 2$ for $q < 1$, while for $q > 1$, $D \to 2 + 3ln(q)/ln(4N)$. To get an estimate, suppose that the number of closed domains is $N \approx 4$, and $\xi \approx 1/e$. The value of the parameter q can be easily calculated, $q \approx 0.5$. Hence, the fractal dimension of the system of closed walls is $D \approx 2$.

So, if quantum fluctuations lead to the formation of spatial areas with the field taking a value near a potential maximum, its further evolution results in a system of enclosing walls. The characteristic size of the next generations of walls differs from the previous one approximately by a factor of e. The fractal dimension of such system is $D \approx 2$.

According to this scenario, it is interesting to find the number of walls inside an arbitrary sphere of radius R given by

$$N_{tot} = \sum_{i=1}^{n} N^i = N\frac{N^n - 1}{N - 1} \approx \frac{N^{n+1}}{N - 1}. \tag{4.65}$$

By analogy with the previous calculations and using Eq. 4.65, one obtains the distribution of pointlike dense objects with fractal dimension $D' \approx ln N/ln(1/\xi)$. For realistic values $N \approx 4$, $\xi \approx 1/e$ we find $D' \approx 1.4$ which differs somewhat from the value $D \approx 2$ previously obtained. This is not surprising because in the first case we measure the area of surfaces of walls within a certain volume while in the second case we measure the number of walls.

Of course, other mechanisms at later stages contribute to the distribution and change the fractal dimension somewhat, but the model discussed gives a primordial reason of fractality in the galaxy and star distribution. Observational data indicate that the distribution of stars and galaxies really carries fractal character. So, the number of galaxies inside a sphere of radius R is $N(R) \sim (R)^{2.2\pm0.2}$ up to the sizes of 200 Mpc [267]. The distribution of stars inside galaxies also carries fractal character. In Ref. [268] this fractal dimension was determined by averaging observational data of ten galaxies and was found to be equal to $D \sim 2.3$.

The mechanism of fractal structure production discussed here is of rather general nature and can find applications far beyond the case of the considered model. The production of massive black holes due to Gaussian fluctuation in a framework of a model of chaotic new inflation was discussed in [252]. Another example is based on hybrid inflation, one of the most promising models of inflation [269, 270, 271]. In the standard version of hybrid inflation the potential contains two fields

$$V = V_0 + \frac{1}{2}m_\varphi^2\sigma^2 + \frac{1}{2}\lambda_1\sigma^2\chi^2 - \frac{1}{2}m_\psi^2\chi^2 + \frac{1}{2}\lambda_2\chi^4. \tag{4.66}$$

During inflation, the field σ rolls down along a valley $\chi = 0$. In the meantime field fluctuations around the critical line $\chi = 0$ lead to a formation of fractal structure of domains. This critical line plays the same role as the critical point π in previous discussion. Just after passing the critical point $\sigma = m_\chi^2/\lambda_1$ the state $\chi = 0$ becomes unstable and field χ moves (in average) to one of the new stable minima. These minima are separated by potential maximum and we again come inevitably to the fractal structure of domain walls. The latter are converted into black holes after the end of inflation.

c. Correlation in space distribution of secondary black holes. Megaparsec scale.

One could guess that the same mechanism of the massive BH formation provides the production BHs of smaller masses. Below, we prove this

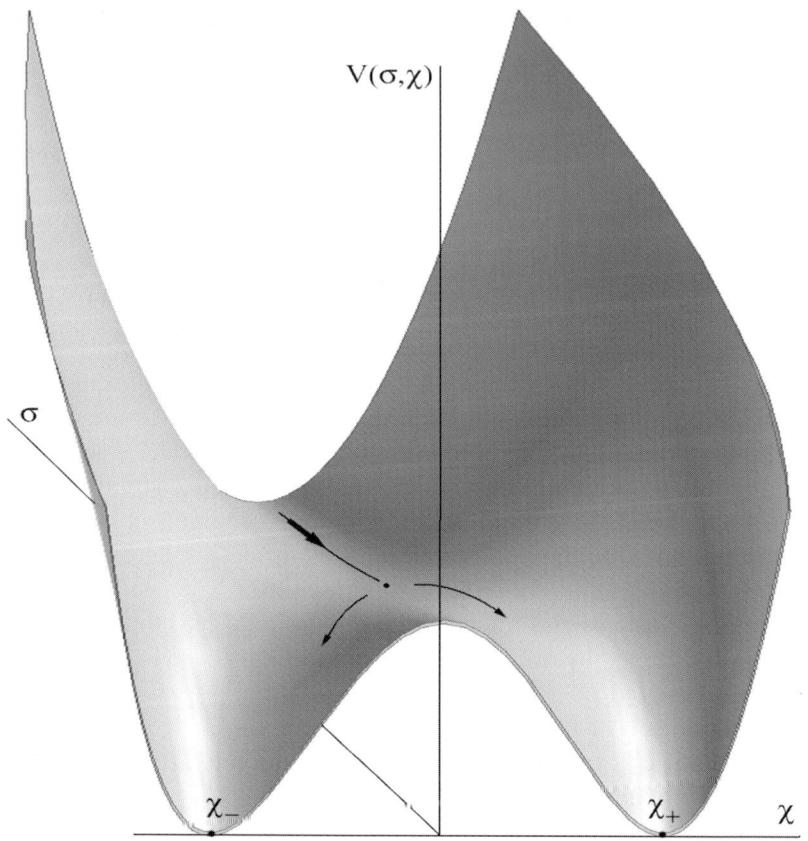

Figure 4.2.   The form of hybrid potential. The field σ moves in the vicinity of critical line. The critical point marks the beginning of classical motion to the specific minima.

statement and consider following [82] space distribution of intermediate BHs around a massive one in the scale of galaxy cluster ($\sim 10 Mpc$).

Let M_0 be the mass of most massive BH; M is the mass of the BH inside the sphere. There are three scales in our problem: R_0 is a size of closed wall that is converted into the most massive BH in future; r is a radius of the sphere with the most massive BH in its center and R is a size of closed wall that is converted into a BH of intermediate mass inside the sphere. These scales correspond to e-folding numbers N_0, N_r and N at the inflation stage and each of them can be expressed in terms of others. The domain with size $r(t_0)$ which will be filled by a part of galaxy cluster was formed N_r e-folds before the end of inflation. If the closed wall gives small contribution to total energy density during early times it expands along with the Universe expansion and we can approximately admit

$$r = r(N_r, t) = r_U(t)e^{N_r - N_U}, \tag{4.67}$$

where $r_U(t)$ is the size of the Universe. The connections $R \longleftrightarrow N$ and $R_0 \longleftrightarrow N_0$ are derived below.

The value important for our consideration is the phase ϑ_r of the domain of the size $r(t)$ at the moment of its formation at the stage of inflation, i.e. at the e-folding number N_r. It can be determined from the following arguments.

To the period characterized by e-folds number N_0 we have already $e^{3(N_r - N_0)}$ causally disconnected areas inside the volume in question. By assumption, only one of them contains the phase $\vartheta_m = \pi$ which leads to formation of most massive BHs. So in the considered case the probability to have the phase equal to π is equal to $1/e^{3(N_r - N_0)}$. One can find the value of the phase ϑ_r by equating this probability to the general expression for the probability to find specific value of the phase $\vartheta_m (= \pi)$ at the given averaged value of the phase ϑ_r

$$\frac{1}{e^{3(N_r - N_0)}} = \frac{\exp\left[-\frac{(\pi - \vartheta_r)^2}{2\delta\vartheta^2(N_r - N_0)}\right]}{\sqrt{2\pi(N_r - N_0)}}, \qquad \delta\vartheta = \frac{H_e}{2\pi f}. \qquad (4.68)$$

Here we supposed Gaussian form for the probability distribution of the phase ϑ

$$P(\vartheta) = \frac{\exp\left[-\frac{(\vartheta_r - \vartheta)^2}{2\delta\vartheta^2(N_r - N)}\right]}{\sqrt{2\pi(N_r - N)}}. \qquad (4.69)$$

From Eq. (4.68) we obtain the phase ϑ_r

$$\vartheta_r = \pi - \delta\vartheta\sqrt{(N_r - N_0)\left[6(N_r - N_0) - \ln(2\pi(N_r - N_0))\right]}. \qquad (4.70)$$

This method of evaluation of the initial phase ϑ_r is correct if

$$P(\vartheta_r - \delta\vartheta) \ll 1.$$

It is supposed from the beginning that the fluctuations of the phase are small, $\delta\vartheta \ll \vartheta_r$. This condition facilitates the estimation. The result can be expressed as a limit on e-folding

$$|N_r - N_0| < \frac{\pi - \vartheta_r}{\delta\vartheta} = 2\pi(\pi - \vartheta_r)\frac{f}{H}.$$

If, for example, $\vartheta_r \sim \pi/2$ and $f/H \sim 5$ the estimation gives $|N_r - N_0| < 50$. Thus we can use the expression (4.70) in a wide range of parameters.

Having this value, the probability of formation of walls with smaller masses at e-folding number N can be obtained from expression (4.69) as follows

$$W(N) = P(\pi) = \frac{[2\pi(N_r - N_0)]^{\frac{N_r - N_0}{2(N_r - N)}}}{\sqrt{2\pi(N_r - N_0)}} \exp\left[-\frac{3(N_r - N_0)^2}{N_r - N}\right]. \qquad (4.71)$$

Total number of the domains at this moment is $e^{3(N_r-N)}$ and hence the total number of closed walls which were originated at e-folding number N is equal to

$$K(N) = W(N)e^{3(N_r-N)} \tag{4.72}$$

The mass M of a closed wall is supposed to be connected with the size of a wall

$$M \approx S(N)\,\sigma. \tag{4.73}$$

The next problem concerns the squaring of a wall surface just after the wall formation. The fact is that up to now we restricted our consideration to spherical walls only, which is too rough. The following discussion is devoted to the problem of nonspherical walls.

By definition, a wall is determined by the phase value $\vartheta = \pi$. Let us have an approximately flat surface of size $1/H_e^2$ at the inflation stage with e-folding number N. This surface will be divided onto $1/e^2$ parts to next e-folding number $N-1$. Total surface thus increases by a factor ς. Our estimations indicate that $\varsigma \approx 1.5 \div 2$. If this process lasts for, say, $\tilde{N}$ e-foldings, the surface is increased by $\varsigma^{\tilde{N}}$ times in addition to ordinary factor of expansion $e^{\tilde{N}}$ during inflation. This fractal-like process stops, when the size of the wall turns to be comparable with the wall width $d \approx 1/m_\chi$. The crinkles of sizes less than d will be smoothed out in the period of wall formation and their effect could be ignored in the analysis of successive wall evolution. Hence our direct aim is to determine the e-folding N_d at which crinkles of the size d are produced.

The scale of the crinkles produced at e-folding number N_d is approximately equal to the wall width $d(\approx 1/m_\chi)$

$$d \approx l_{crinkle} = r_U(t^*)e^{N_d-N_U} = L_U\frac{T_0}{T^*}e^{N_d-N_U}. \tag{4.74}$$

Here T_0 and $L_U \approx 10^{28}cm$ are temperature and size of the Universe at modern epoch and temperature T^* is determined in expression (4.38). Combining these formulas we obtain the last e-folding, N_d, important for crinkles formation

$$N_d = N_U + \ln\left[\frac{T^*d}{T_0L_U}\right]. \tag{4.75}$$

We conclude that the mass of a wall just after its formation can be written in the form

$$E_w \approx \sigma 4\pi R(N)^2 \varsigma^{N-N_d}, \tag{4.76}$$

where $\sigma = 4f\Lambda^2$ is energy density of the wall.

Internal tension in the wall leads to its local flattening while the horizon grows at FRW stage of expansion. Released energy is converted into kinetic energy of the wall and into an energy of outgoing waves which are predominantly the waves of those scalar fields that produce the domain walls. In our case it is the phase ϑ that acquires dynamical sense as the field $\chi = f\vartheta$.

A typical momentum k of the waves is of the order of inverse wall width $1/d$, hence $k \sim m_\chi$ and the waves are necessarily semirelativistic. Their kinetic energy decreases due to the cosmological expansion, leaving particles of mass m_χ in a vicinity of the wall. These particles can further contribute to the process of BH formation. The situation can become more complex if the particles are unstable. In this case, products of their decay are relativistic and they mostly escape capture by the gravitational field of the wall. The main immediate effect of these relativistic products is the local heating of the medium, surrounding the region of BH formation. Their concentration in this region and successive derelativization can later provide their possible contribution into BH and BH cluster formation.

The kinetic energy of the wall that is originated from the local flattening of crinkles also decreases due to cosmological expansion.

Owing to expansion the total size of the wall grows as $R(N) \propto a$. So the current value of the wall energy changes both due to the growth of the total wall size and due to the local flattening of the wall within the horizon. To account for the latter effect one should substitute into (4.76) instead of N_d the value of e-fold N_c, $N_d < N_c < N$ that corresponds to the current size of the horizon. Both factors lead to the respective change in the expression (4.76).

For the estimated values of ς the increase of the wall size $R(N)$ dominates over the decrease of the crinkle contribution into the current value of E_w. The latter contribution can be significant only in the period when the wall enters the horizon and its successive evolution proceeds so quickly that the effects of wall flattening are not suppressed by cosmological expansion. So, when the wall enters the horizon, the effects of crinkles increase the total energy of the wall, as compared with the spherical case, by the factor $A \sim \varsigma^\alpha$, $\alpha \geq 1$, where $\varsigma \approx 1.5 \div 2$, as we estimated earlier.

The size of the wall which has crossed critical point π at the e-folding number N can be estimated in the same manner as it was in the previous subsection. The only difference is that we have treated the surface of the wall more accurately by accounting for crinkles of different scales.

The first restriction coincides with those derived in the previous section so that the size of wall could not be larger than

$$R_1(N) = \frac{1}{2}\left(\frac{L_U T_0}{C}\right)^2 e^{2(N-N_U)}, \quad C = \left(\frac{45}{4\pi^3 g_*}\right)^{1/4}\sqrt{M_P}. \qquad (4.77)$$

The second restriction must be modified with accounting for the crinkle effects. Note that in the previous section this constraint followed from the condition that the energy of the wall should not dominate the energy of plasma within the horizon. This formulation assumes that if the energy of the wall, determined by the intrinsic negative pressure medium inside the wall, exceeds the energy of plasma, the local expansion would come to superluminal regime. Then the further wall evolution would be elusive for the outer observer.

Accounting for crinkles made us re-define the total energy of the wall. The wall flattening and crinkle decays transform the energy, stored in the smaller scale crinkles, into the energy of scalar field waves and the kinetic energy of the wall. These products of wall flattening contribute to the non-negative pressure part of the local energy density, thus preventing the superluminal expansion of the considered region.

To obtain the condition, under which superluminal expansion does not come before the wall enters the horizon, we must now add the effect of wall flattening $\sigma 4\pi R^2 (A - 1)$ to the energy of plasma $E_V \approx \rho \frac{4\pi}{3} R^3$ and to compare this sum with the energy of the flattened wall $\sigma 4\pi R^2$ within the current horizon. We obtain connection of the energy density and the scale of the wall as

$$R = \frac{3\sigma}{\rho} (2 - A), \qquad (4.78)$$

if $A < 2$. The value $R = L_U \frac{T_0}{T} e^{N - N_U}$ depends on the temperature T.

Combining formulas written above one obtains maximal size of the wall

$$R_{max} = R_2(N) \approx \left(\frac{\pi^2}{90\sigma} g_* \right)^{1/3} (L_U T_0)^{4/3} \, e^{\frac{4}{3}(N - N_U)} (2 - A)^{-\frac{1}{3}}. \qquad (4.79)$$

Maximal size of a wall with a phase which has crossed π at e-folding number N is determined from the condition

$$R(N) = \min(R_1(N), R_2(N)).$$

If A approaches 2, the value of R_{max}, given by Eq. (4.79), formally tends to infinity. It means that for sufficiently strong effect of crinkles, corresponding to $A \geq 2$, wall dominance cannot come, since the energy release of flattening always dominates the negative pressure energy of a flattened wall within the horizon (in the Eq. (4.78) it formally corresponds to the tendency of R to zero, when A approaches 2, and to the absence of solution for R at $A \geq 2$). The RD (or MD) regime of dominance of the energy, released in flattening, implies the difference of the expansion law within the considered region from the general expansion, but, being subluminal, it continues until the whole wall enters the horizon.

So, in the case of strong crinkle effects, corresponding to $A \geq 2$, the second restriction has no place and the maximal size of the wall is

$$R(N) = R_1(N).$$

It proves the values of maximal mass of BHs that we consider in the next subsection.

The longer the period of crinkle–decay–products dominance, the larger the over-density in this region, as compared with the mean cosmological density, $(\delta\rho/\rho) \gg 1$, when the wall enters the horizon. It provides the separation of this region from the general expansion and effective BH formation in it, similar to PBH formation on RD stage, considered by [69] (see for review [71, 3]). The account for this effect needs special study, however, it can only slightly modify (within a factor of A) the numerical values in BH distributions, presented in the following subsection.

d. Primordial black hole cluster seeds for galaxy formation

The clusters of BHs considered in previous sections could be a trigger mechanism for initial baryon fluctuations. In this section we discuss a new model of galaxy formation, offered in [80, 81]. It is based on the prediction of closed wall defects from the succession of symmetry breaking phase transitions in the inflationary Universe and on the formation of massive primordial black holes in the result of collapse of such closed walls. The natural consequence of this mechanism is the spatial correlation between the black holes. Such correlation leads to formation of primordial black hole clusters. The energy of scalar field oscillation from small-scale wall decays is shown to be localized around the black hole cluster. The complex of black hole clusters and concentrated energy of scalar field oscillations provides the effective primordial seeds for matter condensation and galaxy and/or cluster of galaxy formation.

The necessary ingredient of our mechanism is the existence of potential with several minima. As a working example we have chosen the same 'mexican hat' potential with small tilt (4.23), (4.24).

Absolute minima are disposed in points $\varphi = f, \vartheta = 0, 2\pi, 4\pi....$ Local maxima are in points $\varphi = f, \vartheta = \pi, 3\pi....$ Values of the parameters $f \sim 10^{13} GeV, \Lambda = 1 \div 10^8 GeV$ do not contradict any cosmological or physical constraint, providing that the sufficiently effective mechanism of field oscillations decay is present and that the large-scale fluctuations are compatible with the observed level of CMB fluctuations (see the discussion of the latter point in the next chapter). During inflation, quantum fluctuations force the phase ϑ inside some space domains to overcome local maxima. After the end of inflation these domains will be surrounded by closed walls which could collapse into BHs.

Mass distribution of such BHs for specific values of the parameters is represented in Figures 4.3, 4.4, 4.5.

Total mass of such a BH equals $\sim 5 \cdot 10^{53} g$ which is comparable with the baryon mass in the visible part of the Universe. In the preceding sections, we considered only the principal possibility of the formation of domain walls connecting neighboring vacuum states. The numerical calculations were performed for the following values of parameters (which are consistent with the

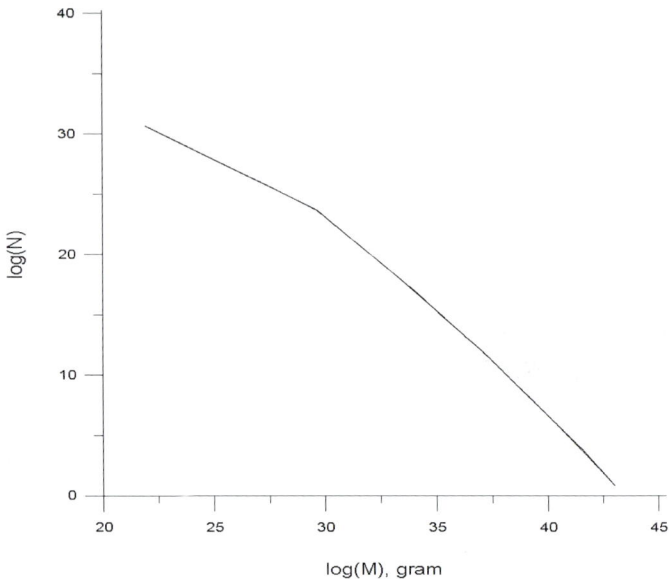

Figure 4.3.   Mass distribution of black holes in the visible part of the Universe. Total mass of BHs equals $5.9 \cdot 10^{53} gram$. Parameters of the model: $\Lambda = 10^3$ GeV, $f = 5 \cdot 10^{13}$ GeV, $\vartheta_U = 0.55\pi$.

observed anisotropy in the cosmic microwave radiation): the Hubble constant at the end of inflation, $H = 10^{13} GeV$; Lagrangian parameters, $f \sim 5H$ and Λ varies from 1 to 10^7 Gev. The initial phase, at which the visible part of the Universe is formed by the time $t_U \approx 60 H^{-1}$ to the end of inflation, controls the number of domains and, accordingly, the number of closed walls formed after inflation. This random value, not related to the Lagrangian parameters, must be selected taking into account the observational restrictions on the abundance of black holes in the Universe. We will use that numerical value ϑ_U, which ensures a sufficiently large number of massive black holes, while the presence of numerous smaller black holes does not contradict experimental restrictions.

As it is seen, the PBH masses fall within the range from 10^{16}–$10^{42} g$ depending on chosen parameters. The total mass of black holes amounts to $\sim 1\% \div 100\%$ of the contemporary baryonic contribution.

The results of calculations are sensitive to the value of the parameters f, Λ and the initial phase ϑ_U. As the Λ value decreases to $\sim 1GeV$, still greater PBHs appear with a mass of up to $\sim 10^{42}$. A change in the initial phase leads to sharp variations in the total number of black holes.

Sharp bends of the curves in the Figures are conditioned due to different mechanisms responsible for small and large wall formation and discussed above.

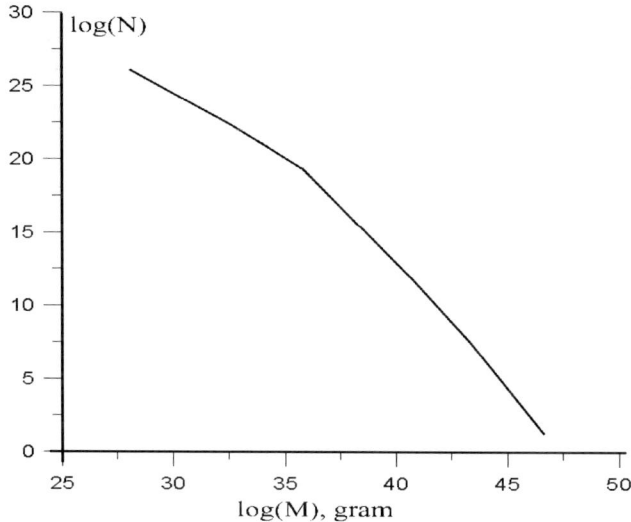

Figure 4.4. Mass distribution of black holes in the visible part of the Universe. Total mass of BHs equals $3.5 \cdot 10^{55} gram$. Parameters of the model: $\Lambda = 1$ GeV, $f = 5.3 \cdot 10^{13}$ GeV, $\vartheta_U = 0.67\pi$.

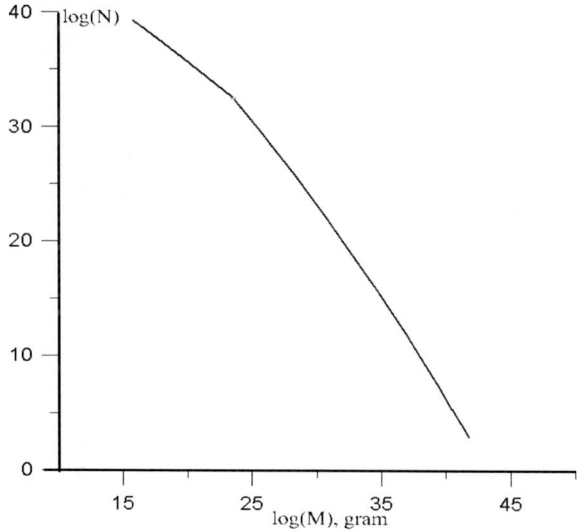

Figure 4.5. Mass distribution of black holes in the visible part of the Universe. Total mass of BHs equals $3.6 \cdot 10^{56} gram$. Parameters of the model: $\Lambda = 10^7$ GeV, $f = 6 \cdot 10^{13}$ GeV, $\vartheta_U = 0.55\pi$.

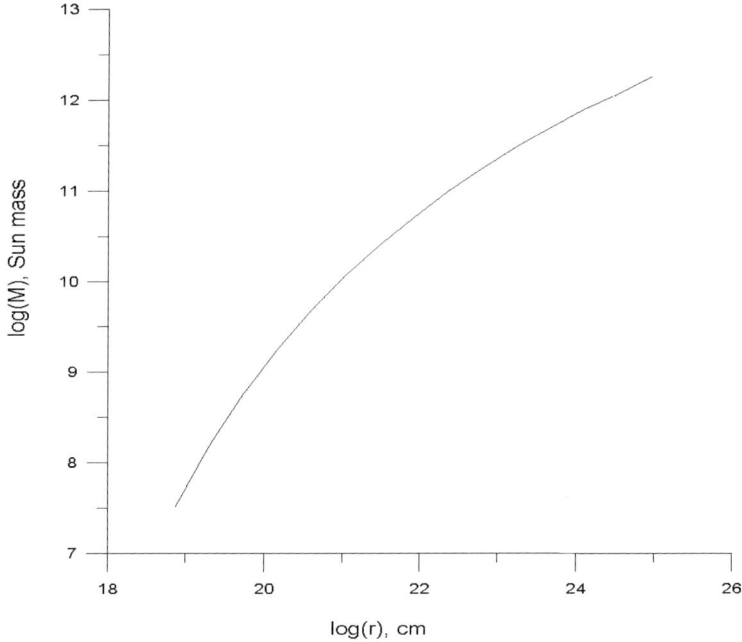

Figure 4.6.   Total mass of black holes within the sphere of radius r and the most massive BH in the center. Parameters of the model: $\Lambda = 10^3$ GeV, $f = 5 \cdot 10^{13}$ GeV.

   Thus, we have BH with different masses which are initially distributed in the space of the Universe. The parameters of the model in Figure 4.4 are chosen in such a way that average distance between the most massive BH, $(10^3–10^4)M_\odot$, is about several Mpc. The latter is a scale of galaxy clusters.

   Figures 4.3, 4.4, 4.5 represent examples of mass distribution of uncorrelated BHs. Each of such a BH being formed gives opportunity for production of smaller BHs. Using formulas (4.67), (4.72) and (4.73) one can calculate desired distribution of intermediate BHs in the vicinity of the most massive one. The distribution of such is represented in Figure 4.6. This BH cluster contains from the beginning $\sim 10^{22}$ BH of smallest masses about $\sim 10^{-11}M_\odot$. There are massive BHs with mass $\sim 10^6 M_\odot$. Total mass of the BH in the cluster is $\sim 5 \cdot 10^{13} M_\odot$ and is comparable with a mass of a baryon component in galaxy clusters. The successive processes of merging could increase the masses of remaining BHs and decrease their number. It is important to note that closed walls of smaller sizes are vanished by radiating energy into surrounding media. This energy contributes to the total mass of the galaxy clusters if the field is stable and heats surrounding media by decay products in the case of unstable field.

We investigated in detail primordial spatial distribution of black holes. It was shown that intermediate mass BHs (in the range $10^{-11} \div 10^{6}\ M_{\odot}$) could concentrate around a massive one in the scale of several Mpc. The total mass of a group of these primordial BHs is comparable with a mass of ordinary galaxy clusters. The amount of such groups coincides approximately with an amount of galaxy clusters in the modern epoch for chosen parameters. Their formation begins at a very early stage which is characterized by the temperature $T^{*} \sim \sqrt{M_{P}m_{\chi}} \sim 10^{5}\,GeV$ and for the biggest walls may continue even after the period of Big Bang nucleosynthesis. Being compatible with the constraints on the average primordial light element abundance and with the averaged restrictions on the distortions of the CMB black body spectrum, the formation of PBH clusters around the most massive BH may lead to local peculiarities of pre-galactic chemical composition and to specific effects in the spectrum, angular distribution and polarization of small-scale CMB fluctuations, accessible to the observational test. After recombination of hydrogen at the modern matter-dominated stage baryons are influenced by gravitational potential of a primordial BH group which facilitates galaxy formation. The proposed model can offer the alternative to the standard picture of slow growth of baryon density fluctuations to produce galaxies. In any case it brings the new element into the theory of large-scale structure formation.

Thus, our calculations confirm the possibility of formation of the clusters of massive PBHs ($\sim 10^{3}M_{\odot}$ and above) in the early pre-galactic stages of cosmological evolution. These clusters represent the fractal structure of localized strong nonlinear energy density fluctuations around which increased baryonic and (cold or warm) dark matter density may concentrate in the subsequent stages, followed by the evolution into galaxies.

3. Discussion

This chapter was devoted to the new phenomena in the theory of large-scale structure and galaxy formation, arising as a cosmological impact of particle theory. The new mechanism is offered for the formation of protogalaxies, which is based on the cosmological inferences of the elementary particle models predicting nonequilibrium second-order phase transition in the inflation stage period and the domain wall formation upon the end of inflation. The presence of closed domain walls with the size markedly exceeding the cosmological horizon in the period of their formation leads to the wall collapse in the post-inflation epoch (when the wall size becomes comparable with the cosmological horizon), which results in the formation of massive black hole clusters that can serve as nuclei for the future galaxies.

Results of calculation of the black hole mass distribution are compatible with the available observational data. The number of black holes with $M \sim 10^{3} \div 10^{4}M_{\odot}$ (and above) is comparable with the number of galaxies in the visible part of the Universe. A mechanism of deceleration of the wall motion is considered and it is shown that this process does not affect the dynamics of collapse of supermassive walls.

Development of the proposed approach gives ground for a principally new scenario of the galaxy formation in the model of the Big Bang Universe. Traditionally, the hot Universe model assumes a homogeneous distribution of matter on all scales, whereas the appearance of observed inhomogeneities is related to the growth of small initial density perturbations. However, an analysis of the cosmological inferences of the theory of elementary particles indicates the possible existence of strongly inhomogeneous primordial structures in the distribution of both the dark matter and baryons. These primordial structures represent a new factor in the theory of galaxy formation.

Topological defects such as the cosmological walls and filaments, primordial black holes, archioles in the models of axion cold dark matter [77, 78, 3], and essentially inhomogeneous baryosynthesis (leading to the formation of antimatter domains in the baryon–asymmetric Universe) [83, 79, 273, 3] by no means offer a complete list of possible primary inhomogeneities inferred from the existing elementary particle models. The proposed approach discloses a number of interesting aspects in this direction.

Indeed, most of the above-mentioned phenomena can be naturally described with the use of a simple Nambu–Goldstone model, applied to different conditions in the inflationary Universe.

Thermal and non-thermal post-inflationary $U(1)$ breaking phase transitions lead in this model to primordial large-scale nonhomogeneity in the distribution of energy density of coherent oscillations of a stable scalar field – to the archiole structure. Non-thermal phase transitions on the inflationary stage can result in spikes in the spectrum of density fluctuations. Fluctuations of phase on the inflationary stage after such transition can result in closed wall defects.

In the latter case, the model provides for a possibility of the quantitative analysis of correlations in the formation of massive PBHs and the primary inhomogeneity of the dark matter and baryons. Originally inherent in this mechanism is the inhomogeneous phase distribution which eventually acquires (similar to what takes place in the invisible axion cosmology) a dynamical sense of the initial amplitude of the coherent oscillations of a scalar field. Irrespective of the efficiency of dissipation of the energy of these oscillations, the regions of closed wall formation must be correlated with the regions of maximum energy density of the dark matter. If these oscillations are not decaying, their energy density may provide for the contemporary dark matter density. Inhomogeneity in the initial amplitude of these oscillations would then imply an inhomogeneity in the initial energy density and, hence, the regions of black hole formation will become the regions of increased dark matter density. A qualitatively similar effect (albeit not as pronounced) takes place in the dissipation of coherent oscillations at the expense of particle production. An increase in the oscillation energy density transforms into a local increase in the density of dark matter particles produced in this region.

Thus, development of the proposed approach may lead to a number of interesting scenarios of initial stages in the formation of protogalaxies, depending on the selection of particular elementary particle models and their parameters. We continue to discuss these models in the following chapters.

Chapter 5

BARYON ASYMMETRICAL UNIVERSE WITH ANTIMATTER REGIONS

The statement that our Universe is baryon asymmetrical is a quite firmly established observational fact, being one of the cornerstones of contemporary cosmology (see review in [3]). Indeed, if large regions of matter and antimatter co-exist now, the annihilation would take place at the borders of these regions being the source of enormous gamma radiation that is not observed. If the typical size of such a domain structure was small enough, domains would be annihilated completely. Then the energy released by the annihilation would result, depending on the period of annihilation, in diffuse γ –ray background, in distortions of the spectrum of the cosmic microwave radiation, or in peculiarities of light element abundance, neither of which is observed (see [313] for review). Recent analysis of this problem [272] for a baryon symmetric Universe claimed that the size of domain regions should exceed 1000 Mpc, being comparable with the modern cosmological horizon. It therefore seems more plausible that the Universe is fundamentally matter–antimatter asymmetric.

However the arguments used in [272] do not exclude the case when the Universe is composed almost entirely of matter with relatively small insertions of primordial antimatter. Thus, we may expect the existence of macroscopically large antimatter regions in the Universe which differs drastically from the case of the baryon symmetric Universe. We call the region filled with antimatter in the baryon asymmetrical Universe, antimatter island. Of course the existence of antimatter islands is not the rigorous requirement of baryosynthesis, but natural modification of baryogenesis scenarios will result in formation of domains with different signs of baryon charge (see for review [273] and [3]). The principal possibility of appearance of antimatter domains as the profound signature of nonhomogeneous baryosynthesis was first put forward in [83].

The only condition, which is necessary to satisfy, is that the amount of antibaryons within antimatter islands must be small compared to the total baryon number of the Universe. At first glance it is not difficult to obtain some amount of domains with antimatter, if we simply suppose that the C- and CP-violation have different signs in different space regions. This may be achieved, for example, in models with two different sources of CP-violation, explicit and spontaneous [274]. However, the processes, involving spontaneous CP–violation, appear as the result of first- or second-order phase transition in the early Universe. This implies that any primordial antimatter islands should be too small [273]. For example, if they are formed in the second-order phase transition their size at the moment of formation is determined by so-called Ginzburg temperature $l_i \simeq 1/(\lambda T_c)$, where T_c is the critical temperature at which the phase transition happens and λ is the self-interaction coupling constant of a field which breaks CP symmetry [274]. In this case different domains would expand together with the Universe and now their sizes would be $l_0 \simeq l_i(T_c/T_0) = 1/(\lambda T_0) \simeq 10^{-21} pc/\lambda$, where T_0 is the present temperature of the background radiation.

On the other hand it has been revealed [275] that the average displacement of the antimatter domain's boundary caused by annihilation with surrounding matter is about $0.5pc$ at the end of the radiation-dominated (RD) epoch. Therefore only a primordial antimatter island, having initial size up to $0.5pc$ or more at the end of the RD stage, survives to the contemporary epoch and in the case of successive homogeneous expansion has the modern size $\simeq 1kpc$ or more. Any primordial antimatter islands with scale less then critical survival size $l_c \simeq 1kpc$ at the contemporary epoch must be eaten up by annihilation processes. Thus, it is a serious problem for models with thermal phase transition to provide the size of primordial antimatter islands exceeding the critical survival size l_c in order to avoid total annihilation.

It was first shown in [276] that the problem of formation of sufficiently large antimatter islands in the result of nonhomogeneous baryosynthesis implies with necessity some reflection of the inflationary stage. In this chapter we present the issue for nonhomogeneous baryosynthesis in the inflationary Universe in which the relationship between the nonhomogeneity of baryosynthesis and inflation has the manifest form.

The proposed approach is based on the mechanism of spontaneous baryogenesis [273]. This mechanism implies the co-existence with inflaton of a complex scalar field carrying baryonic charge. The phase of this additional field has initially arbitrary value along the valley of its potential. The baryon/antibaryon number excess is produced, when due to a small tilt of potential, the phase moves to its minimum.

It is supposed that the vacuum energy responsible for inflation is driven by a scalar inflaton field, and an additional complex field coexists with the inflaton. Due to the fact that vacuum energy during inflation period is too large, the tilt of potential is vanishing. This implies that the phase of the field behaves as an ordinary massless Nambu–Goldstone (NG) boson. Owing to quantum fluctuations of massless field at the de Sitter background [260, 51, 265, 263, 264] the phase is varied in different regions of the Universe. When the vacuum energy decreases the tilt of potential becomes topical, and the pseudo-Nambu–

Goldstone (PNG) field starts to oscillate. As the field rolls in one direction during the first oscillation, it preferentially creates baryons over antibaryons, while the opposite is true as it rolls in the opposite direction. Thus, to have a baryon asymmetric Universe as a whole one must have the phase sited in the point, corresponding to the positive baryon excess generation, just at the beginning of inflation (when the modern Universe size is leaving the horizon). Then subsequent quantum fluctuations may adjust the phase at the appropriate position causing the antibaryon excess production. If it takes place not too late after the inflation begins, the size of antimatter islands may exceed the critical surviving size l_c.

1. Phase distribution for NG field at the inflation period

We start our consideration with the discussion of evolution of a $U(1)$ symmetric scalar field at the inflation epoch with a pseudo Nambu–Goldstone tilt emerging after the end of exponential expansion of the Universe. The $U(1)$ symmetry is supposed to be associated with baryon charge. It is shown that quantum isocurvature fluctuations lead in a natural way to a baryon asymmetrical Universe with antibaryon excess regions. The range of parameters is calculated at which the fraction of the Universe occupied by antimatter and the size of antimatter regions satisfy the observational constraints and lead to effects, accessible to experimental search for antimatter. The $U(1)$ symmetric scalar field coexists with inflaton at the inflation epoch. The quantum fluctuations of such field during the inflation stage cause the isocurvature perturbations for the phase marking the Nambu–Goldstone vacuum. The size distribution of domains containing the appropriate phase values, caused by isocurvature fluctuations, coincide with the size distribution of antimatter islands.

Thus, to estimate the number density of antimatter regions with sizes exceeding the critical survival size l_c in the baryogenesis model under consideration, we have to deal with long-wave quantum fluctuations of the NG boson field in the period of inflation. Various aspects of this question have been examined in the numerous papers [277, 256, 278, 279, 269, 57, 280, 281, 265, 264, 263] in connection with cosmology of invisible axion. Such quantum fluctuations could be a reason of axionic topological defects or could be reprocessed into isocurvature density perturbations.

The effective potential of the complex field is taken in the usual form

$$V(\chi) = -m_\chi^2 \chi^* \chi + \lambda_\chi (\chi^* \chi)^2 + V_0, \qquad (5.1)$$

where constant V_0 was added to make the potential (5.1) non-negative. The field χ can be represented in the form

$$\chi(\vartheta) = \frac{f}{\sqrt{2}} \exp\left(\frac{i\vartheta}{f}\right). \qquad (5.2)$$

The $U(1)$ symmetry breaking implies that the radial component of the field χ acquires a nonvanishing classical part,

$$f = m_\chi/\sqrt{\lambda_\chi},$$

and the angular field ϑ in Eq. (5.2) becomes a massless NG scalar field with a vanishing effective potential, $V(\vartheta) = 0$. In this case, χ has the familiar Mexican hat potential, and the vacuum is placed in a circle of radius f. Throughout the present chapter we will work with dimensionless angular field $\theta = \vartheta/f$.

We are concerned here with the possibility to store appropriate phase value in the domain of size exceeding the critical survival size. Such value of phase plays the role of starting point for clockwise movement, which is going to generate antibaryon excess when the tilt of potential (5.1), breaking $U(1)$ explicitly, will turn to be topical.

We assume that the Hubble constant varies slowly during inflation. Also we use well-established behavior of quantum fluctuations on the de Sitter background [264], [265], [263] (see Chapters 3 and 4). It implies that spatial size of vacuum fluctuations of every scalar field grows exponentially in the inflating Universe. When the wavelength of a particular fluctuation becomes greater than H^{-1}, the average amplitude of this fluctuation freezes at some nonzero value because of the large friction term in the equation of motion of the scalar field, whereas its wavelength grows exponentially. In other words such a frozen fluctuation is equivalent to the appearance of a classical field that does not vanish after averaging over macroscopic space intervals. Because the vacuum must contain fluctuations of every wavelength, inflation leads to the creation of more and more new regions containing the classical field of different amplitudes with scale greater than H^{-1}. The average amplitude of such fluctuations for the NG field generated during each time interval H^{-1} is equal to [260, 51] (see also the previous chapter)

$$\delta\vartheta = \frac{H}{2\pi}. \tag{5.3}$$

During this time interval the Universe expands by a factor of e. Since the NG field (the PNG tilt is still vanishing) is massless during the inflation period, one can see that the amplitude of each frozen fluctuation is not changed in time at all and the phases of each wave are random. Thus the quantum evolution of the NG field looks like one-dimensional Brownian motion [57, 282] along the circle valley corresponding to the bottom of the NG potential. It means that the values of the phase θ in different regions become different, and the corresponding variance grows as [265, 263, 264]

$$\langle(\delta\theta)^2\rangle = \frac{H^3 t}{4\pi^2 f^2}. \tag{5.4}$$

As a result, the dispersion grows as $\sqrt{\langle(\delta\theta)^2\rangle} = \frac{H}{2\pi f}\sqrt{N}$, where N is the number of e-folds. In other words the phase θ makes quantum step $\frac{H}{2\pi f}$ at each e-fold, and the total number of steps during time interval Δt is given by $N = H\Delta t$.

Let us consider the scale $k^{-1} = H_0^{-1} = 3000h^{-1}Mpc$ which is the biggest cosmological scale of interest. The Universe is assumed to be baryon asymmetric in the scale which leaves the horizon at the corresponding e-fold $N = N_{max}$. On the other side this scale is the one entering the horizon now, namely $a_{max}H_{max} = a_0 H_0$, where the subscript "0" indicates the contemporary epoch. This implies that:

$$N_{max} = \ln\frac{a_{end}H_{end}}{a_0 H_0} - \ln\frac{H_{end}}{H_{max}}. \tag{5.5}$$

Here the subscript "*end*" denotes the epoch at the end of the inflation period. The slow-roll paradigm assumes slow variation of H during inflation, so that the last term of (5.5) is usually ≤ 1. The first term depends on the evolution of scale factor a between the end of slow-roll inflation and the present epoch. Assuming that inflation ends by a short matter-dominated period, which is followed by the RD stage lasting until the present matter-dominated era begins, one has [283]

$$N_{max} = 62 - \ln\frac{10^{16}GeV}{\sqrt{H_{end}M_P}} - \frac{1}{3}\left(\frac{\sqrt{H_{end}M_P}}{\rho_{reh}^{1/4}}\right), \tag{5.6}$$

where ρ_{reh} is energy density determined at the reheating temperature T_{reh}, when the RD stage is established. With $H_{end} \simeq 10^{13}GeV$ and instant reheating this gives $N_{max} \approx 62$, the largest possible value.

In local supersymmetric models to avoid gravitino overproduction [284, 285] T_{reh} should be much smaller, even as small as $T_{reh} < 4 \cdot 10^6 GeV$, as it follows from the analysis [286] of ^6Li production by gravitino decay products (see [3] for review). There should have been a long MD stage after the end of inflation until the heating of the Universe. Then the value of N_{max} should be even less than 58. On the other hand, the constraints on T_{reh}, following from the analysis of primordial gravitino effects, imply the realization of a model of supergravity, coupled with matter, being, thus, model-dependent. Moreover, even in the case of low T_{reh}, N_{max} can still be 62, if the inflaton field oscillations provide the RD stage after the end of inflation (see Chapter 2). With all these reservations we will use further $N_{max} = 60$.

The smallest cosmological scale of antimatter islands that survived after annihilation is $k_c^{-1} = l_c \approx 8h^2 kpc$ [275]. It is 9 order of magnitude smaller than H_0^{-1}, that corresponds to

$$N_c \approx N_{max} - 13 - 3\ln h \approx 45. \tag{5.7}$$

Thus, the l_c should have left the horizon at 45 e-folds before the end of inflation to provide the survival of the corresponding domain in future.

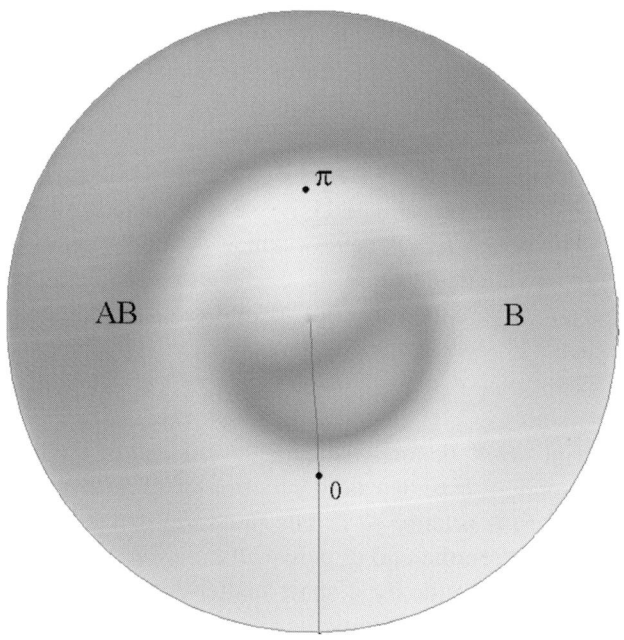

Figure 5.1.   Mexican hat potential, the view from above.

Let us imagine that the phase value $\theta = 0$ corresponds to the North Pole of the NG field circle valley, and $\theta = \pi$ corresponds to the South Pole (see Figures 4.1, 5.1). The phase grows along the anti-clockwise direction, and the bottom of the PNG potential is located at the North Pole of the circle. It will be shown below that the antibaryon production corresponds to the regions (marked as 'AB' in Figure 5.1) that would contain phase values that correspond to anticlockwise rolling of the PNG field ϑ during the first half-period of oscillation. Baryon production should take place, if the field ϑ rolls clockwise toward the North Pole of tilted potential just after the first oscillation began. This area is marked as 'B' in Figure 5.1

Now we are able to perform a preliminary estimation of the fraction of volume of the Universe containing antimatter islands. According to the previous paragraph to ensure that the Universe would be baryon asymmetric as a whole with the net baryon charge being positive, it is necessary to suppose that the phase average value $\theta = \theta_U$ within the biggest cosmological scale of interest, emerging at the $N_{max} = 60$ e-folds before the end of inflation, is located in the range $[0, \pi]$. The θ_U is the starting point for Brownian motion of the phase value along the circle valley during inflation. As has been mentioned above, the phase makes Brownian step $\delta\theta = \frac{H}{2\pi f}$ at each e-fold. Because the typical wavelength of the fluctuation $\delta\theta$, generated during such step, is equal to H^{-1}, the whole domain H^{-1}, containing θ_U, after one e-fold effectively becomes divided onto e^3 separate, causally disconnected, domains of radius H^{-1}. Each domain contains almost homogeneous phase value $\theta_{59} = \theta_U \pm \delta\theta$. In half of these domains the phase evolves toward π (the South Pole) and in the other domains it moves toward zero (the North Pole). One should require that the phase value reaches the values π or 0 during the first 15 steps. Only in this case the antimatter islands would have the size larger than l_c, and they would survive up to the modern era. It means that at least one of the two following conditions must be satisfied

$$\theta_{60} \geq \pi - \frac{15H}{2\pi f}, \tag{5.8}$$

$$\theta_{60} \leq \frac{15H}{2\pi f}.$$

Consider initially the case of exact equalities in the expression (5.8), when the main part of antimatter is contained in the domains with antimatter of size l_c. The number of domains containing the equal values of phase at the 45 e-folds before the end of inflation satisfies the following expression

$$n_{45} \approx (e^3/2)^{15} \approx 10^{15}. \tag{5.9}$$

In the case of the second inequality in (5.8) this value represents the number of islands with extremely dilute antimatter content because of the phase θ being almost zero. To obtain more or less dense antimatter islands we have to select those domains which move away from the phase $\theta = 0$. The probability that every domain of size l_c would not be separated at the next e-fold into e^3 domains with size one order of magnitude lower than l_c, and with different signs of the phase θ is given by $P_s \approx (1/2)^{e^3} \approx 10^{-6}$. Thus, the number of domains serving as the prototypes for antimatter islands of size l_c can be estimated as follows

$$\bar{n} = n_{45} P_s \approx 10^9. \tag{5.10}$$

The same estimation is valid in the case of the first inequality in (5.8). The probability P_s accounts in this case for the non-division of high density antimatter domain with the size l_c onto smaller domains of the different baryon charge. Otherwise, the annihilation of such smaller size domains with the opposite baryon charge can prevent the domain's survival.

There are about 10^{11} galaxies in the observed part of the Universe. Thus, according to (5.10) we reveal that 1% of all galaxies contain the region of size l_c filled with antimatter of highest possible antibaryon density, if the θ_{60} coincides with the first inequality (5.8), or of the lowest one in the case, if the second inequality is held.

We can also find the size distribution for antimatter islands. To this end it is necessary to study the inhomogeneities of phase induced by (5.3). It has been well established that for any given scale $l = k^{-1}$ large-scale component of the phase value θ is distributed in accordance with Gauss's law [260, 51, 57, 282]. The quantity which will be especially interesting for us is the dispersion (5.4) for quantum fluctuations of phase in the period from $k = H^{-1}$ to $k_{min} = l_{max}^{-1}$ (where the biggest cosmological scale l_{max} corresponds to 60 e-folds). This quantity can be expressed in the following manner

$$\sigma_l^2 = \frac{H^2}{4\pi^2} \int_{k_{min}}^{k} d\ln k = \frac{H^2}{4\pi^2} \ln \frac{l_{max}}{l} = \frac{H^2}{4\pi^2 f^2}(60 - N_l), \qquad (5.11)$$

where N_l is the number of e-folds for any scale l. This means that the distribution of phase has the Gaussian form

$$P(\theta_l, l) = \frac{1}{\sqrt{2\pi}\sigma_l} \exp\left\{-\frac{(\theta_{60} - \theta_l)^2}{2\sigma_l^2}\right\}. \qquad (5.12)$$

Suppose that at e-fold N_t before the end of inflation the volume $V(\bar{\theta}, N_t)$ has been filled with phase value $\bar{\theta}$. Then at the e-fold $N_{t+\Delta t} = N_t - \Delta N$ the volume filled with average phase $\bar{\theta}$ satisfies the following iterative expression

$$V(\bar{\theta}, N_{t+\Delta t}) = e^3 V(\bar{\theta}, N_t) + (V_U(N_t) - e^3 V(\bar{\theta}, N_t))P(\bar{\theta}, N_{t+\Delta t})\sigma_{N_{t+\Delta t}}, \quad (5.13)$$

here the $V_U(N_t) \approx e^{N_t} H^{-1}$ is the volume of the Universe at N_t e-fold. With the use of expression (5.13) one can calculate the size distributions of domains filled with appropriate value of phase numerically. In order to illustrate quantitatively the number distribution of domains, we present here the numerical results for specific values of $\theta_U \equiv \theta_{60} = \pi/6$ and $h = \frac{H}{2\pi f} = 0.026$. The table contains the results as concerns the number of domains containing antimatter with average phase $\bar{\theta} \le 0$ at e-fold number N,

Number of e-fold	Number of domains	Size of domain
59	0	$1103 Mpc$
55	$5.005 \cdot 10^{-14}$	$37.7 Mpc$
54	$7.91 \cdot 10^{-10}$	$13.9 Mpc$
52	$1.291 \cdot 10^{-3}$	$1.9 Mpc$
51	0.499	$630 kpc$
50	74.099	$255 kpc$
49	$8.966 \cdot 10^3$	$94 kpc$
48	$8.012 \cdot 10^5$	$35 kpc$
47	$5.672 \cdot 10^7$	$12 kpc$
46	$3.345 \cdot 10^9$	$4.7 kpc$
45	$1.705 \cdot 10^{11}$	$1.7 kpc$

The fraction of the Universe filled with phase $\bar{\theta}$ appears to be equal to $7.694 \cdot 10^{-9}$.

One can see that the size distribution of domains is peaked at smallest value. Assuming that h is the free parameter we can easily obtain several regions with a negative phase $\bar{\theta}$ and sizes larger or equal to the critical surviving size. In spite of a sufficiently large total number of antimatter islands, only the small part of our Universe could be occupied by them.

Up to now we did not distinguish two rather different cases when the phase crosses South and North Poles. The only similarity is that both cases lead inevitably to domains with antimatter. Meantime, the final picture looks rather different in these cases.

If we came to negative phase from the minimum of the potential (North Pole), a boundary between baryon and antibaryon areas is mostly very wide. It takes place because the phase $\theta = 0$ corresponds to zero baryon charge. So the average antibaryon density in domains, originated from the crossing of the North Pole, is very low. Only small amount of more dense antimatter domains (with the internal density approaching the average baryon density) are formed due to Gaussian tail in the amplitude distribution of fluctuations.

Another situation takes place if the phase reaches the values $\theta > \pi$. First of all, the matter-antimatter boundary will be very sharp and strong annihilation would take place after the end of inflation. Secondly, this boundary must contain a field wall according to discussion in previous Chapter. Estimations revealed that if the antimatter domain is large enough for to be survived in future, the wall appears to be unacceptably massive.

2. The problem with large-scale fluctuations

This mechanism suffers also from another shortage. Indeed, fluctuations of the amplitude of the phase θ must not be smaller than ~ 0.01 to obtain a substantial amount of antimatter domains. In this case the first several phase fluctuations, corresponding to the modern scales from galaxy superclusters to the modern horizon, are too large to be unobservable. They should have created fluctuations of CMB temperature at the level exceeding the observed one. A possible way to suppress these large-scale fluctuations and hence to improve the situation is to take into account the interaction of the χ-field with inflaton.

Consider this interaction in the form (φ – inflaton, χ – complex field)

$$V(\varphi, \chi) = \lambda(|\chi|^2 - f^2/2)^2 - g|\chi|^2 (\varphi - cM_P)^2, \qquad (5.14)$$

where λ, g and c are parameters of the potential. This potential having the same form of the Mexican hat possesses the minimum at

$$|\chi| \equiv f_{eff}(\varphi) = \sqrt{f^2 + \frac{g}{\lambda}(\varphi - cM_P)^2}. \qquad (5.15)$$

Position of the minimum is not a constant now, but it is strongly dependent on the classical value φ of the inflaton. This value is ruled by classical equations of motion and varies in the range $(M_P \div 10M_P)$. Average amplitude of fluctuations of the phase θ of the field χ is inversely proportional to the scale $f_{eff}(\varphi)$.

Inflaton varies with time as

$$\varphi(t) = \varphi_U - \frac{m_\varphi M_P}{2\sqrt{3\pi}} t \qquad (5.16)$$

for the quadratic potential $U(\varphi) = m_\varphi^2 \varphi^2 / 2$. In this case we have $m_\varphi \cong H$ at the end of inflation. Thus, N e-foldings after our Universe was born inflaton has the value

$$\varphi = \varphi_N = \varphi_U - \frac{M_P}{2\sqrt{3\pi}} N. \qquad (5.17)$$

Now effective scale f_{eff} (5.15) can be represented in the form

$$f_{eff}(N) = f \sqrt{1 + \frac{g}{\lambda} \frac{M_P^2}{f^2} \left[(\frac{\varphi_U}{M_p} - c) - \frac{N}{2\sqrt{3\pi}} \right]^2}. \qquad (5.18)$$

Denoting $(\frac{\varphi_U}{M_p} - c) \equiv \frac{N_f}{2\sqrt{3\pi}}$ one comes to final expression for effective energy scale

$$f_{eff}(N) = f \sqrt{1 + \frac{g}{12\pi\lambda} \frac{M_P^2}{f^2} (N_f - N)^2}. \qquad (5.19)$$

This expression has an important property. Indeed, there is an *a priori* large parameter $M_P^2 / f^2 \sim 10^{10}$ and for reasonable relation between the parameters g and λ one can reveal that the value $f_{eff}(N)$ has a very sharp minimum at $f_{eff}(N_f) = f$. The immediate consequence is that the amplitude of the phase θ fluctuations

$$\langle \delta\theta \rangle = \frac{H}{2\pi f_{eff}(N)} \qquad (5.20)$$

increases sharply in the vicinity of the e-fold number $N = N_f$.

In this way the problem of large-scale fluctuations could be resolved. In addition, we acquire an interesting feature of the fluctuations – their amplitude is large only at e-folds around e-fold number N_f. As a result, we come to very tight size distribution of the antimatter domains. To give some idea about the values, let us choose numerical values of parameters as follows: initial phase $\theta_U = \pi/8$; parameter $A = 5$ $(A \equiv \frac{g}{12\pi\lambda} \frac{M_P^2}{f^2})$ and $N_f = 15$. Using iteration procedure (5.13), one can easily find that there are $1.3 \cdot 10^{10}$ antimatter domains with antibaryon density equal to the average baryon density in our Universe and with the size of about $10^{21} cm$. The distribution is very sharp so that actually the domains of other sizes are virtually absent. Antibaryon islands with lower antibaryon density have a peak at the same size, but their abundance is about 10^{16} in the visible part of the Universe. If this situation takes place, our Universe contains one dense antimatter domain per ten galaxies. Such domain is able to form globular clusters of stars made from antimatter. Low density antibaryon domains could not participate in galaxy formation and should be spread in the intergalactic space. The strategy of the experimental search for both low and high density antimatter areas in the modern Universe is discussed in Chapter 6.

3.   Spontaneous baryogenesis mechanism

Another important element of our scenario of inhomogeneous baryogenesis contains the conversion of the phase θ into baryon/antibaryon excess in the considered mechanism of spontaneous baryogenesis [287, 288]. The basic feature of this mechanism is that the sign of baryon charge created by relaxation of energy of the PNG field critically depends on the direction that the phase is rotated toward the minimum in the bottom of slightly tilted Mexican hat potential. By this reason domains containing the phase values $\theta < 0$ or $\theta > \pi$ convert into the domains with antimatter, when the potential gets the tilt and the PNG field moves to its minimum.

One reasonable issue to the spontaneous baryogenesis has been considered in the work [84]. Let us briefly discuss it. It was assumed that in the early Universe a complex scalar field χ coexists with inflaton ϕ responsible for inflation. This field χ has non vanishing baryon number. The possible interaction of χ that violates lepton number can be described by the following Lagrangian density

$$L = -\partial_\mu \chi^* \partial^\mu \chi - V(\chi) + i\bar{Q}\gamma^\mu \partial_\mu Q + i\bar{L}\gamma^\mu \partial_\mu L -$$

$$\tag{5.21}$$

$$m_Q \bar{Q}Q - m_L \bar{L}L + (g\chi\bar{Q}L + h.c.).$$

The fields Q and L could represent heavy quark and lepton coupled to the ordinary quark and lepton matter fields. Since fields χ and Q possess baryon number, while the field L does not, the couplings in Lagrangian (5.21) violate lepton number [84]. The $U(1)$ symmetry that corresponds to baryon number conservation is expressed by the following transformations

$$\chi \rightarrow \exp{(i\beta)}\chi, \qquad Q \rightarrow \exp{(i\beta)}Q, \qquad L \rightarrow L. \tag{5.22}$$

The effective Lagrangian density for θ, Q and L eventually has the following form after symmetry breaking

$$L = -\frac{f^2}{2}\partial_\mu \theta \partial^\mu \theta + i\bar{Q}\gamma^\mu \partial_\mu Q + i\bar{L}\gamma^\mu \partial_\mu L \tag{5.23}$$

$$-m_Q \bar{Q}Q - m_L \bar{L}L + (\frac{g}{\sqrt{2}}f\bar{Q}L + h.c.) + \partial_\mu \theta \bar{Q}\gamma^\mu Q.$$

At the energy scale $\Lambda \ll f$, the symmetry (5.22) is explicitly broken and the Mexican hat circle gets a little PNG tilt described by the following potential

$$V(\alpha) = \Lambda^4(1 - \cos\theta). \tag{5.24}$$

This potential, of height $2\Lambda^4$, has minima at $\alpha = \theta f = 2\pi N f$ with N integer or zero, and so it has the unique minimum $\alpha = \theta f = 0$ at small amplitudes of the field α. Of course, in most cases, the potential (5.24) is the lowest-order approximation to a more complicated expression, emerging from particle physics models (see, e.g., [289] and Refs. therein).

The important parameter for spontaneous baryogenesis is the curvature of potential (5.24) in the vicinity of its minimum, which determines the mass of the PNG field

$$m_\theta^2 = \frac{\Lambda^4}{f^2}.$$ (5.25)

As it was mentioned above the field α is an additional field with nondominant energy density contribution into the total density on the de Sitter stage. Also we assume that the field α behaves as a massless NG field during inflation implying that the condition

$$m_\theta \ll H$$ (5.26)

is valid, where H is the Hubble constant during inflation. After the end of inflation, when condition (5.26) is violated, the oscillations of field θ around the minimum of potential (5.24) are started. The energy density $\rho_\theta \simeq \theta_i^2 m_\theta^2 f^2$ of the PNG field which has been created by quantum fluctuations of θ during the inflation converts to baryons and antibaryons [287, 84]. The sign of baryon charge depends on the local initial value of phase from which the oscillations are started.

Let us estimate the number of baryons and antibaryons produced by classical oscillations of phase θ with an arbitrary initial phase θ_i. The appropriate expression for the density of produced baryons (antibaryons) $n_{B(\bar{B})}$ is represented in [84] in the limit of small θ_i

$$n_{B(\bar{B})} = \frac{g^2}{\pi^2} \int_{m_Q+m_L}^{\infty} \omega d\omega \left| \int_{-\infty}^{\infty} dt \chi(t) e^{\pm 2i\omega t} \right|^2,$$ (5.27)

that is valid if $\chi(t \to -\infty) = \chi(t \to +\infty) = 0$.

The general case with arbitrary initial phase can be obtained in the limits $\chi(t \to -\infty) \neq 0; \chi(t \to +\infty) = 0$ without loss of generality. After integration by part expression (5.27) has the form

$$N_{B(\bar{B})} = \frac{g^2}{4\pi^2} \Omega_{\theta_i} \int d\omega \left| \int_{-\infty}^{\infty} d\tau \dot{\chi}(\tau) e^{\pm 2i\omega\tau} \right|^2,$$ (5.28)

where Ω_{θ_i} is the volume containing the phase value θ_i. Here the surface terms appear to be zero at $t = \infty$ due to asymptotes of field χ and at $t = -\infty$ due to Feynman radiation conditions.

For our estimations it is enough to accept that the phase changes as

$$\theta(t) \approx \theta_i(1 - m_\theta t)$$ (5.29)

during first oscillation. More correct formulae lead to more complicated expressions that could be calculated numerically, which is not necessary for our estimations made below. Substituting (5.29) and (5.2) into (5.28) and after some algebra, we come to

$$N_{B(\bar{B})} \approx \frac{g^2 f^2 m_\theta}{8\pi^2} \Omega_{\theta_i} k(\theta_i), \quad k(\theta_i) = \theta_i^2 \int_{\mp\frac{\theta_i}{2}}^{\infty} d\omega \frac{\sin^2 \omega}{\omega^2}, \qquad (5.30)$$

where the sign in the lower limit of integral corresponds to baryon or antibaryon net excess production, respectively.

To compare our result with the result of [84], let us calculate the integral in the limit $\theta_i \ll 1$. We find

$$N_B - N_{\bar{B}} = \frac{g^2 f^2 m_\theta}{8\pi^2} \Omega_{\theta_i} \theta_i^3. \qquad (5.31)$$

The comparison shows good agreement between the results of both approaches in the limit of small amplitude oscillations around the minimum, corresponding to $\theta = 0$. Using for spatially homogeneous field $\chi = \frac{f}{\sqrt{2}} e^{i\theta}$ the following formula for the created baryon charge

$$Q = i(\chi^* d\chi/dt - d\chi^*/dt\chi) = -f d\theta/dt, \qquad (5.32)$$

one can easily conclude that the baryon charge $Q > 0$ if $\theta > 0$ during classical movement of phase θ to zero. Thus, the clockwise rotation gives rise to origin of baryon excess while the anti-clockwise rotation leads to the antibaryon excess.

During reheating, the inflaton energy density converts into the one of radiation. It is assumed that reheating takes place, when the Mexican hat potential does not yet feel the PNG tilt. This implies that the total decay width of inflaton into light degrees of freedom should be rather quick, as compared with the period of the PNG field oscillations, so that $\Gamma_{tot} \gg m_\theta$. The oscillations of θ field start, when $H \approx m_\theta$. The time variation of the phase leads to creation of baryons or antibaryons according to the above consideration. The entropy density after thermalization is given by

$$s = \frac{2\pi^2}{45} g_* T^3, \qquad (5.33)$$

where g_* is a total effective massless degrees of freedom. Here we are concerned with the temperature above the electroweak symmetry breaking scale. At this temperature all the degrees of freedom of the Standard Model are in equilibrium and g_* is at least equal to 106.75. The temperature is connected with the expansion rate as follows

$$T = \sqrt{\frac{M_P H}{1.66 g_*^{1/2}}} \approx \frac{\sqrt{M_P m_\theta}}{g_*^{1/4}}. \qquad (5.34)$$

The last part of the expression (5.34) takes into account that the relaxation starts at the condition $H \approx m_\theta$. Using the formulas (5.30), (5.33), (5.34) we get the baryon/antibaryon asymmetry

$$\frac{n_{B(\bar{B})}}{s} = \frac{45g^2}{16\pi^4 g_*^{1/4}} \left(\frac{f}{M_P}\right)^{3/2} \frac{f}{\Lambda} k(\theta_i).$$ (5.35)

The function $k(\theta_i)$ accounts for the dependence of the amplitude of baryon asymmetry and of its sign on the initial phase value, which appeared in the different space regions during inflation. The behavior of this function could be easily calculated numerically using its definition by the expression (5.30).

Expression (5.35) allows us to get the observable baryon asymmetry of the Universe as a whole $n_B/s \approx 3 \cdot 10^{-10}$. In the model under consideration we have supposed initially $f \geq H \simeq 10^{-6} M_P$. The natural value of coupling constant is $g \leq 10^{-2}$ and the observed baryon asymmetry is obtained at quite reasonable condition $f/\Lambda \geq 10^5$ (see, e.g., [289]).

4.   Radiation of baryon charge during wall shrinking

As we have discussed, there are two ways to obtain domains with antimatter in the baryon-dominated Universe. Up to now we dominantly considered the case, when the phase θ crosses the value $\theta = 0$ due to the fluctuations at inflation stage. There is another case, when the phase θ crosses the value $\theta = \pi$. This case leads to almost inevitable closed wall formation.

Our analysis performed above concerned (anti-) baryon abundance as the result of oscillation of the phase θ in a volume as a whole. As one can see from Eq. (5.28), the particle production is, in fact, the result of any process of radiation of the phase θ. After the phase reaches its vacuum value $\theta = 0, 2\pi...$ the particle production is terminated. However, the crossing of the phase $\theta = \pi$ during the inflationary stage, leads to islands of vacuum with $\theta_{domain} = 2\pi$ in the sea of the phase $\theta_{sea} = 0$. The phase continuously changes from 0 to 2π across the wall that is placed at the phase value $\theta_{wall} = \pi$. The classical motion of the wall represents some kind of phase variation with time. As it was shown in Chapter 4, closed walls accelerate rapidly during shrinking, and hence their radiation of (anti-) baryons could contribute significantly in some conditions.

Let us estimate the possible effect of wall radiation. To proceed, suppose that the created baryon/antibaryon excess is proportional to the total energy of the wall, whereas such excess in the whole volume inside the wall is proportional to the total energy of phase oscillations. The energy of the wall is $E_w \sim \Lambda^2 f \cdot R_w^2$. The energy of the phase oscillations, when they start inside the wall is $E_V \sim \Lambda^4 \cdot R_w^3$ by an order of magnitude. Consequently, the ratio of baryon excess B_w, originated from wall radiation to the one B_V in the whole volume is less than unity in that period

$$\frac{B_w}{B_V} \sim \frac{E_w}{E_V} \sim \frac{d}{R} < 1. \tag{5.36}$$

In the course of successive expansion the size of the wall grows as the scale factor a, so that the total energy of the wall and the baryon excess created by its motion grows correspondingly as a^2, whereas the baryon excess in the volume inside the wall does not change. It leads to violation of the condition (5.36), and the contribution of wall radiation could be important. Additional argument to investigate this process is that the wall motion produces baryon content of the sign opposite to that produced by the phase oscillation in the volume, which could be important in the case of small walls. Let Lagrangian of interaction of the complex field with fermions have the form

$$L_{\text{int}} = g\varphi(x)\overline{Q}(x)L(x) + h.c. \tag{5.37}$$

Standard result of quantum mechanics gives the expression for number of particles radiated by the wall

$$\sum_{s_Q, s_L} \int |A_{k_Q, k_L}|^2 \frac{d^3 k_Q}{2\varepsilon_Q (2\pi)^3} \frac{d^3 k_L}{2\varepsilon_L (2\pi)^3}, \tag{5.38}$$

where A_{k_Q, k_L} is the amplitude of creation of Q – particle with momentum k_Q and L – particle with momentum k_L

$$\left\langle Q, L \left| \int d^4 x g\varphi(x)\overline{Q}(x)L(x) \right| 0 \right\rangle. \tag{5.39}$$

To be more specific, consider complex field φ with potential (4.24), (4.26). As usual, radial component of the field is placed at its minimum $\varphi = f/\sqrt{2}$. As it was discussed above, field configuration of the angular component represents the wall, which separates the vacuum with $\theta = 0$ from the other with $\theta = \pi$. The vacua are degenerated and the motion of the wall is governed by initial conditions and by its surface tension.

Straightforward calculations of the matrix element and summation over spins leads to the expression

$$|A_{k_Q, k_L}|^2 = g^2 \frac{f^2}{2} Tr \left(\hat{k}_L + m_L \right) \left(\hat{k}_Q - m_Q \right) \cdot$$
$$\cdot \left| \int d^4 x \exp\{i\theta + iEt - i(\mathbf{P r})\} \right|^2, \tag{5.40}$$
$$E = \varepsilon_Q + \varepsilon_L, \quad \mathbf{P} = \mathbf{k}_Q + \mathbf{k}_L.$$

Calculation of integral (5.40) represents nontrivial (though solvable) problem due to space and time dependence of the phase θ. Integration by part of the integral slightly facilitates this problem. Surface terms contribute only to renormalization of wave function and must be omitted at the calculation of radiation [294]. One could apprehend it more easily by considering a wall at rest. Such wall does not radiate in spite of the presence of surface terms. Some calculations lead to the matrix element of the form

$$\left|A_{k_Q,k_L}\right|^2 = 2g^2 f^2 \left[(k_Q k_L) - m_L m_Q\right] |I|^2,$$

$$I \equiv \frac{1}{E} \int d^4 x \dot\theta \exp\left\{i\theta + iEt - i(\mathbf{P}\,\mathbf{r})\right\}.$$

Now we have to specify the form of the classical configuration of the field θ. It would be convenient to choose a motion of a spherical wall which is collapsing due to its internal pressure. Back reaction of radiated fermions on the total energy of the wall is neglected in the following. The wall is chosen being at rest in the infinite past and in the infinite future. If the Lagrangian of the field θ has the form

$$L = \frac{1}{2}\left(\partial_\mu \theta\right) + \frac{\Lambda^2}{f}\left(\cos\theta - 1\right), \tag{5.41}$$

its classical equation of motion possesses the known solution for a plane wall at rest

$$\theta(x,t) = -4arctg\left[\exp\left(\frac{\Lambda^2}{f}\left(x - x_0\right)\right)\right] \tag{5.42}$$

The spherical wall is characterized by its radius, being much greater, than the wall width $d \approx f/\Lambda^2$. Hence, a small mistake will be made, if we choose the solution in the form

$$\theta(r,t) = -4arctg\left[\exp\left(\frac{\gamma}{d}\left(r - R(t)\right)\right)\right], \tag{5.43}$$

where $R(t) = R_0 - u(t)\cdot t$ stands by the wall radius and $\gamma(t) = [1 - u(t)^2]^{-1/2}$. In this case the integral is represented by the formula

$$I = \frac{2\pi}{Ed}\int dt\, r^2 dr\, d\cos\chi\,\gamma\,\theta'(\xi)\dot R(t)\exp(i\theta\,(\xi) + iEt - iPr\cos\chi),$$

where $\zeta = \gamma/d((r - R(t))$. Integrating out angle χ, we obtain

$$I = \frac{4\pi}{EPd}\int dt\,\gamma(t)R(t)\dot R(t)e^{iEt}\int dr\,\theta'(\xi)\sin(Pr)e^{i\theta(\xi)}. \tag{5.44}$$

In this formula it was taken into account that the function ζ is varied in a tight area $\sim d/\gamma$ at $r = R(t)$. To proceed it is necessary to know a time dependence of the wall radius $R(t)$, that can be easily obtained from the energy conservation. The energy of the wall at rest is $W_0 = 4\pi R_0^2\sigma$. Equating it to the wall energy at an arbitrary instant of time $W(t) = 4\pi R(t)^2 \gamma(t)\sigma$ and taking into account the expression for Lorentz factor with wall velocity $u(t) = \dot{R}(t)$, one comes to the equation

$$\dot{R}^2 = 1 - R^4/R_0^4. \tag{5.45}$$

This equation is used below to change the variable $t \to R$:

$$dt = dR/\left[1 - (R/R_0)^4\right].$$

In this case integral (5.44) is expressed in the form

$$I = \frac{4\pi R_0^2}{EPd} \int\limits_{d/2}^{R_0} dR \, R^{-1} \, e^{iEt(R)} \int dr \, \theta'(\xi) \sin(Pr) e^{i\theta(\xi)}, \tag{5.46}$$

where the relationship $\gamma = R_0^2/R_2$ was kept in mind. The final formula for the number of radiated particles can be obtained after the integration out of the angles in the expression (5.40)

$$N = \frac{g^2 f^2 R_0^4}{8\pi^6 d^2} \int d^3 k_Q d^3 k_L \frac{(k_Q k_L) - m_L m_Q}{\varepsilon_Q \varepsilon_L E^2 P^2} J,$$

$$J = \left| \int\limits_{d/2}^{R_0} dR \, R^{-1} \, e^{iEt(R)} \int\limits_0^{\infty} dr \, \theta'(\xi) \sin(Pr) e^{i\theta(\xi)} \right|^2. \tag{5.47}$$

Numerical calculation of this expression is possible but it is not simple. Fortunately, we need only its estimation. For this we neglect masses of the radiated particles and estimate their momenta by order of magnitude as $1/d$. It leads to the estimation of baryon (or more definitely, Q-quanta) number emitted by the wall during its shrinking

$$N_{B(\bar{B})} \approx \frac{g^2 f^2 R_0^4}{8\pi^6 d^2} J_{\pm}, \tag{5.48}$$

$$J_{\pm} = \left| \int\limits_{d}^{R_0} dR \, R^{-1} \, e^{it(R)/d} \int\limits_{-\infty}^{\infty} d\xi \, \theta'(\xi) \sin(\xi + R/d) e^{\pm i\theta(\xi)} \right|^2.$$

Numerical calculations indicate that $J_+ \gg J_-$ provided that $R_0 \gg d$. For example, $J_+ = 24.5$, $J_- = 0.15$ at initial radius $R_0 = 40d$. Thus, if the closed wall, with the phase $\theta = 2\pi$ inside it, is shrinking, baryon generation dominates. It is interesting to note that in such domains antibaryons were produced at first stage. This stage is characterized by classical motion of the phase in the whole volume to its stationary value which was equal 2π.

The comparison of the analytical estimations indicates that baryon production by collapsing walls is very effective. Namely, a baryon–antibaryon ratio equals approximately R_0/d. Nevertheless a caution is necessary. Indeed, our estimation is valid if the back reaction is neglected. It means that the total energy of radiated particles must in any case be smaller than the total wall energy. Estimating with the use of Eq. (5.48) the total energy of radiated semirelativistic particles as $E_p \sim \varepsilon_Q \cdot N_B \sim \frac{1}{d} \frac{g^2 f^2 R_0^4}{8\pi^6 d^2} J_+$ and comparing it with the energy of the wall $E_w \sim f\Lambda^2 R_0^2 \sim \frac{f^2}{d} R_0^2$, one obtains that the whole energy of the wall can be radiated for walls with the size

$$R_0 > R_c \equiv \frac{2\pi^3}{g} \cdot \sqrt{\frac{2}{J_+}} \cdot d. \qquad (5.49)$$

On the other hand, the calculations made in Chapter 4 show that the number of walls increases sharply with decreasing of their size. Thus, we come to an interesting conclusion – the large walls produce antibaryon islands, meantime a large number of small walls contribute to baryon background.

Note that the possibility of energy dissipation due to radiation for large moving walls can slow them down more effectively than the radiation friction, considered in Chapter 4.

5. Discussion

In this chapter we have considered the particular example of an inflationary model with inhomogeneous baryogenesis. This is a successful model for generation of antimatter islands with appropriate sizes exceeding the critical surviving size. The antibaryon density relative to background baryon density in the resulting antimatter islands and the number of these islands depends on the incidental value of phase that has been established at the 60 *e*-folds and on the parameters of the PNG field potential. It is possible to obtain one or several antimatter domains in our galaxy depending on the values of these parameters. The observational consequences of existence of such domains and the restrictions on their number and sizes have been analyzed in papers [275, 85] and will be discussed in the next chapter.

As we have mentioned, one of the additional problems for most models of inhomogeneous baryogenesis, invoked by the phase transitions at the inflation epoch, is the prediction of large-scale topological defects. Our scheme also contains the premise for existence of domain walls. When the PNG tilt is significant and if initial phase is close to π, domain walls are formed along the closed surfaces with $\theta = \pi, 3\pi, \ldots$ [295, 265]. In other words every antimatter island with high relative antibaryon density will be surrounded by a domain wall. The wall energy per unit surface is

$$\Delta \approx 8f\Lambda^2. \qquad (5.50)$$

This stress energy responds to the oscillation of the wall bag. During the oscillations, the energy, stored in the walls, is released in the form of quanta of the PNG field and gravitational waves.

We would like to note that the regions with antimatter in baryon asymmetrical Universe arise naturally in the various models, using in different ways the similar idea of baryon excess generation by primordial scalar field, as in the model of spontaneous baryosynthesis. The main issue, that is needed, is the existence of a valley of potential for this field. It is the valleys that are responsible for formation of causally separated regions with different values of a field, which in turn lead to antimatter domains. Many models based on supersymmetry possess this property.

To be more specific, consider briefly another model of baryogenesis by baryonic charge condensate [296], [273] with the potential of the form

$$U(\varphi) = m^2 |\varphi|^2 + \frac{1}{2}\lambda_1 |\varphi|^4 + \frac{1}{4}\lambda_2 \left(\varphi^4 + \varphi^{*4}\right).$$

Baryonic charge of field φ is not conserved due to the last term:

$$\partial_\mu j_B^\mu \equiv i\partial_\mu \left(\varphi^* \cdot \partial^\mu \varphi - \partial^\mu \varphi^* \cdot \varphi\right) = i\lambda_2 \left(\varphi^{*4} - \varphi^4\right),$$

where $\lambda_1 = -\lambda_2 \equiv \lambda > 0$ is supposed for simplicity. After the end of inflation the field is governed by classical equations of motion

$$\ddot{\varphi}_1 + 3H\dot{\varphi}_1 + \left(m^2 + 4\lambda\varphi_2^2\right)\varphi_1 = 0,$$
$$\ddot{\varphi}_2 + 3H\dot{\varphi}_2 + \left(m^2 + 4\lambda\varphi_1^2\right)\varphi_2 = 0.$$

Near the bottom of the potential at $\varphi_1 = \varphi_2 = 0$ the field rotates with almost constant momentum that gives rise to production of baryon charge. The direction of rotation is determined by an arbitrary initial condition at inflation stage. If the conditions are slightly changed, it could easily result in another direction of rotation after inflation. But it surely happens due to fluctuations in the case of domains of smaller sizes and some of them acquire antibaryon excess.

In the multi-field picture the possibility of unstable walls arises. The succession of phase transitions can lead to appearance of walls at certain temperatures and to their disappearance at smaller temperatures [216].

The set of theoretical arguments for possible existence of antimatter domains, surviving to the present time in the Universe with globally positive baryon excess, is, as we discuss further, even much wider. It provides serious grounds to experimental search for antimatter in the Universe. The theoretical analysis of possible forms of antimatter objects and their signature is the important component of such search. The results of this analysis are presented in the next chapter.

Chapter 6

ANTIMATTER IN THE MODERN UNIVERSE

The use of travelling is to regulate imagination by reality, and instead of thinking how things may be, to see them as they are.

Samuel Johnson

It was shown in [85], [276], [83] and discussed in the previous chapter that the existence of antimatter domains in the baryon-dominated Universe is a profound signature for the origin and evolution of primordial baryon matter inhomogeneity. Depending on its parameters the mechanism of inhomogeneous baryosynthesis can lead to both high and low antibaryon density domains. According to [85] high density domains can evolve into antimatter stellar objects so that a globular cluster of antimatter stars can exist in our Galaxy, that may be tested in the cosmic searches for antimatter planned for the near future. Such searches involve both direct search for pieces of antimatter – for antinuclei or antimeteorites, or use indirect probes by gamma radiation that may be originated from antimatter annihilation.

1. Introduction

In the baryon asymmetric Universe the Big Bang theory predicts the exponentially small fraction of primordial antimatter and practically excludes the existence of primordial antinuclei. The secondary antiprotons may appear as a result of cosmic ray interaction with the matter. In such interaction it is impossible to produce any sizeable amount of secondary antinuclei. Thus, non exponentially small amounts of antiprotons in the Universe in the period from 10^{-3} to 10^{16} s and antinuclei in the modern Universe are the profound signature for new phenomena, related to the cosmological consequences of particle theory.

The inhomogeneity of baryon excess generation and antibaryon excess generation as the reflection of this inhomogeneity represents one of the most important examples of such consequences. It turned out [85], [276], [83], [3] that practically all the existing mechanisms of baryogenesis can lead to generation of antibaryon excess in some places; the baryon excess, averaged over the whole space, being positive. So domains of antimatter in baryon asymmetric Universe provide a probe for the physical mechanism of the matter generation.

The original Sakharov's scenario of baryosynthesis [59, 60] has found physical grounds in GUT models. It assumes CP-violating effects in out-of-equilibrium B-non-conserving processes, which generate baryon excess proportional to CP-violating phase. If sign and magnitude of this phase varies in space, the same out-of-equilibrium B-non-conserving processes, leading to baryon asymmetry, result in $B < 0$ in the regions where the phase is negative. The same argument is appropriate for the models of baryosynthesis, based on electroweak baryon charge non-conservation at high temperatures as well as on its combination with lepton number violation processes, related to the physics of Majorana mass of neutrino. In all these approaches to baryogenesis independent of the physical nature of B-non-conservation the inhomogeneity of baryon excess and generation of antibaryon excess is determined by the spatial dependence of the CP-violating phase.

Spatial distribution of this phase is predicted in models of spontaneous CP violation, modified [216] to escape the supermassive domain wall problem (see rev. in [83, 276] and Refs. therein).

In this type of model the CP-violating phase acquires discrete values $\phi_+ = \phi_0 + \phi_{sp}$ and $\phi_- = \phi_0 - \phi_{sp}$, where ϕ_0 and ϕ_{sp} are, respectively, constant and spontaneously broken CP phase, and antibaryon domains appear in the regions with $\phi_- < 0$, provided that $\phi_{sp} > \phi_0$.

In models where the CP-violating phase is associated with the amplitude of invisible axion field, spatially-variable phase ϕ_{vr} changes continuously from $-\pi$ to $+\pi$. As was shown in Chapter 4, the axion-induced antibaryon excess forms the Brownian structure looking like an infinite ribbon along the infinite axion string (see [77, 78]). This structure is smoothed by the annihilation at the border of matter and antimatter domains. When the antibaryon diffusion scale exceeds the minimal width of the ribbon $l_h(T)$, given by Eq. (4.17), the infinite structure decays on separated domains. The distribution on domain sizes turns to be strongly model-dependent and was calculated in [85].

The size and amount of antimatter in domains, generated in the result of local baryon-non-conserving out-of-equilibrium processes, is related to the parameters of models of CP violation and/or invisible axion (see rev. in [276, 18, 3]). SUSY GUT-motivated mechanisms of baryon asymmetry imply flatness of superpotential relative to existence of squark condensate. Such a condensate, being formed with $B > 0$, induces baryon asymmetry, after squarks decay on quarks and gluinos. The mechanism does not fix the value and sign of B in the condensate, opening the possibilities for inhomogeneous baryon charge distribution and antibaryon domains [18]. The size and amount of antimatter in such domains is determined by the initial distribution of squark condensate.

So antimatter domains in baryon asymmetric Universe are related to practically all the mechanisms of baryosynthesis, and serve as the probe for the mechanisms of CP violation and primordial baryon charge inhomogeneity. The size of domains depends on the parameters of these mechanisms. In the previous chapter we gave a quantitative estimation of possible domain size distribution in the mechanism of spontaneous baryogenesis

With account of all possible mechanisms for inhomogeneous baryosynthesis, predicted on the basis of various and generally independent extensions of the Standard Model, the general analysis of possible domain distributions is rather complicated. Fortunately, the test for the possibility of the existence of antistars in our Galaxy, offered in [85], turns to be practically model independent and as we show here may be accessible to cosmic ray experiments, and to AMS experiment, in particular.

EGRET data [218] on diffuse gamma background show a visible peak around $E_\gamma \approx 70$ MeV in gamma spectrum, which fact can be naturally explained by the decays of π^0-mesons, produced in nuclear reactions. Interactions of the protons with gaseous matter in the Galaxy shift the position of such a peak to higher values of gamma energy due to 4-momentum conservation. At the same time the secondary antiprotons, produced in the cosmic ray interactions with interstellar gas, are too energetic [219] and their annihilation also cannot explain the observational data.

The above consideration draws attention to the model with an antimatter globular cluster existing in our Galaxy, which cluster can serve as a permanent source of antimatter due to (anti)stellar wind or (anti)Supernova explosions.

On the other hand, as was mentioned in [85], low antibaryon density domains cannot evolve into gravitationally bound objects. With the case of such "diffused antiworld" we begin the discussion of the possible forms and signatures for antimatter in the modern Universe to be considered in the present chapter.

2. Diffused antiworld

There are several reasons for the possibility of low antibaryon density in antimatter domains. In models of inhomogeneous baryosynthesis with spontaneous CP violation (see [3, 290, 84, 276] for review) both constant φ_c and spontaneously broken φ_s , CP-violating phases are involved in baryosynthesis. Provided that $\varphi_s > \varphi_c$ in CP domains with $\varphi = \varphi_c - \varphi_s$ the antibaryon excess is generated. If both phases are of the same order of magnitude $\varphi_s \sim \varphi_c$, so that $\varphi_s - \varphi_c << \varphi_s \sim \varphi_c$, the antibaryon excess density within the antimatter domain is much lower than the baryon excess density.

Another possibility for a low antibaryon density can appear in the model of axion-dominated CP violation in baryosynthesis [85]. Small-scale domain annihilation in the region with a local antibaryon excess results in spreading over the larger volume, thus reducing antibaryon density. The same is generally true for a stochastic small-scale baryon–antibaryon domain structure of any origin.

At the density of antimatter ρ_{ab} within a domain by 3 order of magnitude less, than the baryon density ρ_b (which we assume in the further discussion corresponding to $\Omega_b = 0.1$) cosmological nucleosynthesis in the period t $\sim 1 - 10^3 s$ results in the nontrivial chemical composition [275]. For $10^{-4} < \rho_{ab}/\rho_b < 10^{-3}$ antideuterium is the dominant product. For smaller densities of antibaryons within domains no antinuclei are formed. At the densities $\rho_{ab}/\rho_b < 10^{-4}$ owing to low antimatter densities inside the domain no recombination takes place at $z \sim 1500$ and the antimatter domain remains ionized after recombination in the Universe [275]. The radiation pressure and the radiation dominance in the energy density within the domain suppress then the development of gravitational instability, so that antimatter domains of sufficiently large size should now be clouds of ionized positron–antiproton plasma presumably situated in voids.

Below we give a quantitative estimation for the surviving size and observational effects of domains of diffused antiworld [275].

Let us assume that the Universe contains regions of a very small antibaryon excess density. At temperatures above some MeV these regions cannot be strongly affected by the diffusion of surrounding particles, because their mean free path is small enough. Therefore, we will consider the evolution of these antibaryon domains at the temperatures $4 \cdot 10^3 K = T_{rec} < T < T_{nucl} = 10^9 K$, when the Universe is filled with plasma of electrons, protons, photons and neutrinos and matter diffuse inside the antimatter domains. Separate consideration will be done below for temperatures below the period of recombination at $T < T_{rec}$, when neutral atoms move almost freely.

If the size of the antibaryon region is much greater, than the mean free path of surrounding particles, we can solve a one-dimensional problem assuming that the "initial" baryon density at $T_{in} = 10^9 K$ is given by

$$n_b(\mathbf{R}, t_{in}) = n_0 \theta(-x), \qquad (6.1)$$

where θ is step function.

Note that since the antibaryon component is very small we neglect it for the moment. In this case the diffusion equation for baryons is

$$\partial n_b / \partial t = D(t) \partial^2 n_b / \partial x^2 - \alpha n_b. \qquad (6.2)$$

The last term in Eq. (6.2) takes into account the expansion of the Universe. Diffusion coefficient $D(t)$ is expressed as follows

$$D(t) \approx \frac{3 T_\gamma c}{2 \rho_\gamma \sigma_T} \approx 0.61 \cdot 10^{32} Z^{-3}, cm^2/s, \qquad (6.3)$$

where T_γ and ρ_γ are the temperature and the energy density of the radiation, respectively, c is the velocity of light, σ_T is the Thomson cross-section, Z is the red shift which is related with the time t on the RD stage by $t \approx 2.6 \cdot 10^{19} s/Z^2$.

Let us introduce a new variable r defining baryon to photon ratio $r = n_b/n_\gamma$. Since the evolution of the photon density is given by the equation $\partial n_\gamma/\partial t = -\alpha n_\gamma$, we can rewrite Eq. (6.2) in terms of r as

$$\partial r/\partial t = D(t)\partial^2 r/\partial x^2, \qquad (6.4)$$

where

$$r(\mathbf{R}, t_0) = r_0 \theta(-x). \qquad (6.5)$$

To solve this equation, it is suitable to introduce new variable u instead of time t :

$$u = \int_{t_{in}}^t D(t')dt'.$$

Eq. (6.4) acquires the simple form

$$\partial r/\partial u = \partial^2 r/\partial x^2,$$

which has the solution

$$r(x, u) = \frac{1}{2} n_0 \left[1 + \Phi \left(\frac{x}{2\sqrt{u}} \right) \right].$$

Due to the properties of the error integral Φ one can conclude that the boundary between matter and antimatter regions is determined as

$$x_b = 2\sqrt{u}.$$

The diffusion coefficient is connected with the temperature in the following manner

$$D(T) = D_{nucl} \left(\frac{T_{nucl}}{T} \right)^3.$$

Its time dependence has different forms at RD and MD periods. Our aim is to calculate the value $u_{rec} \equiv u(t_{rec})$. There are two different periods $t_{nucl} < t < t_{eq}$ and $t_{eq} < t < t_{rec}$. The first, radiation-dominated period is characterized by the time–temperature relationship

$$T = T_{eq} \left(\frac{t_{eq}}{t} \right)^{1/2},$$

where t_{eq} is the time, corresponding to the transition from RD to MD stage, and T_{eq} is the temperature at this time.

Such relationship at the second period of matter dominance may be expressed in the form

$$T = T_{eq} \left(\frac{t_{eq}}{t} \right)^{2/3}.$$

Now we are ready to calculate the value u_{rec},

$$u_{rec} = \int_{t_{nucl}}^{t_{rec}} D(t')dt' = \int_{t_{nucl}}^{t_{eq}} D(t')dt' + \int_{t_{eq}}^{t_{rec}} D(t')dt' = \qquad (6.6)$$

$$= \int_{t_{nucl}}^{t_{eq}} D_{nucl} \left(\frac{T_{nucl}}{T} \right)^3 dt' + \int_{t_{eq}}^{t_{rec}} D_{nucl} \left(\frac{T_{nucl}}{T} \right)^3 dt' =$$

$$= D_{nucl} T_{nucl}^3 \left[2t_{eq} T_{eq}^2 \int_{T_{eq}}^{T_{nucl}} \frac{dT'}{T'^6} + \frac{3}{2} t_{eq} T_{eq}^{3/2} \int_{T_{rec}}^{T_{eq}} \frac{dT'}{T'^{11/2}} \right] \simeq$$

$$\simeq D_{nucl} t_{eq} \left[\frac{2}{5} \left(\frac{T_{nucl}}{T_{eq}} \right)^3 + \frac{1}{3} \left(\frac{T_{nucl}}{T_{rec}} \right)^3 \left(\frac{T_{eq}}{T_{rec}} \right)^{3/2} \right].$$

For the initial time t_{nucl}, corresponding to the temperature $T_{nucl} = 10^9 K$, $D(t_{nucl})$ is given, according to Eq. (6.3), by

$$D(t_{nucl}) \equiv D_{nucl} \approx 1.24 \cdot 10^6 cm^2/s. \qquad (6.7)$$

The boundary shift during the whole period can be easily calculated now

$$x_b = 2\sqrt{u_{rec}}.$$

It follows from Eq. (6.6) that the average displacement of the boundary between baryon and antibaryon domains is about $\Delta x \sim 0.2pc$. Therefore, the primordial antibaryon regions of low density, which grow up to $1pc$ or more to the period of recombination, have to be conserved in spite of the diffusion of ordinary matter.

Note that according to Eq. (6.6) the motion of boundary is mainly determined by the second term which is responsible for the matter-dominated epoch. This is due to the fact that the radiation friction is less effective for plasma on the MD stage and as a consequence the mean free path increases, giving the important contribution to the boundary motion.

Below the temperature $T_{rec} = 4000K$ atoms are formed in baryon domains. Since the antibaryon density is assumed to be small enough ($\rho_{ab}/\rho_b < 10^{-4}$) in our approach, we can consider the flow of hydrogen atoms into antibaryon region as a motion of free-streaming atoms into empty space. The physical distance travelled by the atoms after recombination until the present time t_p is given by

$$d \approx a_p \int_{t_{rec}}^{t_p} \frac{v(t)dt}{a(t)}, \qquad (6.8)$$

where $v(t)$ is an average velocity of atoms, $a(t)$ is the scale factor of the Universe. It is convenient to set $a = ya_p$, where $y = 1/(1 + Z) = T_p/T$ and to rewrite Eq. (6.8) as

$$d \approx \int_{a_{rec}}^{a_p} \frac{v(a)da}{a\dot{a}} = \int_{a_{rec}/a_p}^{1} \frac{v(y)dy}{y\dot{y}} \tag{6.9}$$

At the recombination, the velocity of atoms is given by the thermal value $v_{rec} \approx c\sqrt{T_{rec}/m}$, where m is the mass of atom. After $4000K$, the typical velocity $v \approx p/m$ of atoms is red-shifted down like $1/a \sim T$. Therefore, below 4000 K we have $v \approx c\frac{T_p}{m}\sqrt{\frac{m}{T_{rec}}}\frac{1}{y}$. Taking the equation

$$(\dot{y})^2 = \frac{8\pi}{3}G\rho(y)y^2 \tag{6.10}$$

where G is Newton's constant and substituting a density in the form $\rho(y) = \rho_0/y^3$ in the case of the matter-dominated Universe we find

$$d \approx \frac{c}{\sqrt{\frac{8\pi}{3}G\rho_0}}\frac{T_p}{m}\sqrt{\frac{m}{T_{rec}}}\int_{T_p/T_{rec}}^{1}\frac{dy}{y^{3/2}} = \frac{2c}{\sqrt{\frac{8\pi}{3}G\rho_0}}\sqrt{\frac{T_p}{m}} \tag{6.11}$$

Substituting $\rho_0 = \rho_c$, where $\rho_c = 1.88 \cdot 10^{-29}h^2g/cm^3$ is the critical density of the Universe, we obtain $d \sim 3/h$ kpc. For value of Hubble constant h between 0.4 and 1 the free-streaming length of atoms will be of the order of several kpc. Therefore antibaryon regions of the same size will be filled at present by hydrogen atoms. Note that mutual penetration of antiprotons in the matter regions and of matter atoms into the antimatter regions is not equivalent in the considered asymmetric case. For the matter gas the antimatter domain is transparent due to the low density of antiprotons in the domain, making $n_{ab}\langle\sigma v\rangle t < 1$. On the other hand, matter is opaque for antiprotons, since $n_b\langle\sigma v\rangle t > 1$ even now.

A key observation to test the model of diffused world could be the search for gamma rays from a boundary annihilation of antimatter and hydrogen atoms. Let us consider first the possibility of the annihilation in antimatter domains filled with hydrogen atoms (the annihilation of matter–antimatter domains in baryon symmetric Universe was considered in [272]). Taking into account the annihilation and expansion of the Universe the number density of antiprotons is described by the equation

$$dn_{ab}/dt = - <\sigma v> n_b n_{ab} - \alpha n_{ab}. \tag{6.12}$$

In the limit $n_b \gg n_{ab}$ we can neglect the variation of n_b due to annihilation. Then introducing $r = n_b/n_\gamma$, $\hat{r} = n_{ab}/n_\gamma$ and solving Eq. (6.12) we find that at present time in antimatter domains filled with hydrogen atoms

$$\hat{r}_p = \hat{r}_{rec} \exp[-\int_{t_{rec}}^{t_p} \langle \sigma v \rangle r n_\gamma dt]. \tag{6.13}$$

Here indexes p and rec denote present period and recombination. Since according to [291] at energies below $10eV$ the cross-section of $\bar{p}H$ annihilation is given by

$$< \sigma v > \approx 2.7 \cdot 10^{-9} cm^3/s \tag{6.14}$$

we find that the integral in Eq. (6.13) is much greater than unity. So, no gamma radiation takes place at present from such regions because practically all antiprotons have been already annihilated.

Therefore the radiation is possible only from the narrow region of the boundary between matter and antimatter domains, where antibaryons have not been annihilated yet. The width of this region $d \sim v\Delta t$, where Δt is defined from the condition

$$\int_{t_p-\Delta t}^{t_p} \langle \sigma v \rangle n_b dt \approx \langle \sigma v \rangle n_b \Delta t \sim 1 \tag{6.15}$$

where the velocity is expressed in the form

$$v \approx c \frac{T}{m} \sqrt{\frac{3m}{T_{rec}}} \tag{6.16}$$

Substituting all necessary numbers we find the width of the region

$$d \approx c \sqrt{\frac{T_p}{T_{rec}} \frac{T_p}{m}} \frac{1}{\langle \sigma v \rangle n_b} \approx 0.86 \left(\frac{10^{-7} cm^{-3}}{n_b} \right), pc. \tag{6.17}$$

The gamma flux at the Earth in this case is given by

$$\frac{d\Phi}{d\omega d\Omega} \approx \frac{dn}{dt} \frac{dN_\gamma}{d\omega} \frac{V}{4\pi r_A^2}, \tag{6.18}$$

where

$$dn/dt = < \sigma v > n_b n_{ab} \tag{6.19}$$

is the rate of annihilation per unit volume per unit time; $dN_\gamma/d\omega$ is the differential cross-section for an inclusive gamma production; $V = 4\pi R^2 d$ is a volume of the annihilating part of the diffused world at present; R is the size of the diffused world; r_A is a distance between the Earth and the diffused world. Integrating Eq. (6.18) over photon energy we obtain

$$\frac{d\Phi}{d\Omega} \approx \langle \sigma v \rangle \beta n_b^2 \langle N_\gamma \rangle d \left(\frac{R}{r_A} \right)^2, \tag{6.20}$$

At small energies the cross-section must be proportional to the inverse power of the antiproton velocity. To find this dependence we have to match the available experimental data on σ_{ann} with this expected behavior. As it follows from data [220, 221], obtained at CERN–LEAR, the dependence $\sigma_{ann} \sim v^{-1}$ is valid already for laboratory antiproton momenta $p_{lab} \leq 1000$ MeV/c. The annihilation cross-section is the difference between total and inelastic ones, $\sigma_{ann} \approx \sigma_{tot} - \sigma_{el}$. Thus, at $P_{lab} \geq 300$ MeV/c the data from [223] for the total and elastic cross- sections were used [86] and at momenta less than 300 MeV/c the dependence

$$\sigma_{ann}(P < 300 \text{ MeV/c}) = \sigma_0 \, C(v^*)/v^*$$
$$\sigma_{el} = const \,, \tag{6.25}$$

was used [86] for annihilation and elastic cross-sections, respectively, where v^* is the velocity of the antiproton in the $\bar{p}p$ center-of-mass system. Additional Coulomb factor $C(v^*)$ gives a large increase for the annihilation cross-section at small velocities of the antiproton and is defined by the expression [215]:

$$C(v^*) = \frac{2 \pi v_c/v^*}{1 - \exp\left(-2 \pi v_c/v^*\right)} \,, \tag{6.26}$$

where, $v_c = \alpha c$, with α and c being the fine structure constant and the speed of light, respectively.

Using the experimental data on the $\bar{p}p$ annihilation cross-section [220, 221] it was found in [86] that value σ_0 in Eq. (6.25) is equal to:

$$\sigma_0 = \sigma_{ann}^{exp}(P = 300 \text{ MeV/c}) = 160 \text{ mb}\,.$$

Consider following [86] the spherical model for halo with z axis directed to the North Pole and x axis directed to the Solar system. The number density distribution of interstellar hydrogen gas $n_H(r, z)$ along z direction was parameterized as:

$$n_H(z) = n_H^{halo} + \Delta_H(z)\,,$$
$$\Delta_H(z) = \frac{n_H^{disk}}{1 + (z/D)^2}\,, \tag{6.27}$$

with $n_H^{halo} = 5 \cdot 10^{-4}$ cm^{-3} being the hydrogen number density in the halo, $n_H^{disk} = 1$ cm^{-3} being the hydrogen number density in the disk and $D = 100pc$ being the half-width of the gaseous disk. Here the hydrogen number density in the halo is chosen in suggestion that $\sim 90\%$ of the halo mass is a non-baryonic dark matter. Such a distribution of the matter gas is to a large extent the worst case for our aims since the matter density along z axis falls slowly and visible fraction of the antiprotons will annihilate sufficiently far from the galactic disk plane. Nevertheless, as we shall see, even in this case the picture is still quasi-stationary and the antiproton number density in the halo is practically not disturbed by the annihilation in the dense regions.

The validity of the stationary approximation depends on the interplay of the lifetime of the antiprotons relative to the annihilation and their confinement time in the Galaxy. To evaluate the antiproton confinement time the results of the "two–zone" leaky box model (LBM) [219] were used in [86]. The authors of [219] considered the spectra of secondary antiprotons produced in collisions of the cosmic ray protons with interstellar gas. If we compare the antiproton spectrum, obtained in [219], one easily observes that shape of the spectrum beautifully reproduces the observational data on $\bar{p}/p$ ratio. But the predicted total normalization is lower by factor $2 \div 3$, than the data. Owing to the fact that confinement time enters as a common factor in the predicted $\bar{p}/p$ ratio, we find necessary factor, performing the fit to the observational data. Experimental points have been taken in [86] from [222], where references on the data can be found.

The data on the cosmic ray $\bar{p}/p$ ratio, used in [86], have been collected in balloon experiments. The region of low kinetic energies, $E_{kin} \leq 100$ MeV, is strongly affected by the heliosphere [224]. To avoid this influence two of the most left points in Figure 6.1 were removed in [86] from the fit.

The solid curve in Figure 6.1(a), taken from [86], represents the "two–zone" LBM predictions for the $\bar{p}/p$ ratio, multiplied by the fitted factor $K = 2.58$, which factor increases the confinement time for slow antiprotons in the Galaxy up to $5.5 \cdot 10^8$ years. The dashed curve is the phenomenological fit in the form $R(E) = a E^{b+c \lg E}$, which is plotted for comparison. The shapes of both curves match fairly. Figure 6.1(b) shows the resulting antiproton confinement times for Galaxy as whole (solid) and for disk only (dashed).

Figure 6.2, taken from [86], shows the antiproton lifetime to the annihilation (a) and the free path length of the antiprotons (b) versus their distance from the galactic plane, z for three values of the antiproton velocity. In the stationary case to compensate the annihilation of antiprotons with matter gas the number density of antiprotons must satisfy the equation:

$$\frac{d^2 n_{\bar{p}}}{dE\, dt} = I_{\bar{p}}(E) - v\,\sigma(v)\, n_H\, \frac{dn_{\bar{p}}}{dE}. \tag{6.28}$$

The solution of this equation is:

$$\frac{dn_{\bar{p}}}{dE} = I_{\bar{p}}(E)\, t_{ann}(E) \left(1 - e^{-t/t_{ann}}\right), \tag{6.29}$$

with $t_{ann} = [v\,\sigma(v)\, n_H]^{-1}$ being the lifetime of antiprotons relative to the annihilation.

From Figure 6.2(a) we can conclude that for antiprotons with velocities 10^3 km/s (stellar wind) the confinement time in the halo, starting from distances $z \sim 2$ kpc, is less than their annihilation time. Thus, from Eq. (6.29) we obtain for the halo:

$$n(E) \approx I_{\bar{p}}(E)\, T_{conf}. \tag{6.30}$$

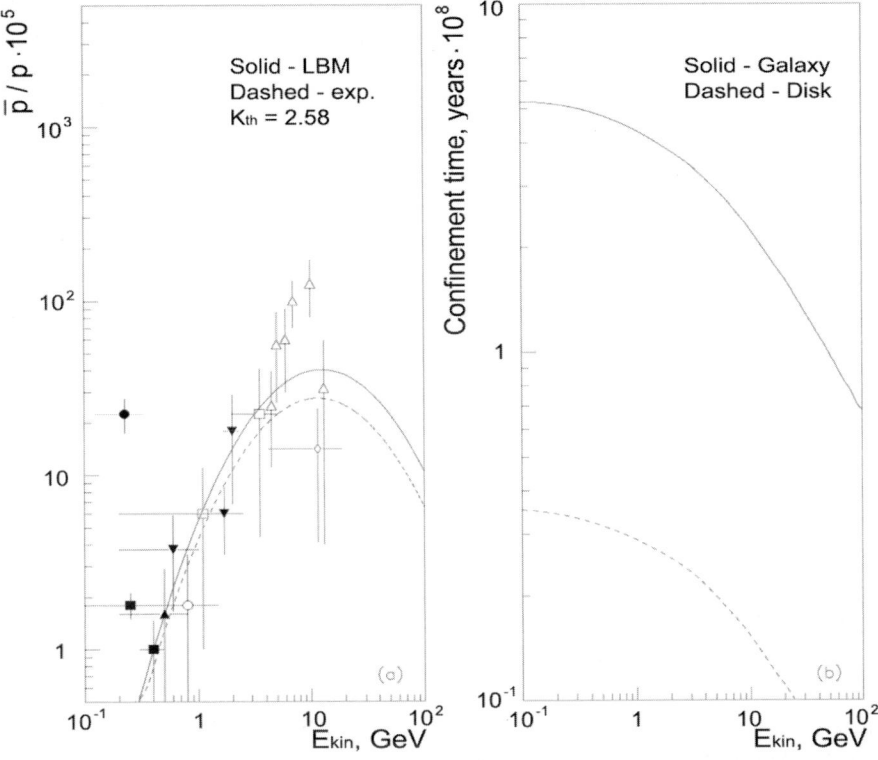

Figure 6.1. (a) Fit of the $\bar{p}/p$ ratio to experimental data. Solid line shows predictions of the two–zone leaky box model [219], increased by factor $K \approx 2.6$. The dashed curve is the phenomenological fit, described in the text. (b) The respective confinement times for the antiprotons in the Galaxy (solid) and in the disk (dashed). The curves are taken from [219] and multiplied by factor K.

In the gaseous disk the situation is just the opposite. Antiprotons annihilate with high rate and their lifetime relative to the annihilation is much less than the time necessary for them to escape from the volume of the Galaxy.

In other words, antiprotons are stored in the halo during the confinement time $\approx 5 \cdot 10^8$ yrs increasing the gamma flux by factor T_{conf}. We can also conclude that during large confinement time antiprotons are being spread over the halo with constant number density independent of the position of the antistar cluster. Under the assumption of the universal acceleration mechanism in the halo their energy spectrum comes to the stationary form. Additionally from Figure 6.2(a) we see that the "storaging" volume is of the order of the volume of the halo $V_{halo} = 4\pi R_{halo}^3/3$, when the region with $T_{conf} \gg T_{ann}$ is restricted by $|z| \leq 2$ kpc. Thus, intensive annihilation takes place within the volume $V_{ann} \approx \pi R_{halo}^2 \, 4 \, kpc$. The ratio of these two volumes is of the

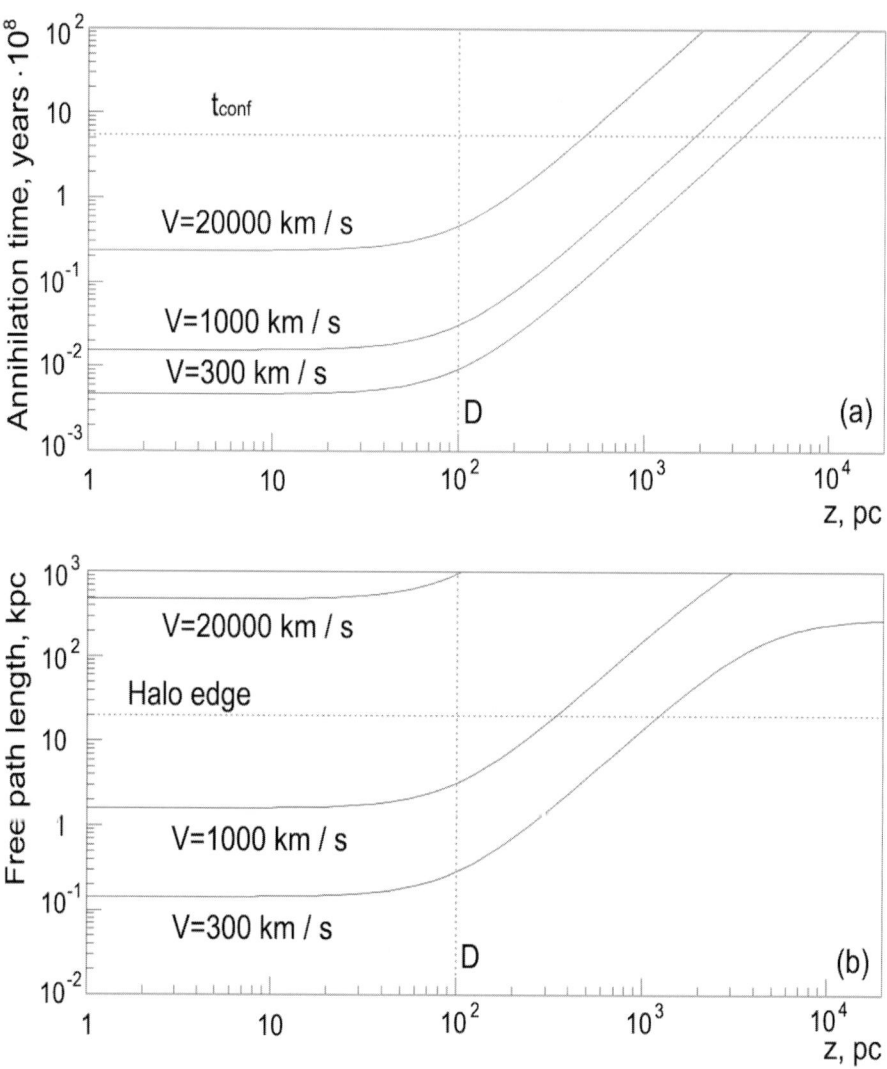

Figure 6.2. (a) The dependence of the antiprotons annihilation time on z coordinate. The horizontal dashed line is the antiproton confinement time in the Galaxy. (b) The dependence of free path length of the antiprotons. The horizontal dashed line is the halo edge $z = 20$ kpc. The curves are calculated for three values of the antiproton velocity: 300 km/s, 10^3 km/s and $2 \cdot 10^4$ km/s. The vertical dashed line shows the half-width of the disk $D = 100$ pc.

order of

$$\frac{V_{ann}}{V_{halo}} \sim \frac{4\,kpc}{4/3\,R_{halo}} \leq 20\%$$

and the annihilation of antiprotons in the gaseous disk practically does not affect the number density of antiprotons in the Galaxy as a whole.

The above consideration provides quasi-stationary distribution of antimatter in the halo and, as a result, leads to the constant number density of the antiprotons in the galactic halo. Figure 6.2(b) shows z dependence of free path length of antiprotons at three values of their velocity [86].

4.2   Diffuse gamma flux

The gamma flux arriving from the given direction is defined by the well-known expression:

$$J_\gamma(E_\gamma) = \int_0^L dl\, \psi(E_\gamma, r, z). \qquad (6.31)$$

The integration must be performed up to the edge of the halo $L = -\alpha_x R_\odot + \sqrt{R_{halo}^2 - R_\odot^2 (1 - \alpha_x^2)}$ with α_x being the cosine of the angle between the line-of-sight and the x axis, directed from the Galaxy center to the Sun and lying in the plane of the Solar orbit.

Function $\psi(E_\gamma)$ in Eq. (6.31) is the intensity of gamma sources along the observation direction l in assumption of isotropic distribution of gamma emission. This function is defined as:

$$\psi(E_\gamma, r, z) = \int_{E_{min}}^\infty dE\, v(E)\, \sigma_{ann}(E)\, n_H(r, z)\, n_{\bar{p}}(E, r, z)\, W(E_\gamma; E)$$

$$(6.32)$$

$$W(E_\gamma; E) = \frac{dn_\gamma(E_\gamma; E)}{dE_\gamma\, dO}.$$

To simulate the gamma energy spectrum and angular distribution $W(E_\gamma; E)$ the Monte Carlo technique was used in [86]. The experimental data [225] on the $\bar{p}p$ annihilation at rest (see Table) have been used to simulate the probabilities of different final states. In practice, the approximation of the annihilation at rest is valid with very good accuracy up to laboratory momenta of the incoming antiprotons about 0.5 GeV because at these laboratory momenta the kinetic energy of the antiproton is still by order of magnitude less than the twice antiproton mass. The simulation of particle distribution in the final state has been performed according to phase space in the center-of-mass of the $\bar{p}p$ system. PYTHIA 6.127 package [226] has been used in [86] to perform the subsequent decays of all unstable particles. Momenta of stable particles ($e^\pm$, $p/\bar{p}$, $\mu^\pm$, γ and neutrinos) have been boosted to the laboratory reference frame. The resulting average number of γs per annihilation is [86]

$$< n_\gamma > = \int d\Omega\, dE_\gamma\, W(E_\gamma; E) = 3.93 \pm 0.24$$

and agrees with the experimental data.

COSMOLOGICAL PATTERN OF MICROPHYSICS

Table 6.1. Relative probabilities of $\bar{p}p$ annihilation channels.

Channel	Rel. prob., %	Channel	Rel. prob., %
$\pi^+\pi^-\pi^0$	3.70	$2\pi^+2\pi^-\eta$	0.60
$\rho^-\pi^+$	1.35	$\pi^0\rho^0$	1.40
$\rho^+\pi^-$	1.35	$\eta\rho^0$	0.22
$\pi^+\pi^-2\pi^0$	9.30	$4.99\,\pi^0$	3.20
$\pi^+\pi^-3\pi^0$	23.30	$\pi^+\pi^-$	0.40
$\pi^+\pi^-4\pi^0$	2.80	$2\pi^+2\pi^-$	6.90
$\omega\pi^+\pi^-$	3.80	$3\pi^+3\pi^-$	2.10
$\rho^0\pi^0\pi^+\pi^-$	7.30	$K\bar{K}\,0.95\pi^0$	6.82
$\rho^+\pi^-\pi^+\pi^-$	3.20	$\pi^0\eta'$	0.30
$\rho^-\pi^+\pi^+\pi^-$	3.20	$\pi^0\omega$	3.45
$2\pi^+2\pi^-2\pi^0$	16.60	$\pi^0\eta$	0.84
$2\pi^+2\pi^-3\pi^0$	4.20	$\pi^0\gamma$	0.015
$3\pi^+3\pi^-\pi^0$	1.30	$\pi^0\pi^0$	0.06
$\pi^+\pi^-\eta$	1.20		

In the stationary case we can put the annihilation rate in the halo being constantly compensated by the permanent source of the antiprotons. But, owing to the fact that the antiproton's annihilation rate in the gaseous disk is much greater than in the halo, we must take into account the dependence of the antiproton density on z coordinate. Figure 6.2(b) demonstrates that free path length of the slowest antiprotons is comparable with half-width of the disk D. To take this effect into account we have to consider antiproton annihilation with the gas in the disk. For given value of z we have:

$$\frac{dn_{\bar{p}}(z, E)}{dz} = \sigma_{ann}(E)\,\Delta_H(z)\,n_{\bar{p}}(z, E). \qquad (6.33)$$

The differential equation Eq. (6.33) can be easily solved and results in the following antiproton number density distribution along the z axis [86]:

$$n_{\bar{p}}(z, E) = n_0 \exp\left\{-\sigma_{ann}(E)\int_z^{z_{max}} dz'\,\Delta_H(z')\right\}, \qquad (6.34)$$

where, $z_{max} = L\,\alpha_z$ is the maximal value of z coordinate, defined by the edge of the halo, and n_0 is the antiproton number density far from the disk.

The next point we need to consider is the antiproton energy spectrum. As it will be shown further, the stellar wind from antistars has to give the most significant contribution in the antimatter pollution from the anticluster. The original distribution of the stellar wind particles has a Gaussian form peaking at velocities $v \approx 500$ km/s [227]. The interplanetary shocks accelerate emitted particles and the resulting stellar cosmic rays flux becomes proportional to $J_{SW} \sim v E_{kin}^{-2}$ in the range of kinetic energies up to ~ 100 MeV [227]. Additional acceleration occurs in the interstellar plasma and, as we believe, produces the observable spectrum of the galactic cosmic rays $\sim v E_{kin}^{-2.7}$. The acceleration mechanisms are defined by collisionless shocks in interplanetary or Galaxy plasma and are charge-independent. One has to take into account also the relative movement of the hypothetical antistar cluster with velocity ~ 300 km/s as well as the similar velocities of the matter gas defined by the gravitational field of the Galaxy. Thus, one can expect that the minimal velocity of antiprotons from (anti)stellar wind relative to the matter gas is something about $v_{min} \approx 600 - 700$ km/s. Following the above consideration, we chose the antiproton spectrum in the halo (far from the regions with high matter gas density) to be similar to the galactic cosmic rays proton spectrum in the whole range of the antiproton energies:

$$ n_{\bar{p}}(E, z >> D) \sim \left(\frac{1\,\text{GeV}}{E_{kin}} \right)^{2.7} , \tag{6.35} $$

with the normalization:

$$ \int_{E_{min}}^{\infty} n_{\bar{p}}(E, z >> D)\, dE = n_0 . $$

Actually, reasonable variation of the form of the antiproton flux does not affect significantly the total normalization and changes only the gamma spectrum at higher energies. The main contribution in the integrated antiproton number density comes from the slowest antiprotons owing to fast rise of the annihilation cross-section with the decrease of the velocity. We do not consider here the contribution in the gamma flux from the annihilation of the secondary antiprotons produced in the collisions of the cosmic ray protons with interstellar gas. This effect must give the main contribution at higher energies of gammas and needs careful investigation of the deceleration mechanisms in the halo.

If we assume that all the gamma background at high galactic latitudes is defined by the antiproton annihilation, we have the only free parameter in our model – the minimal velocity of the antiprotons v_{min}. Therefore, for given value v_{min} the integrated number density of antiprotons in the halo n_0 can be found from comparison with the observational data on diffuse gamma flux. If we choose the minimal velocity of antiprotons to be of the order of the velocity of the stellar wind, $v_{SW} \approx 1000$ km/s, being equivalent to kinetic energy of the antiprotons $E_{kin}^{SW} \approx 5.2$ keV, we obtain the necessary integral number density of antiprotons n_0, corresponding to the fit of EGRET data by $\bar{p}p$ annihilation, to be equal to [86]:

$$ n_0^{SW} \approx 5.0 \cdot 10^{-12} \text{ cm}^{-3}. \tag{6.36} $$

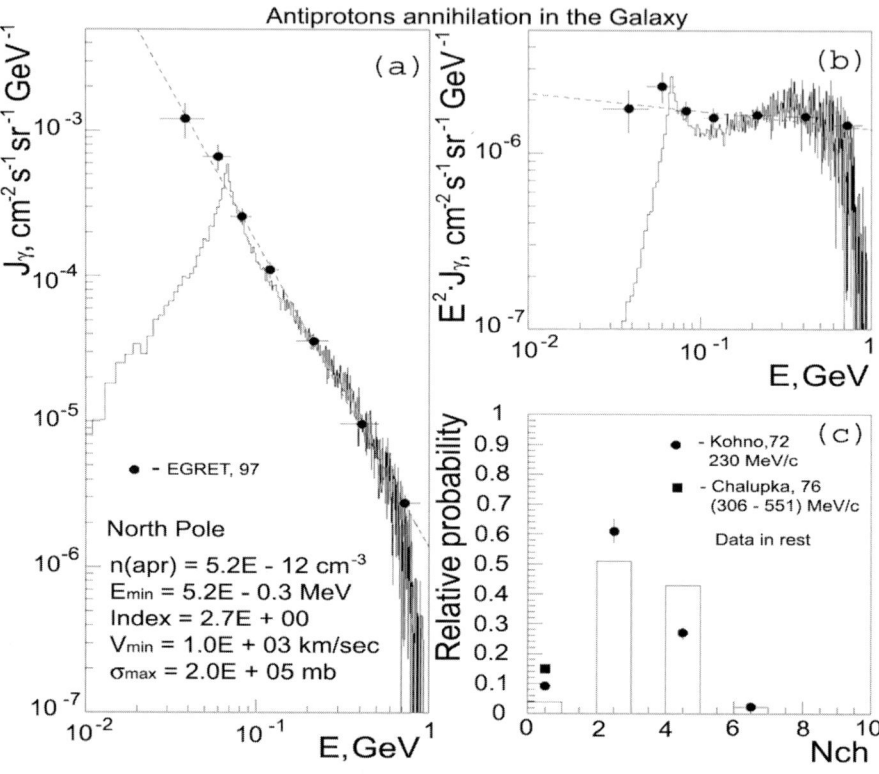

Figure 6.3. Comparison of the calculated differential fluxes of γ quanta from $\bar{p}/p$ annihilation for the minimal antiproton velocity $v_{min} = 10^3$ km/s with experimental data *EGRET* [218] on diffuse gamma background (a,b). The observational direction is to the North Pole of the Galaxy. There is also shown the comparison of the charged multiplicity distribution in the annihilation model described in the text with the existent experimental data (c). Circles – [228], squares – [229].

Figure 6.3(a,b), taken from [86], demonstrates the resulting differential gamma distribution in the Galactic North Pole direction in comparison with EGRET data [218] in the range $10 \leq E_\gamma \leq 1000$ MeV. The peak of π^0 decay is clearly seen both in calculations as well as in experimental distributions. Figure 6.3(c) shows the charged multiplicity distribution in the annihilation model described above. The comparison with the experimental points taken from [228, 229] serves as additional confirmation of the calculations [86].

There were also performed in [86] the calculations for two other values of the minimal velocity of the antiprotons $v_{disp} = 300$ km/s and for the velocity of the (anti)matter thrown out by the Supernovae, $v_{SN} = 2 \cdot 10^4$ km/s. The respective necessary values of the integral antiproton number density are:

$$n_0^{disp} \approx 2.0 \cdot 10^{-12} \, \text{cm}^{-3}$$

$$n_0^{SN} \approx 6.0 \cdot 10^{-11} \, \text{cm}^{-3}. \tag{6.37}$$

Thus, one can see that necessary integral antiproton density in the halo practically linearly depends on minimal velocity of antiprotons in the range $300 \le v \le 10^4$ km/s. Note, that the approximation about annihilation at rest is valid for all the range of above minimal velocities and the resulting gamma spectrum does not change its form at such a variation of v_{min}.

4.3 Gamma ray constraint on antimatter

Let us estimate the intensity of the antiproton source and, as a result, the total mass of the hypothetical globular cluster of antistars for three values of the minimal antiproton velocity: v_{disp}, v_{SW} and v_{SN}. The first case assumes that antiprotons have been decelerated and travel in the halo with velocities equal to the velocity dispersion defined by the galactic gravitational field. The second value of v_{min} is of the order of the speed of the fast stellar wind and the third case is the velocity of particles blown off by the Supernova explosion without possible deceleration.

If we integrate over the volume of the whole halo and take into account the antiproton storage in the halo during the confinement time, we obtain for the integral intensity of the antiproton source $\dot{M} \sim (n_0 \, m_p \, V_{halo}) / t_{conf}$. For the above three variants of the minimal antiproton velocity and $t_{conf} \sim 5 \cdot 10^8$ years we obtain from Eq. (6.36) and Eq. (6.37) the following values of the necessary antiproton source intensity:

$$\dot{M}^{disp} \approx 3.0 \cdot 10^{-9} \, M_\odot/yr$$

$$\dot{M}^{SW} \approx 8.5 \cdot 10^{-9} \, M_\odot/yr \tag{6.38}$$

$$\dot{M}^{SN} \approx 1.0 \cdot 10^{-7} \, M_\odot/yr$$

From the analogy with elliptical galaxies in the case of constant mass loss due to stellar wind one has the mass loss $10^{-12} M_\odot$ per Solar mass per year. In the case of stellar wind we find for the mass of the anticluster:

$$M_{clu}^{SW} \approx 2 \cdot 10^4 \, M_\odot. \tag{6.39}$$

To estimate the frequency of Supernova explosions in the antimatter globular cluster the data on such explosions in the elliptical galaxies were used [85], which gives the mean time interval between Supernova explosions in the antimatter globular cluster $\Delta T_{SN} \sim 1.5 \cdot 10^{15} M_5^{-1}$ s. For $M_5 > 1$ this interval is smaller than the period of the orbital motion of the cluster, and one can use the stationary picture considered above with the change of the stellar wind mass loss by the $\dot{M} \sim f_{SN} \cdot M_{SN}$, where $f_{SN} = 6 \cdot 10^{-16} M_5$ s^{-1} is the frequency of Supernova explosions and $M_{SN} = 1.4 M_\odot$ is the antimatter mass blown off in the explosion. Following the theory of Supernova explosions in old star populations only the supernovae of the type I (SNI) take place, in which no hydrogen is observed in the expanding shells. In strict analogy with the matter SNI the chemical composition of the antimatter Supernova shells should include roughly half of the total ejected mass in the internal anti-iron shell with the velocity dispersion $v_i \leq 8 \cdot 10^8$ cm/s and more rapidly expanding $v_e \sim 2 \cdot 10^9$ cm/s anti-silicon and anti-calcium external shell. The averaged effective mass loss due to Supernova explosions gives the antinucleon flux $\dot{N} \sim 10^{42} M_5$ s^{-1}, but this flux contains initially antinuclei with the atomic number $A \approx 30 - 60$, so that the initial flux of antinuclei is equal to $\dot{A} \sim (2-3) \cdot 10^{40} M_5$ s^{-1}. Due to the factor $\sim Z^2 A^{2/3}$ in the cross-section the annihilation lifetime of such nuclei is smaller than the cosmic ray lifetime, and in the stationary picture the products of their annihilation with $Z < 10$ should be considered. With account of the mean multiplicity $<N> \sim 8$ of annihilation products one obtains the effective flux $\dot{A}_{eff} \sim (1.5 - 2.5) \cdot 10^{41} M_5$ s^{-1}, being an order of magnitude smaller than the antiproton flux from the stellar wind.

If we take the antimatter stellar wind to be as small as the Solar wind $(\dot{M}_\odot - 10^{-14} M_\odot$ yr$^{-1})$ this corresponds to the antiproton flux by two orders of magnitude smaller than the one chosen above in Eq. (6.38), and the antimatter from Supernova should play the dominant role in the formation of galactic gamma background. For the Supernova case we have for the mass of the anticluster the value

$$M_{clu}^{SN} \approx 4.0 \cdot 10^5 M_\odot ,$$

which value agrees with the estimation [85]. If we assume that a significant fraction of the antiprotons from stellar wind is decelerated up to v_{disp} the respective mass of the globular cluster of antistars can be reduced up to

$$M_{clu}^{disp} \approx 7 \cdot 10^3 M_\odot.$$

It is necessary to make a small remark. Namely, in principle, one cannot exclude that the secondary antiprotons produced in pp collisions can be decelerated in the halo magnetic fields up to velocities order of few hundreds km/s. In this case they will also give contribution in the diffuse gamma flux annihilating with the matter gas and the calculations performed above are valid also in this case.

5. Antihelium flux signature for antimatter

The estimation of the previous section puts an upper limit on the total mass fraction of antimatter clusters in our Galaxy. Their integral effect should not contradict the observed gamma ray background.

The uncertainty in the distribution of magnetic fields causes even more problems in the reliable estimation of the expected flux of antinuclei in cosmic rays. It is also accomplished by the uncertainty in the mechanism of cosmic ray acceleration. The relative contribution of disc and halo particles into the cosmic ray spectrum is also unknown.

To have some feeling of the expected effect we may assume that the mechanisms of acceleration of matter and antimatter cosmic rays are similar and that the contribution of antinuclei into the cosmic ray fluxes is proportional to the mass ratio of globular cluster and Galaxy. Putting together the lower limit on the mass of the antimatter globular cluster from the condition of survival of antimatter domain and the upper limit on this mass following from the observed gamma ray background, one obtains [85, 87, 86] the expected flux of antihelium nuclei in the cosmic rays with the energy exceeding 0.5 Gev/nucleon to be $10^{-8} \div 10^{-6}$ of helium nuclei observed in the cosmic rays.

Such estimation assumes that annihilation does not influence the antinuclei composition of cosmic rays, which may take place if the cosmic ray antinuclei are initially relativistic. If the process of acceleration takes place outside the antimatter globular cluster one should take into account the Coulomb effects in the annihilation cross-section of non relativistic antinuclei, which may lead to suppression of their expected flux.

On the other side, antinuclei annihilation invokes a new factor in the problem of their acceleration, which is evidently absent in the case of cosmic ray nuclei. This factor may play a very important role in the account for antimatter Supernovae as the possible source of cosmic ray antinuclei. From the analogy with elliptical galaxies one may expect [85, 87, 86] that in the antimatter globular cluster Supernovae of the I type should explode with the frequency about $2 \cdot 10^{-13}/M_\odot$ per year. As it was discussed in the previous section, on the basis of theoretical models and observational data on SNI (see c.f. [230]) one expects in such explosion the expansion of a shell with the mass of about $1.4 M_\odot$ and velocity distribution up to $2 \cdot 10^9 cm/s$. The internal layers with the velocity $v < 8 \cdot 10^8 cm/s$ contain anti-iron ^{56}Fe and the outer layers with higher velocity contain lighter elements such as anti-calcium or anti-silicon. Another important property of Supernovae of the I type is the absence of hydrogen lines in their spectra. Theoretically it is explained as the absence of hydrogen mantle in Presupernova. In the case of antimatter Supernova it may lead to strong relative enhancement of antinuclei relative to antiprotons in the cosmic ray effect. Note that similar effect is suppressed in the nuclear component of cosmic rays, since Supernovae of the II type are also related to the matter cosmic ray origin in our Galaxy, in which massive hydrogen mantles (with the mass up to few solar masses) are accelerated.

In contrast with the ordinary Supernova the expanding antimatter shell is not decelerated owing to acquiring the interstellar matter gas and is not stopped by its pressure but annihilates with it [85]. In the result of annihilation with hydrogen, of which the matter gas is dominantly composed, semi-relativistic antinuclei fragments are produced. The reliable analysis of such cascade of antinuclei annihilation may be based on the theoretical models and experimental data on antiproton nucleus interaction. The important qualitative result is the possible nontrivial contribution to the fluxes of cosmic ray antinuclei with $Z \leq 14$ and the enhancement of antihelium flux. With account of this argument the estimation of antihelium flux from its direct proportionality to the mass of antimatter globular cluster seems to give the lower limit for the expected flux.

Here we study another important qualitative effect in the expected antinuclear composition of cosmic rays. Cosmic ray annihilation in galactic disc results in the significant fraction of antihelium-3 so that antihelium-3 to antihelium-4 ratio turns to be the additional signature of the antimatter globular cluster.

5.1   Equations for differential fluxes

Cosidering the $\overline{^4He}$ nuclei travelling through the Galactic disk we have to take into account two processes:

(i) the destruction of an antinucleus in the inelastic interactions with the protons of the galactic media and

(ii) the energy losses during the travelling through the Galaxy.

For the $\overline{^3He}$ nuclei we need to take into account also the possibility of the ^3He nuclei production due to the reaction

(iii) $\overline{^4He} + p \rightarrow \overline{^3He} + all.$

The energy losses occur due to four kinds of processes:

(a) the energy losses on ionization and excitation of the hydrogen atoms in the disk matter;

(b) the bremsstrahlung radiation on the galactic hydrogen atoms;

(c) the inverse Compton scattering on the relic photons;

(d) the synchrotron radiation in the galactic magnetic fields.

The processes (b)–(d) are proportional to $(m_e/M_{He})^2$ and can be neglected at not very high energies of the He nuclei. The energy losses due to ionization and excitation of the hydrogen atoms per one collision are described by the expression [5]:

$$æ(\beta, z) = \frac{4\pi \, Z(z\alpha)^2}{m_e \, \beta^2} \left[\ln \frac{2m_e\beta^2}{I(1 - \beta^2)} - \beta^2 \right], \tag{6.40}$$

where I is ionization potential of the hydrogen atom, $I \approx 15 \, eV$; $Z = 1$, $z = 2$ are the electric charges of the hydrogen and helium nuclei, respectively, $\beta = v/c$ is the dimensionless velocity and $\alpha = 1/137$ is the fine structure constant.

The rates of the energy losses and of the ^4He nuclei destruction are:

$$\frac{dE_{3,4}}{dt} = -n_H \, v_{3,4} \, æ_{3,4},$$

$$\frac{dn_{3,4}}{dt} = -n_H \, v_{3,4} \, \sigma_{ann}^{(3,4)} \, n_{3,4}, \tag{6.41}$$

where n_H is the particle density of H atoms in the Galactic disc.

The source of ^3He nuclei can be written in the form:

$$\frac{dn_3^{(+)}(t, E_3)}{dt} = -\int_{E_3}^{\infty} dn_4(t, E_4) \frac{\partial W(E_4; E_3)}{\partial E_3}. \tag{6.42}$$

Here $\partial W(E_4; E_3)/\partial E_3$ describes the probability to produce ^3He in the inelastic collision $^4He + p \rightarrow {}^3He + all$, with the normalization condition:

$$\int_0^{\infty} dE_3 \frac{\partial W(E_4; E_3)}{\partial E_3} = W_3(E_4).$$

If we introduce the differential flux

$$J(t, E) = v \frac{\partial n(t, E)}{\partial E}$$

and the energy per nucleon ($E \rightarrow E/A$), with $A = 4$ being the atomic weight of the antihelium nucleus, we obtain finally the system of integro–differential equations, describing the behavior of ^4He and ^3He nuclei in the Galaxy [87]:

$$\frac{dJ(t, E_4)}{dt} = -n_H \, c \, \beta_4 \left[\sigma_{inel}(p_4) - A \frac{m_p^2}{p_4 \, E_4^2} \frac{dæ(\beta_4)}{d\beta_4} \right] J(t, E_4),$$

$$\frac{dE_4}{dt} = -n_H \, c \, A^{-1} \, \beta_4 \, æ(\beta_4),$$

$$\frac{dJ(t, E_3)}{dt} = -n_H \, c \, \beta_3 \left[\sigma_{inel}(p_3) - (A - 1) \frac{m^2}{p_3 \, E_3^2} \frac{dæ(\beta_3)}{d\beta_3} \right] J(t, E_3)$$

$$+ n_H \, c \, \beta_3 \int_{E_3}^{\infty} dE_4 \, \sigma_4(p_4) \frac{\partial W(E_4; E_3)}{\partial E_3} \, J_4(t, E_4),$$

$$\frac{dE_3}{dt} = -n_H \, c \, (A - 1)^{-1} \beta_3 \, æ(\beta_3). \tag{6.43}$$

It was suggested in [87] that relative contribution to $\overline{^3He}$ does not depend on energy and the above value was used.

For simplicity it was suggested in [87] that the probability $dW(E_4; E_3)/dE_3$ in Eq. (6.42) can be approximated by the δ-function:

$$\frac{\partial W(E_4; E_3)}{\partial E_3} = W_3 \, \delta(E_4 - E_3),$$

with W_3 from Eq. (6.47).

The initial fluxes for ^4He and $\overline{^4He}$ we chose in the form

$$J_4(0, E) = 0.07 \times \frac{1.93 \beta}{E^{2.7}} \times 10^{-6}, \, cm^{-2} \, s^{-1} \, sr^{-1} \, (GeV/nucleon)^{-1},$$

$$J_3(0, E) = 0.$$

$$(6.48)$$

In the galactic disc, where the hydrogen number density is $n_H \approx 1 \, atom/cm^3$, the typical time scale $T_{conf} = 10^7 \, yr$ was chosen in [87] for the confinement time for He nuclei. It was also accounted for the very low density of the matter in the Galactic halo.

Results of the calculations [87] are shown in Figure 6.5. The solid line shows initial He flux, dashed and dot–dashed lines represent final fluxes of ^4He and ^3He, respectively.

The first two equations in (6.43) can be also applied to the ordinary ^4He nuclei component of cosmic rays, if under the σ_{ann} one understands the inelastic interaction cross-section of the ^4He nucleus with the proton, neglecting again the coherent processes. For comparison there is also plotted by the dotted line the final flux of the ^4He, suggesting that the initial flux is the same as for $\overline{^4He}$. vspace*2cm

In Figure 6.6 the ratios of fluxes $\overline{^4He}/^4He$ and $\overline{^3He}/^4He$ are plotted for two cases: upper curves for $M/M_{MW} = 10^{-6}$ and two lower curves for $M/M_{MW} = 10^{-8}$. These results are compared with the expected sensitivity of the AMS experiment to antihelium flux. One finds the AMS experiment accessible to complete test of the hypothesis on the existence of antimatter globular clusters in our Galaxy. The test of this hypothesis can begin even earlier [91], before AMS, provided that the experimental sensitivity reaches the maximal estimated $\overline{^4He}$ flux, as it is, in particular, expected in the PAMELA experiment.

The important result of these calculations is that the substantial contribution of antihelium-3 into the expected antinuclear flux was found in [87]. Even in the case of negligible antihelium-3 flux originated in the halo its contribution into the antinuclear flux in the galactic disc should be comparable with the one of antihelium-4.

The estimations of [85], on which the calculations [87] were based, assumed stationary in-flow of antimatter in the cosmic rays. In case Supernovae play the dominant role in the cosmic ray origin, the in-flow is defined by their fre-

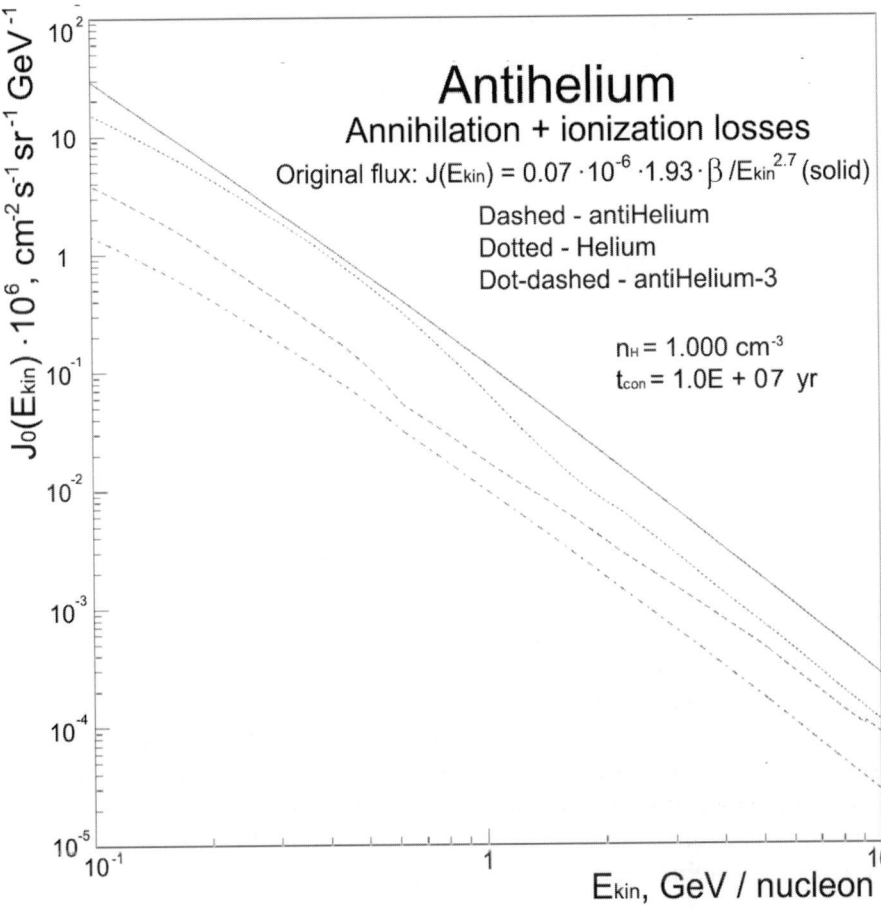

Figure 6.5. Calculated fluxes of $\overline{^4He}$ (dashed), ^4He (dotted) and $\overline{^3He}$ (dash–dotted). The solid line presents initial flux for ^4He nuclei. The confinement time has been chosen equal to 10^7 years.

quency. One may find from [85] that the interval of possible masses of antimatter cluster $3 \cdot 10^3 \div 10^5 M_\odot$ gives the time scale of antimatter in-flow $1.6 \cdot 10^9 \div 5 \cdot 10^7$ years, which exceeds the generally estimated lifetime of cosmic rays in the Galaxy. The succession of antinuclear annihilations may result in this case in the dominant contribution of antihelium and, in particular, antihelium-3 into the expected antinuclear flux. It makes antihelium signature sufficiently reliable even in this case.

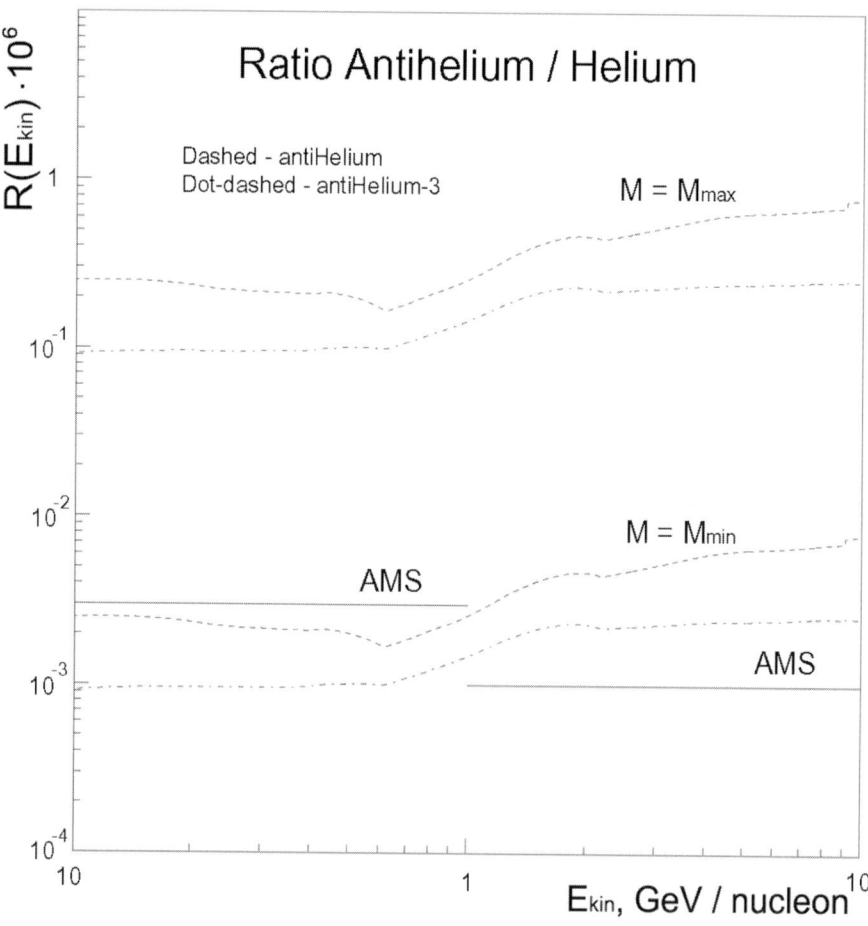

Figure 6.6. Ratios of fluxes $\overline{{}^{4}He}/{}^{4}He$ (dashed) and $\overline{{}^{3}He}/{}^{4}He$ (dash–dotted). Two upper curves correspond to the case of the maximal possible mass of antimatter globular cluster $M_{max} = 10^5\ M_\odot$ and the two lower curves to the case of the minimal possible mass of such cluster $M_{min} = 10^3\ M_\odot$. The results of calculations are compared with the expected sensitivity of the AMS experiment [90] (solid lines).

6.   Anti-Asteroids annihilations on Planets and the Sun

The existence of antimatter stars in the Galaxy as possible signature for inflationary models with non-homogeneous baryosynthesis may leave a trace by antimatter cosmic rays as well as by their secondaries (anti-planets and anti-meteorites) diffused bodies in our galactic halo. The anti-meteorite flux may leave its explosive gamma signature by colliding on lunar soil as well as on terrestrial, Jupiter and Solar atmospheres. However the propagation in the Galaxy and the consequent evaporation in galactic matter gas suppress the lightest

$(m < 10^{-2}g)$ anti-meteorites. Nevertheless the heaviest anti-meteorites ($m >$ 10^{-1} g up to 10^6 g) are unable to be deflected or annihilated by the thin galactic gas surface annihilation; they might hit the Sun (or rarely Jupiter) leading to an explosive gamma event and a spectacular track with a bouncing and even a propelling annihilation on chromosphere and photosphere. Their antinuclei annihilation in pions and their final hard gamma showering may be observable as a "solar flare" at a rate nearly comparable to the observed ones. From their absence we may infer first bounds on antimatter–matter ratio near or below 10^{-9} limit applying already recorded data in gamma BATSE catalog.

6.1 Gamma flashes by anti-meteorite annihilations on the Earth and the Moon

It was shown above that annihilation of antimatter, lost by antimatter stars in the form of stellar wind, can reproduce the observed galactic gamma background in the range tens–hundreds MeV. Still, any source of neutral pions can lead to the same effect and the manifest signature for existence of antimatter stars is the existence of an antinuclear component of cosmic rays, accessible to the future cosmic ray experimental searches, first of all in AMS-II experiment. The other profound signature of antimatter are the pieces of antimatter, coming in the form of antimatter meteorites. We study, following [92], the latter possibility in the present section and find it an interesting tool to probe the origin of matter, related with the creation of antimatter. With all the uncertainties and reservations taken into account, the search for antimatter meteorites can still provide a useful probe for the existence of macroscopic antimatter.

The present flux of meteorites with the mass M observed on the Earth is nearly $10^4 \left(\frac{M}{10kg} \right)^{-1}$ event a year. This power extends for a large range of mass values. It is very possible that most of this matter has a local "solar" origin. However simple argument on nearby stellar encounters and matter exchange imply that up to 1% of the meteorites may be of galactic (extra-solar) origin. Therefore, up to nearly

$$\frac{dN}{dt} = 10^6 \left(\frac{M}{1g} \right)^{-1} \qquad (6.49)$$

of meteorites, hitting the Earth any year, can be of galactic (extra-solar) nature. If the corresponding anti-meteorites rate follows the same power law at any given suppressed ratio, r,

$$r = \left(\frac{N_a}{N_m} \right),$$

where $N_{a(m)}$ the total amount of antibaryons (baryons) in the Galaxy, (let us say a part over a million or a billion or below) its signal will be powerful enough to be (in most cases) observable. Indeed, the amount of energy released during the annihilation follows common special relativity; for any light (milligram unit) anti-meteorites mass M the energy ejected is:

$$E = 10^{18} \left(\frac{M}{1mg} \right) erg. \tag{6.50}$$

Its corresponding "galactic" event rate, following Eq. (6.49) is

$$\frac{dN}{dt} = 10^9 r \left(\frac{M}{1mg} \right)^{-1} year^{-1}. \tag{6.51}$$

The event of the anti-meteorite annihilation on the Earth' atmosphere will give life to an unexpected upward gamma shower that will mimic a mini nuclear atomic test or extreme upward gamma shower. Even for a large suppression ratio $r = 10^{-9}$ this event rate derived from the expression above (one a year) should not escape the accurate BATSE ten-year monitoring. Actually the atmosphere area below BATSE detection is nearly 1% of all Earth leading to a total probability rate of 0.1 in ten years. However the corresponding secondaries gamma flux by consequent nuclei annihilation showering into charged and neutral pion and their decays and degradation in atmosphere, should lead to a huge gamma fluence (the flux integrated over the time of flash) F observable in a near orbit satellite as Beppo–Sax or GRO BATSE:

$$F \simeq 10 \ erg/cm^2 (M/1mg),$$

$$Flux = 100 \ erg \cdot sec^{-1} cm^{-2}.$$

This latter flux is derived assuming a characteristic galactic velocity $v = 300 km/sec$ for the incoming anti-meteorite and a terrestrial atmosphere of nearly 30 km height. Such a signal is nearly 10 order of magnitude above the sensitive BATSE detection threshold. Smaller scale upward gamma flash are indeed known and they are called "Terrestrial Gamma Flashes" . They are corresponding to just 10^8 or 10^9 erg of isotropic fluence energy released at time scale from millisecond up to ten seconds (or even much less energy if originated by beamed upward τ airshowers at $10^{15} eV$ up to nearly horizontal ones at $10^{19} eV$, see review and references in [236]). Therefore such milligram anti-meteorite bangs will be already loudly recorded on data, if they were taking place. Of course such high large event fluence would also not escape other less sensitive astrophysical or military detectors. Therefore it seems that milligram antimatter meteorite rain should be totally excluded at very low level ($r \leq 10^{-9}$). Even more dramatic and sharp gamma signatures should come by their fast Moon annihilation (because of the absence of atmosphere), but at a lesser (Moon surface over Earth) rate. Lunar anti-meteorite annihilation in characteristic nano-second signature, would make very strong signals at lunar orbiting gamma detectors. They provide a complementary tool to exclude very light (microgram) anti-meteorite rains at the same severe bound ($r \leq 10^{-9}$).

6.2 Light anti-meteorite evaporation crossing the Galaxy

However, these results may be alleviated keeping in mind that anti-meteorites can be annihilated or "evaporated" during their propagation in galactic gas. Indeed, the column density of atoms (protons) crossed assuming $n_{disk} = 1 \cdot cm^{-3}$ and a galactic disk height of $h = 100pc$ and a total number of crossing 100 is: $N = 3 \cdot 10^{22}\ cm^{-2}$. Each crossed matter atom annihilates on the surface of the rigid body of the anti-meteorite. Putting the total mass of the crossed matter gas equal to the mass of spherical homogeneous anti-meteorite of radius r and internal density ρ,

$$\pi r^2 N m_H = \frac{4\pi}{3}\rho r^3,$$

one obtains that the anti-meteorite cannot escape complete annihilation, if its radius is smaller, than

$$r_{an} = \frac{3}{4} \cdot \frac{N m_H}{\rho}.$$

The corresponding anti-meteorite mass, given by

$$M_{an} = \frac{9}{16}\pi \cdot \frac{(N m_H)^3}{\rho^2},$$

is (assuming water density) about $2.2 \cdot 10^{-4}$ g. The actual value of minimal mass of the anti-meteorite, surviving annihilation, may be a few orders of magnitude larger. If we take into account the strong (cubic) dependence of M_{an} on N, we find important the increase of N due to effects of annihilation with the gas above the disc. The mass of anti-meteorite, which is completely destroyed by annihilation, can be even larger, if we take into account its atomic composition. To destroy the anti-meteorite, which consists of anti-atoms with atomic number A, it is not necessary to annihilate all the anti-nucleons in all its anti-nuclei, since even the result of one proton anti-nucleus annihilation not only destroys the anti-nucleus, but also causes the successive destructive effects by its fragments. We discuss the effects of energy and momentum transfer due to such processes in the next section, and only estimate here the increase in the minimal mass of anti-meteorite, surviving after annihilation. Putting the total number of matter gas atoms, annihilating on the surface of the anti-meteorite, equal to the total number of anti-atoms with atomic number A in the anti-meteorite, we obtain instead of M_{an} the magnitude

$$M_{surv} = \frac{9}{16}\pi \cdot \frac{(A N m_H)^3}{\rho^2}, \tag{6.52}$$

which is the factor of A^3 larger, than M_{an}. This implies that milligram (and even much heavier, up to 0.3 g for an anti-ice meteorite) anti-meteorites might be suppressed and may be almost absent in the solar system; previous bound by annihilation on the Earth may be considered for heavier (10–100 milligrams or above) anti-meteorites leading to a ratio ($r = 10^{-8}$) of antimatter allowable. Bounds by microgram anti-meteorite annihilation on Moon soil while being very hard and sharp, will be no more effective than the terrestrial bounds. Moreover, there are other processes that may dilute the above anti-meteorite presence in our Solar system.

6.3   Anti-meteorite annihilation and deceleration in gas

Anti-meteorites with a mass heavier than a milligram may survive annihilation: while crossing a gas cloud, their lateral annihilation may heat a meteorite side, leading to a rocket ejection able to decelerate and at large matter gas density gradient even divert and bounce the trajectory. However, for realistic density gradients the latter case cannot be realized and the momentum transfer due to annihilation causes the anti-meteorite deceleration in matter gas, which can be described as follows.

Anti-meteorite of radius r, moving with a velocity v in the central field of gas, distributed around the central mass M isotropically as

$$\rho = \rho_0 \cdot \left(\frac{R_0}{R}\right)^2,$$

experiences the friction force due to annihilation

$$F_f = -\rho(R)\pi r^2 \eta v c,$$

where η is the effectiveness of momentum transfer near unity. Assuming an initial anti-meteorite velocity v_{a_i} and density ρ_a and a normal galactic disk mass density ρ, one finds the characteristic relaxation time τ (for a millimeter anti-meteorite radius) :

$$\tau = \frac{4}{3}\frac{\rho_a}{\rho}\frac{r}{\eta c}$$

$$= 1.3\eta^{-1} \cdot 10^2 \cdot year \frac{r}{mm}\frac{\rho_a}{gcm^{-3}}\frac{10^{-24}gcm^{-3}}{\rho}. \tag{6.53}$$

Therefore, in a short (in galactic scales) time any fast anti-meteorite will be slowed down to a velocity comparable with common galactic gas. Therefore, lightest anti-meteorites will follow a co-moving pattern with matter in the galactic disk. Heavier ones ($m \gg 0.1$ g) will not evaporate and might reach the Earth.

In the presence of any radial gravitational force, near stars or star clusters, the gravitational force is equal to

$$F_g = \frac{4}{3} \cdot \frac{GM\pi\rho_a r^3}{R_0^2},$$

and the friction action leads to a slowed down free fall up to a steady value. The equality of the two forces indeed leads to the constant velocity

$$v = \frac{2}{3\eta} \frac{\rho_a}{\rho_0} \frac{r}{R_0} \frac{R_g}{R_0} c, \tag{6.54}$$

where

$$R_g = \frac{2GM}{c^2}$$

is the Schwarzschild radius of any central body.

The annihilation friction is effective, resulting in anti-meteorite deceleration and successive slow drift and final annihilation towards the star center.

In nearly horizontal motions the fast anti-meteorite may bounce on the star–planet atmosphere and they may escape from the central field. In the case of general motion and matter gas distribution this effect may be estimated by assuming that a fraction of antimatter is annihilated leading to a momentum exchange (See [237]) and a velocity loss $\Delta v \sim v \sim 10^{-3}c$:

$$\Delta v = \eta \cdot E/Mc,$$

where η is the fraction of annihilation energy going into effective anti-asteroid momentum exchange. Being necessary to escape from the galactic plane or from solar atmosphere a $\Delta v > 10^{-3}c$ one finds

$$(\Delta E)/(Mc^2) = (\Delta M)/M \leq 10^{-3}/\eta.$$

This value cannot exceed unity otherwise the anti-meteorite will be totally annihilated; therefore the η efficiency cannot be below 10^{-3} but its value is bounded by the ratio of the interaction length of charged pions on the meteorite volume; the 300 MeV pion crosses nearly 85 cm in water before interacting; the total amount of matter crossed during meteorite lifetime travelling (comparable to galactic age) in the galactic disk is nearly 10^{-2} g or 10^{-2} cm of water. However, in the case of atomic anti-nuclei composition annihilating with hydrogen of galactic gas the main consequence will be a breakdown of anti-nuclei. Its fragments will deposit in a very efficient way (nearly 50%) the energy of annihilation into linear momentum as well as increasing the temperature of the solid antimatter body.

The estimation [92] show that the effective cooling is keeping the temperature below the solid (rock) melting point, while the anti-meteorite moves in the Galaxy and Solar System. The equilibrium temperature is established, provided that the heating rate $2\pi r^2 \kappa \rho c^2 v$ (where κ is the fraction of the total energy, released in the annihilation ($E_{an} = 2m_H c^2$)), that heats the spherically symmetric anti-meteorite of radius r, moving with velocity v in the matter gas of density $\rho = m_H n$, is equal to the rate of radiative cooling $4\pi r^2 \sigma T^4 c$ (where σ is the Stephan–Boltzmann constant). In the considered approximation both heating and cooling are proportional to the surface area, so that the equilibrium temperature is given by $T_e = 168K(n\kappa v)^{1/4}$ for matter gas number density $n = 1cm^{-3}$ and anti-meteorite velocity $v = 300$ km/s. Annihilation of matter gas with anti-nuclei on the anti-meteorite surface leads to its erosion, but its effect, which may deserve special analysis for particular anti-meteorite composition, does not lead to significant change of the above estimation for sufficiently large anti-meteorites.

Nevertheless the "ice" anti-comets might melt efficiently still in the Galaxy and very efficiently near the Solar and Terrestrial atmosphere. The reason is that the estimated value of T_e can easily be factor of 2 larger, but the anti-meteorite, moving with the velocity $v/c \approx 10^{-3}$, with account of all the uncertainties can be hardly heated up to 1000 K due to the annihilation in the low density matter gas (with the number density $n \approx 1cm^{-3}$). The equilibrium condition, rewritten for energy density of radiation ($\epsilon_\gamma = 2.7Tn_\gamma$) and of annihilation products ($\epsilon_{an} = 2nm_Hc^2$) in the form $\epsilon_\gamma c \approx \kappa\epsilon_{an}v$, is reached at $T_e \leq 300$K due to the low values of in-flow velocity $v/c \approx 10^{-3}$ and matter gas density, which compensates the large value of annihilation energy release $\frac{2m_Hc^2}{T_e} \leq 2 \cdot 10^{10}$.

6.4   Annihilation of anti-asteroids on the Sun

The "galactic anti-asteroid" rate on the Sun from Eq. (6.49) is

$$\frac{dN}{dt} = 10^{10}r \left(\frac{1g}{M}\right) year^{-1}. \tag{6.55}$$

The consequent event rate for suppressed anti-asteroids one over a billion is 10 events a year. The fluence F on Earth is $3 \cdot 10^{-7} erg/cm^2$ and comparable to GRB fluence, with a time dilution of nearly 10 seconds. Therefore it may well be missed or misunderstood as a low energy solar flare. The rarest events at 100 g range may mimic observed solar flares. Let us be reminded that present bounds in solar flare activity may be even detectable at a nano-flare intensity. If the above coincidence is not just the hint of the antimatter meteorites in-fall, it provides the present most stringent bound on antimatter. It may be useful to mention that the two anti-meteorite searches undertaken in the USSR in the late 1960s early 1970s, even with no confirmation, exhibited the positive effect, see review in [276]. So not only stringent limits, but even positive discoveries should be in principle considered in the future of such searches.

7.   Conclusions

In conclusion we can say that the hypothesis on the existence of antimatter globular clusters in the halo of our Galaxy does not contradict either modern particle physics models or observational data. Moreover, the Galactic gamma background measured by EGRET can be explained by antimatter annihilation mechanism in the framework of this hypothesis. If the mass of such a globular cluster is of order of $10^4 \div 10^5$ $M_\odot$, we can hope that other signatures of its existence, such as fluxes of anti-nuclei can be reachable for experiments in the near future. The analysis of antinuclear annihilation cascade is important in the realistic estimation of antinuclear cosmic ray composition but seems to be much less important in its contribution into the gamma background as compared with the effect of antimatter stellar wind. This means that the gamma background and the cosmic antinuclei signatures for galactic antimatter are complementary and the detailed test of the galactic antimatter hypothesis is possible in the combination of gamma ray and cosmic ray studies. Antimeteorites annihilations may provide the challenge to search for antimatter in our Galaxy at the same level of sensitivity which is planned to be reached in AMS-II experiment (a part over a billion). With all the uncertainty in the possible relationship between the total mass of antimatter stars and the expected amount of pieces of antimatter to be ejected by antimatter stellar systems and all the possible reservations, our first estimate on Earth and Solar events are showing rather high sensitivity (10^{-8}–10^{-9}) in antimatter search can or even might already be reached.

Chapter 7

ASTRONOMY OF ULTRA HIGH ENERGY COSMIC RAYS

1. Cosmoarcheology of cosmic rays

Ultra High Energy Cosmic Rays (UHECR) is the observed effect of super-high energy physics in the modern Universe, which naturally puts together particle physics and cosmology in the analysis of their possible origin and effects. Even for known particles and interactions UHECR correspond to the energy range at which their properties were not studied experimentally. Particle theory predicts a wide variety of new phenomena in this energy range. One should take into account the possibility of these phenomena in the analysis of the mechanisms of UHECR origin, propagation and detection. Moreover, such phenomena are unavoidable in the modern Big Bang cosmology, based on inflationary models with baryosynthesis and (multicomponent?) nonbaryonic dark matter. The physics of inflation and baryosynthesis, as well as dark matter/energy content implies new particles, fields and mechanisms, predicted in the hidden sector of particle theory. Such particles, fields and mechanisms may play an important role in the problem of UHECR. It makes new physics a necessary component of the analysis of UHECR data.

Methods of cosmoparticle physics [3] offer the way to unbind the complicated knot of the physical, astrophysical and cosmological problems, related with UHECR. They provide the framework to distinguish different types of predicted cosmological effects of new physics and to discriminate them from nontrivial effects of known physics, arising in specific astrophysical conditions. In the present chapter we make some first steps towards this framework for UHECR studies.

In the framework of cosmoarcheology [18] cosmic rays are treated as the source of information on particle processes at different stages of cosmological evolution. In the early Universe such processes in general do not lead directly to fluxes of particles, accessible to cosmic ray detection, and special analysis is needed to relate such processes with their possible reflections in the observed matter spatial distribution, chemical composition or angular distribution and spectrum of CMB [276, 313, 131, 132, 133]. For each component of cosmic particles there exists the period when the Universe loses its opacity for this component, and one can get direct information on these processes from searches for respective nonthermal electromagnetic, proton–antiproton or neutrino backgrounds [134, 135, 136, 137]. After the galaxies are formed, the natural sources of particle acceleration appear in the Universe. Energetic particles from SNs, GRBs, AGNs interact with the matter and these particle processes should contribute to the observed cosmic fluxes.

Even if the observed cosmic ray fluxes are completely explained by these natural mechanisms, the comparison of the observational data with the theoretical prediction for hypothetical sources provides important constraints on such sources [136, 138, 139], thus providing important and in some cases unique information on physics of the very early Universe [137]. However, in some cases we are possibly near positive conjecture on the existence of new physics. So, cosmic fluxes of weakly interacting massive particles (WIMPs), interpreted as stable neutrinos of fourth generation with the mass $\sim 50 GeV$, are probably detected in DAMA experimental search for WIMPs [16], that finds indirect support in the existing gamma background and cosmic positron data and is accessible to test in underground neutrino, cosmic ray and accelerator experiments [62, 25, 63, 64, 65, 140].

The field of physics and astrophysics of cosmic rays is rather huge, so we concentrate here only on some aspects of the possible origin of cosmic ultra high energy particles in their relationship with the effects of new physics in the inflationary Universe. The interference between large-scale structures and microphysics, presented in the preceding chapters, revealed several possible forms of such relationship. It makes us discuss, at least fragmentarily, the corresponding effects in cosmic rays. Namely, it was shown that the microphysical Lagrangian could lead to large-scale structure of primordial massive BH clouds, or to islands of antimatter. In the first case, AGNs are formed, being one of the popular sources of cosmic rays. In the second case, the predicted antinuclear component of cosmic rays provides the direct test of the considered model (in particular, proving or disproving the supposed form of Lagrangian). In this chapter we give another example of such an interference.

The detection [196, 197, 198, 199, 200, 201, 202] of cosmic rays with energy above Greisen–Zatsepin–Kuzmin (GZK) cut-off of $\sim 5 \cdot 10^{19} eV$ presents a serious problem for interpretation. The origin of GZK cut-off [141, 142] is due to resonant photoproduction of pions by protons on cosmic microwave background radiation which leads to a significant degradation of proton energy (about 20% for 6 Mpc) during its propagation in the Universe. Of course, proton energy does not change by many orders of magnitude if high energy protons come from the distances $< 50–100 Mpc$. However, no nearby sources, such as active galactic nuclei have been found up to now in the arrival direction. If there is some correlation with discrete sources, it is claimed to be with very distant BL Lac objects [177]. According to the common belief, BL Lac are QSOs or AGNs of moderate mass, emitting jets along the line of sight directly towards us, which makes the observed effect of their activity so strong.

It is difficult also to relate directly the observed ultra high energy events with the other known particles. For example, in the case of ultra high energy photons due to interaction with cosmic background radiation ($\gamma + \gamma^* \longrightarrow e^+ + e^-$) the photon free mean path should be significantly less than 100 Mpc. A scenario based on direct cosmic neutrinos able to reach the Earth from cosmological distances cannot reproduce the observed signatures of ultra high energy air showers occurring high in the atmosphere.

Different possibilities were considered (see, e.g., [203, 204, 205, 206, 207, 208, 209, 210] and references therein) in order to solve this puzzle. Combining the advantages of different approaches that involve the existence of primordial superheavy particles we consider in the present chapter a nontrivial solution – annihilation of superheavy particles in primordial bound systems as the source of UHECR.

Stability of superheavy particles assumes that they possess some charge. Charge conservation makes these particles be produced in pairs, and the estimated separation of particle and antiparticle in such pair is shown to be in some cases much smaller than the average separation determined by the averaged number density of considered particles. If the new U(1) charge is the source of a long-range field similar to electromagnetic field, the particle and antiparticle, possessing such charge, can form primordial bound system with annihilation time scale, which can satisfy the conditions assumed for this type of UHECR source. These conditions severely constrain the possible properties of considered particles. So, the proposed mechanism of UHECR origin is impossible to realize, if the U(1) charged particles share ordinary weak, strong or electromagnetic interactions. It makes the proposed mechanism of pairing and binding of superheavy U(1) charged particles an effective theoretical tool in the probe of the physics of very early Universe and of the hidden sector of particle theory underlying it.

The necessary decoupling of superheavy particles from the interactions of ordinary particles can be related with physics of neutrino mass, resulting in the dominant annihilation channels to neutrino. It may be importnat for another approach to a possible solution of the GZK paradox that considers the UHECR as secondary products of UHE neutrinos, originated at far cosmic distances, overcoming GZK cut-off, hitting onto relic light neutrino in Hot Dark Halos, leading to resonant Z boson production. A consequent Z-Shower (Z-Burst)(see [145, 146, 147, 148, 149]) takes place, where a boosted ultra-relativistic gauge boson Z (or WW, ZZ pairs) decays in flight and where its UHE nuclear secondaries are the observed UHECR events in terrestrial atmosphere. These ZeV primary UHE neutrinos may be produced either inside compact astrophysical objects (Jets GRBs, AGNs, BL Lac [177]) or by relic topological defects decay [178] or, as in the present chapter, by ultra high heavy particle annihilation. In the first (compact object) case one may easily understand the observed UHECR clustering as well as the possible correlation found recently with BL Lac sources. In the second case one may assume the UHECR clustering toward BL Lac as a pure coincidence; otherwise, one may consider a possible faster induced annihilation of these superheavy particles inside deeper clustered gravitational wells around AGN and BL Lac objects.

The latter possibility can appear in the considered mechanism due to self-adjustment of bound systems in the regions of their high density, where the time scale of collisions of bound systems is less than the cosmological one. Being dominantly disrupted in collisions, the bound systems can also contract in their result, thus increasing the rate of their annihilation.

Assuming the presence of necessary elements of the Z-Shower mechanism (UHE neutrinos and relic neutrinos in the halo of the Galaxy), we show in this chapter that the existence of low-scale gravity at TeV scale could lead to a direct production of photons with energy above 10^{22} eV due to annihilation of ultra high energy neutrinos on relic massive neutrinos of the galactic halo [153]. Air showers initialized in the terrestrial atmosphere by these ultra energetic photons could be collected in the near future by the new generation of cosmic ray experiments.

2.   Primordial bound systems

One of the popular approaches to the problem of UHECR origin is related to decays or annihilation in the Galaxy of primordial superheavy particles [143, 144] (see [150, 151, 152] for review and references there in). The mass of such superheavy particles to be considered here is assumed to be higher than the reheating temperature of the inflationary Universe, so it is assumed that the particles are created in some non-equilibrium processes (see, e.g., [154, 155, 156] and [152] for review), taking place after inflation at the stage of preheating.

The problems related with this approach are as follows. If the source of ultra high energy cosmic rays (UHECR) is related with particle decay in the Galaxy, the time scale of this decay, which is necessary to reproduce the UHECR data, needs special nontrivial explanation. Indeed, the relic unstable particle should survive to the present time, and having the mass m of the order of 10^{14} GeV or larger it should have the lifetime τ, exceeding the age of the Universe. On the other hand, even if particle decay is due to gravitational interaction, and its probability is of the order of (here and further, if not directly indicated otherwise, we use the units $\hbar = c = k = 1$)

$$\frac{1}{\tau} = \left(\frac{m}{M_P} \right)^4 m, \tag{7.1}$$

where $M_P = 10^{19}$ GeV is the Planck mass, the estimated lifetime would be by many orders of magnitude smaller. It implies strong suppression factor in the probability of decay, what needs rather specific physical realization ([143, 150, 151, 152]), e.g. in the model of cryptons [157, 158] (see [159] for review). The general conditions of protection from fast decay can be provided by broken gauge discrete symmetries [160], and the elementary particle models for such long-lived superheavy particles were suggested [161, 162].

If the considered particles are absolutely stable, the source of UHECR is related to their annihilation in the Galaxy. But their averaged number density, constrained by the upper limit on their total density, is so low, that strongly inhomogeneous distribution is needed to enhance the effect of annihilation to the level desired to explain the origin of UHECR by this mechanism.

Such increase of particle concentration can be hardly reached by the simple development of gravitational instability in nearly homogeneous medium, and it appeals to some strong primordial inhomogeneity in particle distribution.

In the present chapter, we consider the solution of the latter problem, offered in [163, 164]. If superheavy particles possess new U(1) gauge charge, related to the hidden sector of particle theory, they are created in pairs. The Coulomb-like attraction (mediated by the massless U(1) gauge boson) between particles and antiparticles in these pairs can lead to their primordial binding, so that the annihilation in the bound system provides the mechanism for UHECR origin. To realize this mechanism the properties of superheavy particles should satisfy a set of conditions, putting severe constraints on the cosmological scenarios and particle models underlying the proposed mechanism.

2.1 Superheavy particles in the inflationary Universe

As we discussed in the previous chapters, the models of inflationary Universe assume that thermodynamic equilibrium conditions of hot Universe (the so-called "reheating") do not take place immediately after the end of inflation, and that there exists a rather long transition period of the so-called "preheating". The non-equilibrium character of superheavy particle production implies strong dependence on the concrete physical processes that can take place at different periods of preheating stage.

It was shown in [154, 155] that the "nonadiabatic" expansion of space-time [156] in the end of inflation at $t \sim 1/H_{end}$, when transition from inflationary to matter- or radiation-dominated post-inflationary stage takes place and preheating begins, can lead to production of superheavy particles with the mass $m \lesssim 10H_{end}$. Here $H_{end} \sim 10^{13}$GeV is the Hubble constant at the end of inflation. The calculations [155] of primordial concentration of superheavy particles produced in this gravitational mechanism exhibit strong dependence on m/H and correspond to a wide range of their modern densities up to $\Omega_X \sim 0.3$.

Superheavy particles can be created at the end of preheating, when reheating takes place at $t \sim 1/H_r$ (H_r being the Hubble constant in the period of reheating), if the quanta of inflaton field contain these particles among the products of decay. The modern density of superheavy particles is then given by

$$\Omega_X = \frac{T_r}{T_{RD}} \frac{2m}{m_\phi} Br(X), \tag{7.2}$$

where $T_r \sim (H_r M_P)^{1/2}$ is the reheating temperature, $T_{RD} \sim 10\text{eV}$ is the temperature at the end of radiation dominance stage and in the beginning of the modern matter-dominated stage, $m_\phi > 2m$ is the mass of the inflaton field quantum and $Br(X)$ is the branching ratio of superheavy particles production in inflaton decay. The condition $\Omega_X \leq 0.3$ constrains the branching ratio as

$$Br(X) \leq 0.1 \frac{T_{RD}}{T_r} \frac{m_\phi}{m}. \tag{7.3}$$

Production of superheavy particles by gravitational mechanism and at the reheating represents two extreme possibilities of the earliest (at the beginning of preheating) and the latest (at the end of preheating) period of superheavy particle production. In the both cases the mass of superheavy particles can not exceed 10^{13}GeV, being restricted either by the value of H_{end} for gravitational production, or for production at reheating by the mass of inflaton that is constrained from the observed level of CMB fluctuations [152].

There are, however, several mechanisms, corresponding to the intermediate value of the Hubble constant H for this period $H_r < H < H_{end}$ and avoiding such severe constraint on mass of superheavy particles.

If inflation ends by the first-order phase transition, bubble wall collisions in the course of true vacuum bubble nucleation can lead to formation of primordial black holes (PBH) with the mass $M \sim M_P \frac{M_P}{H_{end}}$ [316, 305] (See chapter 8). Successive evaporation of such black holes at

$$H_{ev} \sim 1/t_{ev} \sim \frac{M_P^4}{M^3} \sim H_{end} \left(\frac{H_{end}}{M_P}\right)^2 \tag{7.4}$$

is the source of superheavy particles, when the temperature of PBH evaporation, increasing with the loss of mass as $T_{PBH} \sim M_P^2/M$, reaches m.

PBH mass distribution predicted in the mechanism [316, 305] has rather sharp peak at the value $M \approx 1$ g and the period of superheavy particle production corresponds to the value of the Hubble constant $H \approx 10^3$GeV. Taking into account possible interval of parameters of the model [316, 305] one obtains that superheavy particle production by this mechanism takes place at 1GeV$< H < 10^6$GeV.

If α_X is the fraction of PBH mass, evaporated in the form of considered superheavy particles, the relationship between the probability of PBH formation w and Ω_X is given by

$$\Omega_X = \frac{T_r}{T_{RD}} \alpha_X w, \tag{7.5}$$

for dust-like (MD) expansion law at preheating stage and

$$\Omega_X = \frac{(M_P H_{end})^{1/2}}{T_{RD}} \alpha_X w, \qquad (7.6)$$

for relativistic (RD) expansion at the stage of preheating. In the latter case corresponding to the Eq. (7.6), the condition $\Omega_X \leq 0.3$ leads to $w \leq 3 \cdot 10^{-25}/\alpha_X$. Creation of mini black holes with such a low probability does not imply first-order phase transition after inflation, but it is possible even from Gaussian "tails" (see [3] for review) of nearly flat ultraviolet spectra, that are strongly disfavored but still not excluded within the uncertainty of the recent WMAP measurements of CMB anisotropy [74]. It results in a continuous PBH mass distribution, peaking at the minimal value, corresponding to the mass within the cosmological horizon in the end of inflation $M \approx 1$ g. Superheavy particle production in the evaporation of such PBHs should dominantly take place at $H \approx 10^3$ GeV, and continues later with exponentially decreasing probability.

The presence of additional dynamically subdominant fields at the inflationary stage can strongly modify at the small-scales the simple picture of nearly flat power spectrum of density fluctuations, as was shown in the previous chapters. It also leads to the possibility of superheavy particle production in the decay of quanta of such field, ϕ, at the preheating stage. The relationship between Ω_X and the relative contribution r of the field, ρ_ϕ, into the total density ρ_{tot}, $r = \rho_\phi/\rho_{tot}$ in the period of decay, at $\tau \sim 1/H_d$, is given by Eqs. (7.5)–(7.6), in which α_X has the meaning of the branching ratio for superheavy particle production (multiplied by the factor $\sim m/m_\phi$, m_ϕ is the mass of ϕ, in case of relativistic decay products) and H_{end} is substituted by H_d. If ϕ decays due to gravitational interaction, H_d is equal to the probability of decay, Γ, given by Eq. (7.1), $H_d = \Gamma \sim m_\phi(m_\phi/M_P)^4$. In general, for $H_d = \Gamma$, the period of derelativization of the relativistic decay products with the energy $\epsilon \sim m_\phi \gg m$ corresponds to $H \sim (m/m_\phi)^2 H_d$. Pending on the value of $m_\phi \gg m$ the period of superheavy particle production can correspond in this mechanism to any value of H in the interval $H_r < H < H_{end}$.

An interesting realization for production of superheavy particles by an additional field naturally leading to the intermediate values of H in the period of production takes place in the mechanism of instant preheating [165], where the mass of additional particles χ is proportional to the inflaton field. In the end of inflation, at $t \sim 1/H_{end}$, when inflaton goes through the minimum of potential, χ-particles are massless and they are efficiently produced. When in the course of the first oscillation inflaton field increases m_χ increases too and can reach the value close to M_P [165, 152]. Rapid decay of Planckean mass χ-particles gives rise to production of ultrarelativistic superheavy particles with energies $\epsilon \sim m_\chi \gg m$. The period of their derelativization corresponds to $H \sim (m/m_\chi)^2 H_{end}$, what gives for $M_P \geq m_\chi \geq m$ the interval of H: $H_{end} \geq H \geq 10^3$ GeV.

2.2   Pairing of nonthermal particles

Note, first of all, following [163, 164], that in quantum theory particle stability reflects the conservation law, which according to Noether's theorem is related to the existence of a conserved charge, possessed by the considered particle. Charge conservation implies that a particle should be created together with its antiparticle. It means that, being stable, the considered superheavy particles should bear a conserved charge, and such charged particles should be created in pairs with their antiparticles at the stage of preheating, if they possess a local gauge charge. This may not be the case for a global charge and single production of particles bearing a global charge is possible.

Being created in the local process the pair is localized within the cosmological horizon in the period of creation. If the momentum distribution of created particles is peaked below $p \sim mc$, they donot spread beyond the proper region of their original localization, being in the period of creation $l \sim c/H$, where the Hubble constant H at the preheating stage is in the range $H_r \leq H \leq H_{end}$. For relativistic pairs the region of localization is determined by the size of cosmological horizon in the period of their derelativization. In the course of successive expansion the distance l between particles and antiparticles grows with the scale factor, so that after reheating at the temperature T it is equal to

$$l(T) = \left(\frac{M_P}{H}\right)^{1/2} \frac{1}{T}. \tag{7.7}$$

The average number density of superheavy particles n is constrained by the upper limit on their modern density. Say, if we take their maximal possible contribution in the units of critical density, Ω_X, not to exceed 0.3, the modern cosmological average number density should be $n = 10^{-20} \frac{10^{14} GeV}{m} \frac{\Omega_X}{0.3} cm^{-3}$ (being $n = 4 \cdot 10^{-22} \frac{10^{14} GeV}{m} \frac{\Omega_X}{0.3} T^3$ in the units $\hbar = c = k = 1$ at the temperature T). It corresponds at the temperature T to the mean distance ($l_s \sim n^{-1/3}$) equal to

$$l_s \approx 1.6 \cdot 10^7 \left(\frac{m}{10^{14} GeV}\right)^{1/3} \left(\frac{0.3}{\Omega_X}\right)^{1/3} \frac{1}{T}. \tag{7.8}$$

One finds that superheavy nonrelativistic particles, created just after the end of inflation, when $H \sim H_{end} \sim 10^{13} GeV$, are separated from their antiparticles at distances more than 4 orders of magnitude smaller than the average distance between these pairs. On the other hand, if the non-equilibrium processes of superheavy particles creation (such as decay of inflaton) take place at the end of preheating stage, and the reheating temperature is as low as it is constrained from the effects of gravitino decays on ^6Li abundance ($T_{reh} < 4 \cdot 10^6 GeV$ [3, 286]), the primordial separation of pairs, given by Eq. (7.7), can even exceed the value given by Eq. (7.8). It means that the separation between particles and antiparticles can be determined in this case by their averaged

density, if they were created at $H \leq H_s \sim 10^{-15} \cdot M_P(\frac{10^{14}GeV}{m})^{2/3}(\frac{\Omega_X}{0.3})^{2/3} \sim$
$10^4(\frac{10^{14}GeV}{m})^{2/3}(\frac{\Omega_X}{0.3})^{2/3}GeV$.

The value of H_s separates the cases of early, $H \geq H_s$, and late, $H \leq H_s$, particle production. From the discussion in the previous section one easily finds that gravitational mechanism corresponds to very early particle production and production at reheating is very late. For the other mechanisms, pending on their parameters, both early and late particle production can take place.

If the considered charge is the source of a long-range field, similar to the electromagnetic field, which can bind particle and antiparticle into the atomlike system, analogous to positronium, it may have important practical implications for the UHECR problem. The annihilation time scale of such bound system can provide the rate of UHE particle sources, corresponding to UHECR data.

2.3   Formation of bound systems

The pair of particle and antiparticle with opposite gauge charges forms a bound system, when in the course of expansion the absolute magnitude of potential energy of pair $V = \frac{\alpha_y}{l}$ exceeds the kinetic energy of particle relative motion $T_k = \frac{p^2}{2m}$. The mechanism is similar to that proposed in [166] for binding of magnetic monopole-antimonopole pairs. It is not a recombination one. The binding of two opposite charged particles is caused just by their Coulomb-like attraction, once it exceeds the kinetic energy of their relative motion.

If plasma interactions do not heat superheavy particles created with relative momentum $p \leq mc$ in the period corresponding to Hubble constant $H \geq H_s$, their initial separation, being of the order of

$$l(H) = \left(\frac{p}{mH}\right),\tag{7.9}$$

experiences only the effect of general expansion $l \propto a$, where a is the scale factor. Their potential energy is proportional to the inverse first power of the scale factor $V \propto l^{-1} \propto a^{-1}$, while the initial kinetic energy decreases as the square of the scale factor $T \propto p^2 \propto a^{-2}$. It means that the potential energy V, being initially at $a = a_0$ less than the kinetic energy T, $V_0 < T_0$, in the course of successive expansion should inevitably exceed the kinetic energy at $a = a_0(T_0/V_0)$. Thus, the binding condition should be fulfilled in the period corresponding to the Hubble constant H_c, determined by the equation

$$\left(\frac{H}{H_c}\right)^{1/2} = \frac{p^3}{2m^2\alpha_y H}.\tag{7.10}$$

Here H is the Hubble constant in the period of particle creation and α_y is the "running constant" of the long-range interaction, possessed by the superheavy particles. If the local process of pair creation does not involve nonzero orbital momentum, due to the primordial pairing the bound system is formed in the state with zero orbital momentum.

However, the gauge U(1) nature of the charge, possessed by superheavy particles, assumes the existence of massless U(1) gauge bosons (y-photons) mediating this interaction. The inevitable presence of y-photons in the surrounding medium causes interaction with them and induces non-zero orbital momentum. Bound systems with non-zero orbital momentum are also generally formed (see section 5), if the primordial pairing is lost. Moreover, the interaction with plasma not only induces non-zero orbital momentum but even prevents formation of bound system, which we consider in section 6.

In the case of primordial pairing the size of the bound system exhibits strong dependence on the initial momentum distribution

$$l_c = \frac{p^4}{2\alpha_y m^3 H^2} = 2\frac{\alpha_y}{m\beta^2},$$
(7.11)

where

$$\beta = \frac{2\alpha_y m H}{p^2}.$$
(7.12)

which, in principle, facilitates the possibility to fit UHECR data in the framework of hypothesis of bound system annihilation in the halo of our Galaxy. However, there is no such possibility for S-wave annihilation.

Indeed, in the case of gravitational production mechanism [155] both gravity and Coulomb force are radial. Therefore one expects radial motion of particles in the created pair and S-wave annihilation as the dominant mode. The S-wave bound system of particles initially is not in a single eigenstate of their Coulomb-like Hamiltonian and presents a mixture of many states with large principal quantum number n. Correspondingly this state is not stationary state of the Hamiltonian. For large n we may treat the system classically.

If there are two classical particles at rest at the distance l then under impact of their Coulomb-like attraction they will come to contact approximately in time

$$t \sim l_c\left(\frac{l_c m}{\alpha_y}\right)^{1/2} < 6 \cdot 10^{-6}\left(\frac{m}{10^{14}GeV}\right)^4\left(\frac{0.3}{\Omega_X}\right)^2\frac{1}{(50\alpha_y)^2}s.$$
(7.13)

This time is evidently much smaller than the age of the Universe and the annihilation in S-wave would be too fast to provide the origin of UHECR in the modern Universe.

To be effective for UHECR production bound systems should have annihilation time scale, exceeding the age of the Universe, $t_U = 4 \cdot 10^{17}$s. It was found in [143] that to fit the UHECR data by decays in the halo of our Galaxy of superheavy particles with the modern density $\Omega_X \leq 0.3$ and the lifetime τ_X the magnitude

$$r_X = \frac{\Omega_X}{0.3}\frac{t_U}{\tau_X}$$
(7.14)

should have the value $r_X = 2 \cdot 10^{-10}$. As we show in the next Section primordial bound systems with non-zero orbital momentum can satisfy this condition that should be re-formulated accounting for the possible difference between the total density of particles and the density of their bound systems and for the possible change of annihilation time scale of bound systems in the Galaxy.

If late production of superheavy particles takes place at $H \leq H_s$, their initial separation is determined by the $\min\{l(H), l_s\}$, where $l(H)$ is given by Eq. (7.9) and l_s is determined by their mean number density (compare with Eq. (7.8))

$$l_s \approx 3 \cdot 10^7 (\frac{m}{10^{14} GeV})^{1/3} (\frac{0.3}{\Omega_X})^{1/3} (\frac{1}{M_P H})^{1/2} \approx$$

$$\approx 10^{-4} (\frac{m}{10^{14} GeV})^{1/3} (\frac{0.3}{\Omega_X})^{1/3} (\frac{10^4 GeV}{H})^{1/2} cm. \qquad (7.15)$$

The binding condition can retain the form (7.10) for late particle production (i.e. at $H \leq H_s$), if $l(H) \leq l_s$.

2.4 Annihilation of bound systems with non-zero orbital momentum

In the case of late particle production, when $l(H) \geq l_s$, the primordial pairing is lost and even being produced with zero orbital momentum particles and antiparticles originated from different pairs, in general, form bound systems with nonzero orbital momentum. The size of the bound system is in this case obtained from the binding condition for the initial separation, determined by Eq. (7.15), and it is equal to

$$l_c \approx \frac{10^{15}}{2\alpha_y M_P} \left(\frac{m}{10^{14} GeV}\right)^{2/3} \left(\frac{0.3}{\Omega_X}\right)^{2/3} \left(\frac{p}{mc}\right)^2 \left(\frac{m}{H}\right)$$

$$\approx 2 \cdot 10^{-6} cm \left(\frac{m}{10^{14} GeV}\right)^{2/3} \left(\frac{0.3}{\Omega_X}\right)^{2/3} \left(\frac{10^{-12}}{\beta}\right). \qquad (7.16)$$

The orbital momentum of this bound system can be estimated as $M \sim mvl_c$ and the lifetime of such bound system is determined by the time scale of the loss of this orbital momentum. This time scale can be reasonably estimated with the use of the well-known results of the classical problem of falling down the center due to radiation in the bound system of opposite electric charges e_1 and e_2 with masses m_1 and m_2, initial orbital momentum M and absolute value of the initial binding energy E (see, e.g., [167])

$$t_f = \frac{c^3 M^5}{\alpha_y (2E\mu^3)^{1/2}} (\frac{e_1}{m_1} - \frac{e_2}{m_2})^{-2}.$$

$$\cdot ((\mu \alpha_y^2)^{1/2} + (2M^2 E)^{1/2})^{-2}. \qquad (7.17)$$

Here $\mu = \frac{m_1 m_2}{m_1 + m_2}$ is the reduced mass. Putting into Eq. (7.17) $M = \mu v l_c$, $E = \mu v^2 / 2$, and with the account for $2M^2 E \sim \mu \alpha_y^2$ one obtains the lifetime of the bound system as

$$\tau = \frac{l_c^3}{64\pi} \frac{m^2}{\alpha_y^2} = 4 \cdot 10^{20} \left(\frac{l_c}{10^{-6} cm}\right)^3 \left(\frac{m}{10^{14} GeV}\right)^2 (50\alpha_y)^{-2} yr. \quad (7.18)$$

Using the Eq. (7.18) and the condition $l(H) \geq l_s$, one obtains for this case the following restriction

$$r_X = \frac{\Omega_X}{0.3} \frac{t_U}{\tau_X} \leq 3 \cdot 10^{-10} \left(\frac{\Omega_X}{0.3}\right)^5 \left(\frac{10^{14} GeV}{m}\right)^9. \quad (7.19)$$

Note that the condition (7.19) admits $\Omega_X = 0.3$, when superheavy particles dominate in the dark matter of the modern Universe. The strong dependence on m in Eq.(7.19) makes this case rather naturally satisfying the necessary condition for UHECR sources for $m < 10^{14} GeV$.

The above consideration is also valid for the case of early particle production with non-zero orbital momentum. In this case the annihilation time scale is also given by Eq.(7.18), but the value of l_c should be taken from Eq.(7.11). The condition the UHECR sources should satisfy in this case is given by

$$\left(\frac{\Omega_X}{0.3}\right)\left(\frac{1}{50\alpha_y}\right)\left(\frac{m}{10^{14} GeV}\right)\left(\frac{\beta}{10^{-12}}\right)^6 = 10^{-4}. \quad (7.20)$$

The modification of this condition, accounting for dominant invisible channels for bound system annihilation (see Section 3) extends the possibility to satisfy them and even to admit the case $\Omega_X = 0.3$.

So, in the case of non-zero orbital momentum both for early and late particle production the considered mechanism provides conditions, necessary for UHECR sources. The range of the allowed parameters is sufficiently wide to include the case when superheavy particles are the dominant form of the modern dark matter.

Since the considered superheavy particles are the lightest particles bearing this charge, and they are not in thermodynamical equilibrium, one can expect that there should be no thermal background of y-photons and that their non-equilibrium fluxes cannot significantly heat the superheavy particles.

2.5   Particle binding in hot plasma

The situation changes drastically if the superheavy particles possess not only new U(1) charge but also some ordinary (weak, strong or electric) charge [163]. Due to this charge superheavy particles interact with the equilibrium relativistic plasma (with the number density $n \sim T^3$) and for the mass of particles $m \leq \alpha^2 M_P$ the rate of heating

$$n\sigma v \Delta E \sim \alpha^2 \frac{T^3}{m} \quad (7.21)$$

is sufficiently high to bring the particles into thermal equilibrium with this plasma. Here α is the running constant of the considered (weak, strong or electromagnetic) interaction.

Plasma heating causes the thermal motion of superheavy particles. At $T \leq m(\frac{m}{\alpha^2 M_P})^2$ their mean free path relative to scattering with plasma exceeds the free thermal motion path, so it is not diffusion, but free motion with thermal velocity v_T that leads to complete loss of initial pairing, since $v_T t$ formally exceeds l_s at $T \leq 10^{-10} M_P (\frac{\Omega_X}{0.3})^{2/3} (\frac{10^{14} GeV}{m})^{5/3}$.

If the interaction with plasma keeps superheavy particles in thermal equilibrium, potential energy of charge interaction $V = \frac{\alpha_y}{l_s}$ is less than thermal energy T for any $\alpha_y \leq 3 \cdot 10^7 (\frac{0.3}{\Omega_X})^{1/3} (\frac{m}{10^{14} GeV})^{1/3}$. So binding condition $V \geq T_{kin}$ cannot take place when plasma heating of superheavy particles is effective.

For electrically charged particles it is the case until electron positron pairs annihilate at $T_e \sim 100$keV (see [166]) and for colored particles until QCD phase transition at $T_{QCD} \sim 300$MeV. In the latter case colored superheavy particles form superheavy stable hadrons, possessing U(1) charge. For weakly interacting particles after electroweak phase transition, when Eq. (7.20) is not valid, neutrino heating, given by $n \sigma v \Delta E \sim G_F^2 \frac{T^7}{m}$, is sufficiently effective until $T_w \approx 20$GeV. At $T < T_N$, where $N = e, QCD, w$, respectively, plasma heating is suppressed and superheavy particles go out of thermal equilibrium.

In the course of successive expansion kinetic energy of superheavy particles falls down with the scale factor a as $\propto a^2$, and the binding condition is reached at T_c, given by

$$T_c = T_N \alpha_y 3 \cdot 10^{-8} \left(\frac{\Omega_X}{0.3}\right)^{1/3} \left(\frac{10^{14} GeV}{m}\right)^{1/3}. \qquad (7.22)$$

For electrically charged particles, forming after recombination atom-like states with protons and electrons, but still experiencing the Coulomb-like attraction due to non-compensated U(1) charges, the binding in fact does not take place to the present time, since one gets from Eq. (7.22) $T_c \leq 1$K. Bound systems of hadronic and weakly interacting superheavy particles can form, respectively, at $T_c \sim 0.3$eV and $T_c \approx 20$eV.

The size of the bound system is then given by

$$l_c = 10^{15} \left(\frac{0.3}{\Omega_X}\right)^{2/3} \left(\frac{m}{10^{14} GeV}\right)^{2/3} \frac{1}{\alpha_y T_N}, \qquad (7.23)$$

that even for weakly interacting particles approaches a half meter (30m for hadronic particles!). It leads to an extremely long annihilation time scale of these bound systems that cannot fit UHECR data. Moreover, being extremely weakly bound, they should be disrupted almost completely, colliding in the Galaxy. So, for bound systems of weakly interacting superheavy particles, $n \sigma v t_U \sim 10^{14}$ where $n = 3 \cdot 10^{-15} cm^{-3} \frac{10^{14} GeV}{m} \frac{\Omega_X}{0.3}$ is the number density of bound systems, $\sigma \sim \pi l_c^2$ and their relative velocity $v \sim 3 \cdot 10^7$cm/s. It makes it impossible to realize the considered mechanism of UHECR origin if the super-heavy U(1) charged particles share ordinary weak, strong or electromagnetic interactions.

The same is true if superheavy particles are heated by any other means, in particular, by the particles of the same hidden sector, to which they belong. So, not only ordinary, but also hidden sector interactions of considered particles as well as the hidden sector particle content of the Universe turn to be constrained. Such constraints are satisfied, for example, if non-equilibrium y- photons are the only stable form of hidden sector, presents in the Universe after superheavy particles are produced.

3.   Primordial bound systems as the source of UHECR

3.1   Evolution of bound systems in the Galaxy

Superheavy particles, as any other form of CDM, should participate in grav-itational clustering and concentrate to the center in the course of Galaxy form-ation. There are several factors influencing the evolution of bound systems in the Galaxy.

If the size of primordial bound systems $l_c \geq 3 \cdot 10^{-6} cm (\frac{0.3}{\Omega_X} \frac{m}{10^{14} GeV})^{1/2}$ their collision rate in the vicinity of the Solar system $n\sigma v t_U \geq 1$. In general, collision time scale of bound systems with the size l and velocity v is less than the age of the Universe in the regions with superheavy particle density, exceeding

$$n > 10^{-13} cm^{-3} (\frac{10^7 cm/s}{v})(\frac{10^{-6} cm}{l})^2. \qquad (7.24)$$

Since the binding energy of bound systems

$$E_b \leq 5 \cdot 10^{-27} mc^2 (\frac{0.3}{\Omega_X} \frac{10^{14} GeV}{m})^{2/5}$$

is much less than the kinetic energy of their motion in the Galaxy ($T = mv^2/2$), the bound systems should be disrupted in such collisions. On the other hand, large collision cross-section $\sigma \sim \pi l_c^2$ corresponds to the momentum transfer of the order of $\Delta p \sim m\beta c$. Such small momentum transfer leads both to disruption of the bound system on free superheavy particles and with the same order of probability to the reduction of their size down to $l \sim l_c/4$. If the bound systems are within a gravitationally bound cluster (Galaxy, glob-ular cluster, CDM small-scale cluster), both the free particles and contracted bound systems with the size l remain therein and the collisions both between bound systems and between free particles and bound systems continue with the cross-section $\sigma \sim \pi l^2$.

In the vicinity of massive objects with the mass M tidal effects lead to disruption of bound systems with the size l at the distance r, corresponding to tidal force energy $\sim \frac{GMm}{r}\frac{l}{r}$, exceeding the binding energy $\sim \alpha_y/l$. So, in the vicinity of a star with the mass $M = M_\odot$ the bound systems are disrupted at the distances smaller than $r_d = \sqrt{\frac{GMm}{\alpha_y}l}$. Tidal effects disrupt bound systems in the regions of enhanced stellar density, n_M, if $n_M \pi r_d^2 v t_U \geq 1$. Taking for the center of Galaxy $n_M \sim 10^6 M_\odot/pc^3$, one finds that bound systems with the size $l \geq 5 \cdot 10^{-6}$cm should be disrupted there due to tidal effects.

As it was shown in [168], tidal effects strongly influence the formation and mass distribution of small-scale CDM clusters, which should also take place for small-scale clusters of primordial bound systems in the Galaxy. On the other hand, clustering of primordial bound systems may play an important role in the explanation of observed clustering of UHECR events in the framework of the proposed mechanism. It may be easily estimated that if bound systems are clustered around a globular cluster, their disruption due to stellar tidal effects is negligible.

3.2   Space distribution of UHECR events

For the initial number density and size of bound systems, corresponding to $n\sigma v t_U \geq 1$, most bound systems disrupt on the free particles, but a sufficiently large fraction of them $\sim \beta_c/\beta_U \sim (l_U/l_c)^{1/2}$ acquires the size l_U, at which $n\sigma v t_U \approx 1$. The relative amount of bound systems with smaller size $l < l_U$ is of the order of $\sim (l/l_U)^2$, if their annihilation time scale $\tau \geq t_U$ and of the order of $\sim (l/l_U)^2\frac{\tau}{t_U}$ for $\tau \leq t_U$. Annihilation of superheavy particles in bound systems with smaller size is more rapid, which increases the production rate of such UHECR source as compared with the case of superheavy decaying particles with a fixed lifetime. This effect of the self-adjustment of bound systems' annihilation leads to the peculiar space distribution of UHECR sources, corresponding to this mechanism.

The decay rate density, q, of metastable particles with lifetime τ for the number density $n(R)$, depending on the distance R from the center of the Galaxy, is given by

$$q = \left(\frac{n(R)}{\tau}\right). \tag{7.25}$$

Owing to self-adjustment of bound systems' annihilation the density of UHECR production rate for the initial number density $n(R)$ of primordial bound systems with initial annihilation time scale τ has the order of magnitude

$$q = \left(\frac{n(R)}{\tau}\right)\left(\frac{\tau}{\tau_U}\right)^{1/2}\left(\frac{\tau}{t_U}\right)^{1/3}. \tag{7.26}$$

Since $\tau_U \propto l_U^3 \propto n(R)^{-3/2}$ the self-adjustment of bound systems leads to stronger radial dependence $q \propto n(R)^{7/4}$, thus sharpening the concentration of UHECR sources to the center of the Galaxy.

In the case of late particle production with $l(H) \geq l_s$, considered in Subsection 8.2.4, the dependence $l \propto \beta^{-1}$, given by Eq. (7.16), and $\tau \propto l^3 \propto \beta^{-3}$, given by Eq. (7.18), results in the change of Eq. (7.26) by

$$q = \left(\frac{n(R)}{\tau}\right)\left(\frac{\tau}{\tau_U}\right)^{2/3}\left(\frac{\tau}{t_U}\right)^{1/3}.$$

Since $\tau_U \propto l_U^3 \propto n(R)^{-3/2}$ in this case, the radial dependence $q \propto n(R)^{3/2}$ provides the principal possibility to distinguish this case from the case $l(H) \leq l_s$.

It was noticed in [169] that clustering of UHECR events, observed in AGASA experiment [170], can be explained in the model of superheavy metastable particles if such particles with the mass $m \sim 10^{14} GeV$ form clusters in the Galaxy with the mass $M \sim 5 \cdot M_\odot \frac{\tau_X}{10^{10} yr}$.

For the cluster of N metastable particles with lifetime τ the decay rate, P, is given by

$$P = \left(\frac{N}{\tau}\right). \tag{7.27}$$

Owing to self-adjustment of bound systems' annihilation the UHECR production rate for the cluster of N primordial bound systems with initial annihilation time scale τ reaches the order of magnitude

$$P = \left(\frac{N}{\tau}\right)\left(\frac{\tau}{\tau_U}\right)^{1/2}\left(\frac{\tau}{t_U}\right)^{1/3}. \tag{7.28}$$

The enhancement of UHECR production rate due to self-adjustment of bound systems facilitates the possibility to explain clustering of UHECR events in the proposed mechanism. Say, instead of cluster with the mass of $\sim 5 \cdot 10^6 M_\odot$ of particles with the mass $m \sim 10^{14} GeV$ and the lifetime $\tau \sim 10^{16}$yr, corresponding to the considered in [169] possible explanation for the clustering of UHECR events, observed by AGASA, it is sufficient to have the mass of $\sim 10^4 M_\odot$ in the cluster of primordial bounds systems with the superheavy particle number density $n \sim 10^{-10} cm^{-3}$, relative velocity $v \sim 10^7$cm/s and the same initial annihilation time scale $\tau \approx 10^{16}$yr.

In the case of late particle production with $l(H) \geq l_s$, for $p \sim mc$, the condition $H < H_s$ constrains the annihilation time scale

$$\tau > \tau_b = 5 \cdot 10^{19} yr \left(\frac{m}{10^{14} GeV}\right)^9 \left(\frac{0.3}{\Omega_X}\right)^4 (50\alpha_y)^{-5}, \tag{7.29}$$

the size of the bound system l being related with τ by Eq. (7.18). In this case, owing to the effect of self-adjustment, clustering of UHECR events can be also reproduced, e.g., in a cluster with the mass $M \sim 10^5 M_\odot$ and number density $n \sim 10^{-10} cm^{-3}$ of superheavy particles with mass $m \sim 10^{13} GeV$ and initial annihilation time scale of their bound systems $\tau = 4 \cdot 10^{18} yr$ (for metastable particles with the same mass and lifetime the cluster with the mass $4 \cdot 10^8 M_\odot$

is needed). Taking into account the possibility of dominance of superheavy particles in the modern CDM, mentioned above for the considered case, the estimated parameters of such cluster seem to be rather reasonable.

Disruption of bound systems in collisions leads to the decrease of their actual amount in the modern Universe as compared with their primordial abundance. On the other hand, owing to the effect of self-adjustment the initial annihilation time decreases in the dense regions. It leads to the corresponding corrections in the conditions (7.19) and (7.20). Provided that the inequality (7.24) is valid and the collisions of bound systems are significant, the annihilation time scale τ should be corrected by the effect of self-adjustment. In the estimation of this effect we normalize it on the total initial density of primordial bound systems with the annihilation time scale τ in the considered region. One should substitute Ω_X by $\Omega_X(\tau) \leq \Omega_X$, where $\Omega_X(\tau)$ is the averaged initial concentration of bound systems with the annihilation time scale τ. This concentration is less than the total density of superheavy particles Ω_X, if production of superheavy particles or their derelativization was extended in time at the stage of preheating and primordial bound systems of different size and correspondingly different annihilation time scale were produced. The mechanism of superheavy particle production is then constrained by the condition $\Omega_X \leq 0.3$.

3.3   Annihilation into ordinary particles

To be the source of UHECR the products of superheavy particles' annihilation should contain a significant amount of ordinary particles. On the other hand, it was shown above that to be the viable source of UHECR the considered particles should not possess ordinary strong, weak and electromagnetic interactions. Their interaction with ordinary particles, giving rise to UHECR production in their annihilation should be related to the super high energy sector of particle theory and/or physics of inflation and preheating. The self-consistent treatment of this problem should involve the realistic particle physics model, reproducing the desired features of the inflationary scenario and giving detailed predictions for physical properties of U(1) charged superheavy particles. It may be expected that in such models, there can exist superheavy boson (Y), interacting both with superheavy and ordinary particles. If its mass is of the order of $m_Y \geq m$ the annihilation channel into ordinary particles with the cross-section $\sigma \sim \frac{\alpha_Y^2}{m_Y^4} m^2$ will be of the order of the cross-section of the two y-photon annihilation channel.

With account of the invisible yy mode of annihilation, as well as taking into account effects of self-adjustment and destructions of bound systems in collisions, one has to re-define the magnitude r_X, given above by Eq. (7.14), as

$$r_X^{eff} = B_o \frac{\Omega_X(\tau)}{0.3} \frac{t_U}{\tau_{eff}}, \qquad (7.30)$$

where B_o is the branching ratio of annihilation channels to ordinary particles, $\tau_{eff} = (\tau_X \tau_U^3 t_U^2)^{1/6}$ and $\tau_{eff} = (\tau_X \tau_U t_U)^{1/3}$ for the cases of early and late particle production, respectively, provided that the condition (7.24) is valid. If the condition (7.24) is not valid, $\tau_{eff} = \tau_X$. In the Eq. (7.30) $\Omega_X(\tau) \leq \Omega_X$ takes into account possibly non-instant character of superheavy particle production and derelativization, leading to formation of bound systems with different sizes and correspondingly different annihilation time scales.

The total cosmological density of superheavy particles should satisfy the evident condition $\Omega_X \leq 0.3$. It constrains the total amount of bound systems surviving to the present time. For the bound systems, annihilated earlier, the set of astrophysical constraints on the sources of non-equilibrium particles (see review in [3]) should be satisfied. In the case of late particle production the lifetime of bound systems exceeds the age of the Universe and the constraint on the total density of superheavy particles puts restrictions on the mechanism of production.

In the case of early particle production the size of bound system can not exceed

$$l_b = 5 \cdot 10^{-7} cm (\frac{1}{50\alpha_y})(\frac{m}{10^{14} GeV})^{7/3} (\frac{0.3}{\Omega_X})^{4/3}. \qquad (7.31)$$

On that reason their lifetime is less than (7.29). Owing to small size l their self-adjustment can take place only at largest possible $l \leq l_b$ in the center of Galaxy and in very dense dark matter clusters, making $\tau_{eff} = \tau_X$ in the Eq. (7.30) to be the general case. The condition [143] for UHECR sources $r_X^{eff} = 2 \cdot 10^{-10}$ is then given by the Eq. (7.20), in which factor B_o should be added and Ω_X should be changed by $\Omega_X(\tau)$. This condition valid for bound systems with size

$$3 \cdot 10^{-10} cm (\frac{10^{14} GeV}{m})^{2/3} (50\alpha_y)^{2/3} \leq l \leq l_b, \qquad (7.32)$$

and corresponding annihilation timescale between t_U and (7.18). Such bound systems can be formed by superheavy particles, created or derelativized in the period

$$H_s \leq H \leq 2 \cdot 10^5 (50\alpha_y)^{-5/6} (\frac{m}{10^{14} GeV})^{5/6}. \qquad (7.33)$$

Note that at $B_o \sim 10^{-5}$, when annihilation to ordinary particles is strongly suppressed, the case $\Omega_X \sim 0.3$ is possible for early particle production.

Large size of bound systems, created or derelativized at $H \leq H_s$ provides their self-adjustment in the Galaxy. It sharpens the distribution of UHECR sources making them to concentrate within the center of Galaxy (or in dense DM clusters in halo). For the central galactic number density of superheavy particles

$$n = 3 \cdot 10^{-13} cm^{-3} (\frac{\Omega_X}{0.3})(\frac{10^{14} GeV}{m})$$

and their relative velocity $v = 3 \cdot 10^7 \text{cm/s}$ we find

$$l_U = 3 \cdot 10^{-7} cm \left(\frac{m}{10^{14} GeV}\right)^{1/2} \left(\frac{0.3}{\Omega_X}\right)^{1/2},$$

corresponding to

$$\tau_U = 10^{19} yr \left(\frac{m}{10^{14} GeV}\right)^{7/2} \left(\frac{0.3}{\Omega_X}\right)^{3/2} (50\alpha_y)^{-2}.$$

Then condition [143] for $l > l_U$ (and, respectively, $\tau > \tau_U$) takes the form

$$B_o \left(\frac{\Omega_X(\tau)}{0.3}\right)^{17/6} \left(\frac{10^{14} GeV}{m}\right)^{25/6} (50\alpha_y)^{7/3} \left(\frac{H}{H_s}\right) = 3 \cdot 10^{-4}.$$

Even for minimal possible mass $m \sim 10^{13} GeV$ and maximal possible $B_o \sim 1$ and $\Omega_X(\tau) \sim 0.3$ this condition can not be valid for the value $H = H_r$ as low as it follows from constraints on primordial gravitino [286] $H_r < 10^{-6} \text{GeV}$.

Owing to Z- boson "bremsstrahlung" [171] neutrino channel can not strongly dominate in the annihilation to ordinary particles. However, if annihilation of bound systems can not be direct local (galactic) source of UHECR, it can provide the source of UHE neutrinos for the Z-Shower mechanism of UHECR origin.

3.4 Discussion

The combination of the constraints on the conditions of particle creation in the early Universe and on the effective production of UHECR specifies the parameters of the proposed mechanism, which should be considered in the framework of specific models of particle theory underlying the scenarios of very early Universe. The evolution of primordial bound systems should be also analyzed on the base of such models. One, however, can make the general conclusion that the two principal types of bound systems are possible, originated from (i) "Early particle production", when primordial pairing essentially determines the formation of bound systems and from (ii) "Late particle production", when the primordial pairing is not essential for bound system formation. In both cases bound systems with nonzero orbital momentum should be formed. S-wave annihilation is too fast for bound systems to survive to the present time.

To form bound systems with annihilation time scale that can fit UHECR data above GZK cut-off, superheavy particles with mass $10^{13} \text{GeV} < m < 10^{14} \text{GeV}$ should be created or should derelativize in the intermediate period of preheating stage at

$$10^{-4} GeV \leq H \leq 10^6 GeV. \tag{7.34}$$

These data can not be fitted, if particles are produced too early (in the beginning of preheating) as it takes place for gravitational production. Superheavy particle annihilation in bound systems can hardly fit the UHECR data, if the particles are produced too late, in the time of reheating as late, as it follows from the restrictions on primordial gravitino production [286]. These conditions relate the proposed mechanism of UHECRs to proper mechanisms of superheavy particle production and put restrictions on such mechanisms.

If the particles are produced too early, at H exceeding (7.33), their annihilation time scale is so small that either overproduction of UHECRs (for $\tau > t_U$) or too strong effect of annihilation products (for $\tau < t_U$) can take place.

If the particles are produced too late, their annihilation in bound systems can not provide effective source of UHECRs, whereas their presence in the Universe contributes to the total cosmological density and this contribution should not exceed $\Omega_X \sim 0.3$.

The realistic scenario for the considered mechanism of UHECRs favors superheavy particle production by evaporation of PBHs with the mass 0.1–10g, or to derelativization in the period (7.34) of relativistic products of either decay of supermassive additional filed, or of the instant reheating. The restrictions for too early and too late particle production constrain the parameters of these mechanisms. In particular, being the stable relics of mini-PBH evaporation superheavy particles should not over-close the Universe, what puts severe restriction on the possible total amount of such mini-PBHs [316, 305] and on the inhomogeneity of very early Universe on very small scales.

The two cases of superheavy particle production differ by the allowed range of parameters, necessary for UHECR sources.

In the case (i) annihilation of small size (7.32) bound systems with time scale $t_U \leq \tau \leq \tau_b$ takes place. In the case (ii) self-adjustment of bound systems with $l > l_b$ can fit the conditions for UHECRs sources.

In both cases bound systems can dominate in the modern CDM, but the conditions for such dominance are different. In the case (i) annihilation of bound systems with $\Omega_{bs} \sim 0.3$ can reproduce UHECR events, if the branching ratio for annihilation to ordinary particles is small ($B_o \sim 10^{-5}$), whereas in case (ii) this branching ratio should maximally approach to 1.

The possibility of bound system disruption in their collisions in galaxies is specific for the considered mechanism, making it different from the models of decaying and annihilating superheavy particles. If effective, such disruption results in a nontrivial situation, when superheavy particles can dominate in the modern CDM, while the UHECR sources represent a sparse subdominant component of bound systems, surviving after disruption.

Self-adjustment of bound systems' annihilation in the Galaxy sharpens their concentration to the center of the Galaxy and increases their UHECR production rate in clusters. It provides their difference from the case of metastable particles. This property, however, crucially depends on the probability of contraction of bound systems in the course of collisions. More detailed analysis of such collisions is needed to prove this result. If proven to be really specific for the considered type of UHECR sources, it will be principally possible to distinguish them from other possible mechanisms [145, 159, 172] in future AUGER and EUSO experiments.

For suppressed local source of annihilation to quarks, charged leptons, gauge and Higgs bosons, annihilation of superheavy particles in bound systems can provide the effective distant source of UHE neutrinos, thus playing important role in the UHECR production by Z-Shower mechanism.

It also should be mentioned that particle fragmentation from the proposed mechanism, as from any other top-down mechanism can not be responsible for the whole spectrum of UHECRs and offers the possible explanation for its part above GZK cut-off [173], provided that the existence of such UHECRs is confirmed by the future refined data.

If viable, the considered mechanism makes UHECR above the GZK cut-off the unique source of detailed information on the possible properties of the hidden sector of particle theory and on the physics of very early Universe.

4. Possible signature of low-scale gravity in UHECR

It was proposed in [183, 184, 185, 186] that the space is 4+n dimensional, with the Standard Model particles living on a brane. While the weakly, electromagnetically, and strongly interacting particles are confined to the brane in 4 dimensions, gravity can propagate also in extra n dimensions. This approach allows us to avoid the gauge hierarchy problem by introducing a single fundamental mass scale (string scale) M_s of the order of TeV. The usual Planck scale $M_P = 1/\sqrt{G_N} \simeq 1.22 \cdot 10^{19}$ GeV is related to the new mass scale M_s by Gauss's law:

$$M_P^2 \sim R^n M_s^{n+2} \tag{7.35}$$

where G_N is the Newton constant, R is the size of extra dimensions. It follows from (7.35) that

$$R \sim 2 \cdot 10^{-17} \left(\frac{TeV}{M_s} \right) \left(\frac{M_P}{M_s} \right)^{2/n} cm \tag{7.36}$$

gives at $n = 1$ too large a value, which is clearly excluded by present gravitational experiments. On the other hand $n \geq 2$ gives the value $R \lesssim 0.25$ cm, which is below the present experimental limit ~ 1 cm but can be tested for the case $n = 2$ in gravitational experiments in the near future.

It can be shown that the graviton including its excitations in the extra dimensions, so-called Kaluza–Klein (KK) graviton emission, interacts with the Standard Model particles on the brane with an effective amplitude $\sim M_s^{-1}$ instead of M_P^{-1}. Indeed, the graviton coupling to the Standard Model particle $\sim M_P^{-1}$, the rate [187] of the graviton interaction $r \sim (M_P^{-1})^2 N$, where N is a multiplicity of KK states. Since this factor is $\sim (\sqrt{S}R)^n$, where $\sqrt{S}$ is the c.m. energy, then substituting R from (7.36) we get $r \sim M_s^{-2}$. Thus, the graviton interaction becomes comparable in strength with weak interaction at TeV scale.

This leads to the varieties of new signatures in particle physics, astrophysics and cosmology (see, e.g., [187, 188, 189, 190, 191, 192, 193, 194, 195]) which have already been tested in experiments or can be tested in the near future.

In this section we consider the possible signature of the low-scale gravity in ultra high energy cosmic rays.

The origin of cosmic rays with energies, exceeding the GZK cut-off energy [141], is widely discussed as the possible effect of new physics. In particular, it was proposed [145, 211] that ultra high energy neutrinos reaching the Earth from cosmological distances interact with a halo of relic light neutrinos in the Galaxy, producing due to Z, $W^{\pm}$ boson exchange secondaries inside the galactic halo. Photons from π^0 decays and nucleons can easily propagate to the Earth and be the source of the observed ultra high energy air showers. Crucial elements of models [145, 211] are: the existence of neutrino mass in the range 0.1–10 eV and significant clustering of relic neutrinos in the halo up to $10^5 n_v$, where n_v is the cosmological neutrino number density ($n_v \sim 100 cm^{-3}$). Also the existence of ultra high energy ($> 10^{21}\text{-}10^{23}$ eV) neutrino flux is necessary in order to produce multiple secondaries with energies above GZK cut-off.

If the graviton interaction is comparable in strength with weak interaction at TeV scale, then photons can be produced directly [153, 214] in a reaction

$$v + \bar{v} \quad \to g \longrightarrow \gamma + \gamma \tag{7.37}$$

due to virtual graviton exchange (Figure 7.1). In the Standard Model the process (7.37) occurs via loop diagram and therefore is severely suppressed.

At high energies the cross-section for the process (7.37) can be obtained immediately from that for the process $e^+ e^- \longrightarrow \gamma\gamma$ including graviton exchange (see, for example, [188]) by substituting e = 0. Then

$$\frac{d\sigma}{dz} = \frac{\pi}{16} \frac{S^3}{M_s^8} F^2 (1 - z^4) \tag{7.38}$$

where $\sqrt{S}$ is c.m.s. energy, z = $|\cos\theta|$ is the polar angle of the outgoing photon. The factor F depends on the number of extra dimensions:

$$F = \begin{cases} \log(M_s^2/S), & n = 2, \\ 2/(n-2), & n > 2, \end{cases}$$

at $\sqrt{S} \ll M_s$. In Eq. (7.38) it is also taken into account that primary beam of neutrinos is polarized.

Integrating (7.38) over the polar angle and including a symmetry factor for two γ we get

$$\sigma = \frac{\pi}{20} \frac{S^3}{M_s^8} F^2 \approx 7 \cdot 10^{-35} F^2 (\sqrt{S}/TeV)^6 (TeV/M_s)^8 cm^2. \tag{7.39}$$

One can see from (7.39) that at TeV energies the rate of the reaction (7.37) is comparable with the rate of weak processes [145].

Assuming $M_s \sim \sqrt{S} \sim$ TeV we find [153] for example for $n = 3$ the following probability for the interaction of ultra high energy neutrinos inside the galactic halo: $P \approx \sigma n_G L_G \sim 10^{-3}$, where $L_G \sim 100$ Kpc is the size of the galactic neutrino halo, $n_G \sim 10^5 n_\nu$ is the neutrino number density in the galactic halo. This probability is significantly greater than the probability of ultra high energy neutrino interaction in terrestrial atmosphere [212].

Let us note that nearby galaxies also can be sources of additional ultra high energy photons due to neutrino interaction with relic neutrinos of galactic halos [145, 211].

TeV range in c.m.s. corresponds to the energy of extragalactic neutrino flux $E \approx 10^{22}\text{--}10^{23}$ eV since

$$E \approx \frac{S}{2m} \approx 5 \cdot 10^{22} (\sqrt{S}/TeV)^2 (10eV/m) eV \qquad (7.40)$$

where m is neutrino mass.

Photon distribution in reaction (7.37) in laboratory system is given by

$$\frac{d\sigma}{d(\omega/E)} = 8\pi F^2 \frac{m^3 E^3}{M_s^8} \frac{\omega}{E} \left(1 - \frac{\omega}{E}\right) \left[\left(1 - \frac{\omega}{E}\right)^2 + \left(\frac{\omega}{E}\right)^2\right] \qquad (7.41)$$

where $\omega >> m$ is photon energy. This distribution is shown in Figure 7.2. It follows from (7.41) that photons are produced in the reaction (7.37) mainly within the energy range $0.2E \leqslant \omega \leqslant 0.8E$ with an average energy $\approx E/2$.

Therefore existence of low-scale gravity at TeV scale or above could lead to the direct production of photons with energy $\omega > 10^{22}$ eV (at these energies the mean interaction length for pair production for photons in the radio background is $\approx 1 - 10 Mpc$ [213]). Such photons can hardly be produced in standard weak interaction processes because in the last ones photons appear as a result of cascade processes significantly reducing photon energy in comparison with the initial neutrino energy. For example, as it was shown in [145] final energy of photons produced due to cascade processes can be by 10–100 times less than the energy of the initial neutrino flux.

Of course photons with the energy $\sim 10^{23}$ eV could be produced in cascade processes induced by neutrinos of the energy $> 10^{24} - 10^{25}$ eV but from the observations of cosmic rays we know that cosmic ray fluxes decrease with the energy as E^{-3}, and therefore the probability of such events is significantly suppressed.

The effect of prompt gamma production in UHE neutrino collision with light neutrinos in the galactic halo was estimated in [153] in the assumption of a very high local neutrino over-density ($n_G \sim 10^5 n_\nu$). The calculations of neutrino concentration in the galactic neighborhood in the framework of CDM model lead [179, 180], however, to much smaller values of the over-density for the light neutrino masses ($\leqslant 1eV$), estimated from the searches for neutrino oscillations [12, 13, 14, 15].

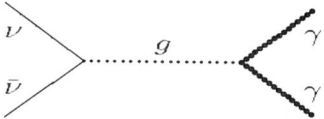

Figure 7.1.   Neutrino annihilation into two photons.

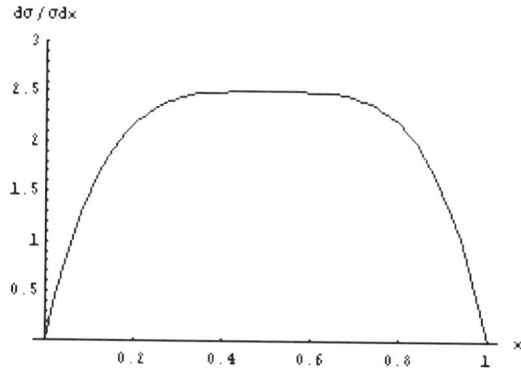

Figure 7.2.   Energy distribution of photons ($x = \omega/E$).

5.   Conclusions

Pair correlation, considered in the present chapter, takes place if the local process of superheavy particle creation preserves charge conservation. This condition has serious grounds in the case of a local U(1) gauge charge, similar to electric charge, but it may not be the case for global charge, say, for mechanisms of R-parity nonconservation due to quantum gravity wormhole effects [143]. The crucial physical condition for the formation of primordial bound systems of superheavy particles is the existence of new strictly conserved local U(1) gauge symmetry, ascribed to the hidden sector of particle theory. Such symmetry can arise in the extended variants of GUT models (see, e.g., [3] for review), in heterotic string phenomenology (see [24] and references therein) and in D-brane phenomenology [26, 27]. Note, that in such models the strictly conserved symmetry of the hidden sector can be also SU(2), which leads to a nontrivial mechanism of primordial binding of superheavy particles due to macroscopic size SU(2) confinement, as was the case for "thetons" [174, 175, 176].

The proposed mechanism is deeply involved in the details of the hidden sector of particle theory. The necessary combination of conditions (superheavy stable particles, possessing new strictly conserved U(1) charge, existence of their superheavy Y-boson interaction with ordinary particles, nontrivial physics of inflation and preheating) can be rather naturally realized in the hidden sector of particle theory. In this aspect, the proposed mechanism offers the link between the observed UHECR and the predictions of particle theory, which cannot be tested by any other means and on which the analysis of primordial pairing and binding can put severe constraints.

Even so, while we may agree that the number and sequence of the assumption of the present scenario may sound artificial and *ad hoc*, (maybe at the same level of topological defect lifetime) we have taken into account a large number of astrophysical and cosmological bounds narrowing the parameter window into a very severe and fragile regime which may soon survive (or not) future theoretical self-consistence and experimental test. Indeed, if HIRES and AGASA data converge to a GZK cut-off with no spectra extensions to Grand Unified Energies or, in a different scenario, in case of more evidence for UHECR clustering to BL Lac sources compatible only to Z-Shower model [145], the model may be considered as an untenable local solution of the UHECR puzzle. Let us be reminded that UHE neutrinos of all flavors will be produced by such heavy particle annihilation leading to important signals in new generation UHE neutrino telescopes based on Horizontal Tau Showering (or Earth Skimming Neutrinos) [181] – see review and some answers to the recent arguments [179] against Z-Shower mechanism in [236].

Fluxes of ultra high energy cosmic rays at the Earth are very small $\Phi \sim 0.03 km^{-2} sr^{-1} yr^{-1}$. Until now only about 60 events were collected with energies above GZK cut-off. However in the near future improved Fly's Eye (7000 $km^2 sr$) [202] will allow us to detect about 20 events/yr. It seems possible that such detector could collect rare ultra energetic photons ($\omega > 10^{22} eV$). The detection of such events could be an indication that these ultra high energy photons were produced in $v=v$ annihilation in the galactic halo due to effects of low-scale gravity at TeV scale. The possibility of search for this effect is, however, strongly conditioned by both the existence of UHE neutrino sources and the sufficiently high local over-density of relic neutrinos.

Chapter 8

HIGH DENSITY REGIONS FROM FIRST-ORDER PHASE TRANSITIONS

In this chapter we consider dynamics of first-order phase transitions in the early Universe. Numerical results indicate that within a certain range of parameters it leads to formation of separate relatively long-lived clots – configurations filled with a scalar field oscillating around the true vacuum state. Energy is perfectly localized, and density is slightly pulsating around its maximum. This process is accompanied by radiation of scalar waves. Under some conditions the localization of energy leads to formation of small black holes with high probability.

Analysis of physical processes in the early Universe on the basis of particle theory is the important way to study physical conditions in the early Universe and physical mechanisms underlying those conditions. As a result of such analysis, the existence of hypothetical relics of early Universe, such as primordial black holes, topologically stable or metastable solitons etc., have been predicted. Confrontation of predicted effects with observational data provides certain conclusions concerning both cosmological evolution and particle physics models [3].

First-order phase transitions as predicted by unified theories can occur at several periods of cosmological evolution. A wide class of models of particle symmetry breaking [297] contain this possibility. Such transitions are also considered as the final stage of inflation in a wide range of inflationary models. Detailed study of nonlinear configurations arising at the first-order phase transitions and their dynamics is helpful not only for cosmology – nonlinear dynamics of field theories describes a lot of phenomena occurring in laboratory physics.

For brief discussion of first-order phase transitions, let us consider real scalar field ϕ with the Lagrangian

$$L = \frac{1}{2}\partial_\mu \phi \partial^\mu \phi - V(\phi). \qquad (8.1)$$

171

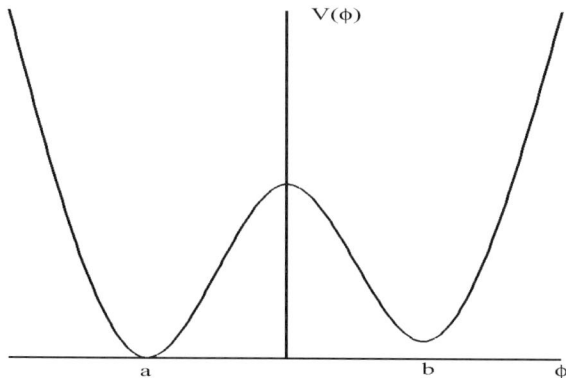

Figure 8.1. Potential with nonequivalent minima. Minimum 'a' is true vacuum of the system, local minimum 'b' is false vacuum.

To compare our results with the results obtained by Hawking, Moss and Stewart [308], and by Watkins and Widrow [307], we choose asymmetric double-well potential of the same form

$$V(\phi) = \tfrac{1}{8}\lambda(\phi^2 - \phi_0^2)^2 + \epsilon\phi_0^3(\phi + \phi_0). \tag{8.2}$$

This potential possesses two nonequivalent minima with different values of vacuum energy density – see Figure 8.1. The minimum with larger vacuum energy density is known as a 'false vacuum' and the other is the 'true vacuum' Both of these are classically stable, but quantum fluctuations are able to destroy false vacuum state, as we know from quantum mechanics. The latter teaches us that probability of the false vacuum decay could be exponentially small provided a width and a height of a barrier between two vacua are large enough. This is true also in the case of field theory, but the picture of the transition is much more complex and nontrivial in this case.

The transition from false vacuum to true consists of decay of a metastable phase by nucleation of bubbles of a new phase [298]. The most probable fluctuation is a spherical bubble nucleated at rest with a certain critical size determined by microphysical processes [298]. The bubbles with true vacuum within are nucleated in different space points and at different instant. Just after their nucleation due to quantum fluctuations, they start their classical motion. The latter looks like expanding of true vacuum regions by quick growth of bubble radii. The false vacuum energy is converted into a kinetic energy of spherical walls that separate both vacua. The picture looks like that on Figure 8.2.

Coleman [298] calculated the bubble nucleation rate in flat space and at zero temperature using the euclidean path-integral formulation of a scalar field

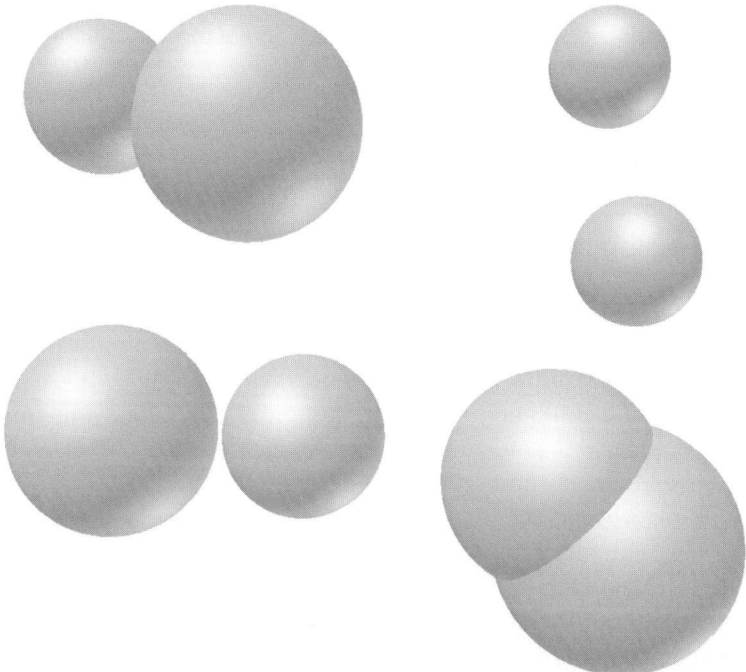

Figure 8.2. Typical picture of false vacuum decay. True vacuum inside spheres with false vacuum around them. The transition from false vacuum state to true is performed by growing of the bubble radii. The elder bubbles are larger. Some of them collide as is shown in the figure.

theory. A nucleated bubble is a true vacuum fluctuation large enough to evolve classically. The nucleation rate in this case is proportional to e^{-S_E}, where S_E is the euclidean action and the solution to the euclidean equation of motion for minimal action is the $O(4)$ symmetric "bounce" solution.

In the very early Universe phase transitions can occur at a finite temperature leading to temperature-dependent form of a scalar field potential, when quantum corrections are taken into account [299]. Generalization of Coleman results to the case of nonzero temperature is based on the remarkable fact that quantum statistics at nonzero temperature is formally equivalent to quantum field theory in the euclidean space, which is periodic in time coordinate with the period T^{-1}. As a result, most probable fluctuations appear to be not $O(4)$ symmetric spherical bubbles but $O(3)$ symmetric (with respect to spatial coordinates) cylindric configurations with certain critical size slightly different from $O(4)$ symmetric case [300, 299].

For a bubble created with a size smaller than the critical one, it could seem that the gain in volume energy cannot compensate for the loss in surface energy and as such the bubbles would have to quickly shrink and disappear. However, detailed analysis discovered that even in this case effects of nonlinearity lead to nontrivial dynamics. The evolution of subcritical bubbles – unstable spherically symmetric solutions of nonlinear Klein–Gordon equation – was first studied numerically by Bogolubsky and Makhankov [301, 302]. Using a quasiplanar initial configuration for the bubbles, they found that for a certain range of initial radii, the bubble, after radiating most of its initial energy, settled into long-lived (as compared with characteristic time-scale) stage and only then disappeared by quickly radiating its remaining energy. Those configurations, called "pulsons", were later rediscovered and revised by Gleiser, who found that their most characteristic feature is not a pulsating mechanism for radiating the initial energy, but the rapid oscillations of the amplitude of a scalar field during long-lived pseudo-stable regime, when almost no energy was radiated away and radial pulsations were rather small [303]. It was shown [303, 304] that those configurations called "oscillons" exist for symmetric and asymmetric double-well potentials, are stable against small radial perturbations, and have lifetimes "far exceeding naive expectations" [304].

Although it is well-known, that three-dimensional nontrivial configurations of a scalar field are unstable, they can be relevant for systems with short dynamical time scales. Detailed study of unstable but long-lived configurations can clarify dynamics of nonlinearities in field theories and their role in a wide class of phenomena ranging from nonlinear optics to phase transitions both in the Universe and in the laboratory [304].

For the bubbles formed with radii large enough (overcritical bubbles) it is classically energetically favorable to grow. The newly formed bubble of true vacuum is separated from the surrounding false vacuum region by the wall at rest. Immediately after nucleation, the wall starts to accelerate outwards absorbing energy stored in a false vacuum region and converting difference of false and true vacuum energy density into kinetic energy of the wall. That way a bubble spreads off converting false vacuum into the true one. This process continues up to the collision with a spherical wall of another bubble. In the first-order phase transitions at the end of inflation the collision of bubbles is considered as the leading mechanism of reheating by converting the wall energy into radiation.

However, the situation with two bubbles appears much more complicated [305], [306]. Even nucleation of two bubbles is not yet studied in the literature in general [307]. Only in the case when bubbles are widely separated at the time of nucleation and thus can be treated as noninteracting (at the stage of nucleation) the generalization of a single bubble solution is straightforward. Two bubble collisions were studied in detail by Hawking, Moss, and Stewart [308] and then by Watkins and Widrow [307], in an elegant approach using symmetry of the problem in a zero temperature case. For zero temperature bubbles produced by quantum tunneling, initial state is $O(4)$ symmetric, as well as euclidean equation of motion, in natural assumption that a scalar field ϕ is invariant under 4-dimensional Euclidean rotations. In analytical continuation to Minkowski space this becomes $O(3,1)$ symmetry. For two bubbles, the line joining their centers is the preferred axis and solution to the euclidean equation of motion is found in the class $O(3)$ ($O(2,1)$ as continued to the Minkowski space) solutions, and field configuration arising in collision belongs to the class of $O(3)$ symmetric solutions.

In this chapter following [305], [306] we investigate two-bubble collision in the case of finite temperature. We are interested not in reheating by two bubble collisions [309] but in evolution of two-bubble configuration during and after collision. As was noted by Hawking, Moss and Stewart [308] and confirmed by Watkins and Widrow [307], collision of two domain walls does not lead to immediate conversion of the wall energy into a burst of radiation. Two walls reflect off one another and move apart creating a new region of false vacuum between them [307]. Our aim is to investigate an evolution of this new false vacuum region to see if it can form a separated object. We connect with such a possibility the hope of formation of metastable relics of the first-order phase transitions such as primordial black holes or self-gravitating particle-like structures with de Sitter-like cores [310, 311]. It appears that a false vacuum configuration evolved into a compact clot filled with an oscillating scalar field.

The fundamental difference of this object from an oscillon is that it arises dynamically as the result of bubble collisions (which increases probability of its production) and that it is made up from an oscillating scalar field at the background of true vacuum. We call it, following [306], a clot (of energy). In numerical simulations [306] non-singular configurations of self-interacting scalar field were observed with asymmetric potential, perfectly localized, but we cannot say that they are non-dissipative, although they are rather long-lived as compared with the characteristic scale for the first-order phase transitions.

1. Temperature transitions vs. quantum transitions

To study mechanism of formation and evolution of false vacuum regions, we shall consider [305], [306] the most favorable regime for their appearance which corresponds to high nucleation rate $\Gamma H^4 \gg 1$, where Γ is the nucleation rate per unit 4-volume and H is the Hubble parameter [308]. We also neglect gravity effects on the process of bubble formation and growth which means that we consider bubbles with the initial size much less than the cosmological horizon, $R(0)H \ll 1$ [298, 307].

Lagrangian (8.1) with potential (8.2) may be considered as effective Lagrangian for a large number of more complex models of the Universe involving the first-order phase transitions (see [299] for more details). In the "thin-wall" approximation, $\epsilon/\lambda \ll 1$, some analytical results are known [298], and we will work in the frame of this approximation. At $T = 0$ the parameters λ, ϕ_0 and ϵ are specified by the particle model. At nonzero temperature they are influenced by temperature corrections. In the case of high nucleation rate, a first-order phase transition is a quick process and we can consider parameters as $\lambda \simeq \lambda(T_c), \phi_0 \simeq \phi_0(T_c), \epsilon \simeq \epsilon(T_c)$, i.e. being constant during the phase transition at the temperature $T = T_c$.

The potential (8.2) has two minima at different values of field ϕ . False vacuum (metastable) state is characterized by the field $\phi = \phi_0(1 - \epsilon/\lambda - 3/2(\epsilon/\lambda)^2)$, whereas the global minimum of the potential $V(\phi)$ represents the true vacuum state $\phi = -\phi_0(1 - \epsilon/\lambda + 3/2(\epsilon/\lambda)^2)$. In our analysis we assume that both mechanisms of the false vacuum decay could take place – tunneling, that is creation of O(4) symmetrical bubbles, and formation of O(3) symmetrical bubbles due to temperature fluctuation. Evidently, if the temperature is low enough, the tunneling mechanism of the false vacuum decay dominates. On the contrary, at high temperatures the decay is realized by the nucleation and growth of the O(3) symmetrical bubbles.

Consider conditions of dominance of the false vacuum decay due to temperature effects. The temperature decay probability was found in [300], [299]:

$$P_{Temp} \propto e^{-S_3/T}, \tag{8.3}$$

where T is the temperature of a phase transition and S_3 is three-dimensional action for O(3) symmetrical bubble. The probability of the vacuum decay due to tunneling, is given by

$$P_{tun} \propto e^{-S_4}, \tag{8.4}$$

where S_4 is the action for O(4) symmetrical bubble. The temperature decay dominates, if $S_3/T < S_4$. The straightforward calculations of the actions S_3 and S_4 give for our potential the condition for the dominance of the temperature decay (the term proportional to $(\epsilon/\lambda)^2$ was omitted):

$$T > \frac{32}{27\pi} \frac{\epsilon}{\lambda} \tag{8.5}$$

in the units $m_\varphi = \hbar = c = 1$ that are used throughout this chapter. Equation of motion of the scalar field in spherical coordinates has the form

$$\frac{\partial^2 \phi}{\partial t^2} - \frac{\partial^2 \phi}{\partial r^2} - \frac{2}{r} \frac{\partial \phi}{\partial r} = -V'(\phi). \tag{8.6}$$

Neglecting terms of order of $O((\epsilon/\lambda)^2$), we obtain the well-known one-dimensional equation

$$\frac{d^2 \phi}{dt^2} - \frac{d^2 \phi}{dr^2} = -V'(\phi) \mid_{\epsilon=0} . \tag{8.7}$$

The properties of this equation have been extensively discussed in the literature since 1975 [312]. The fundamental time-independent solution is defined by

$$r = \int_0^\phi \frac{d\phi}{\sqrt{2V(\phi)}}.$$

It can be easily checked by straight substitution, that for the theory defined by potential (8.2), an approximate solution is represented in the form

$$\phi = \phi_0 \{ th[\frac{\gamma m}{2}(r - R(t))] - \epsilon/\lambda \}, R(t) = vt + R_0, \qquad (8.8)$$

where $\gamma = 1/\sqrt{1 - v^2}$, $v < 1$, $m_\varphi = \sqrt{\lambda}\phi_0$ and $R_0 = 2\lambda/(3\epsilon m_\varphi)$ is critical radius of the nucleated bubble. This approximation is valid, if $R(t) >> 1/m$, which is equivalent to the thin-wall approximation. The initial field configuration can be defined at the moment of the bubble formation $t = 0$ with velocity $v = 0$. But it takes too much computer time and memory to study the development of collision from this initial moment, because the kinetic energy of the walls of the colliding bubbles should be large enough to produce a false vacuum bag (clot of energy (CE)) and hence the initial distance between the centers of colliding bubbles should be large compared with critical radius R_0 also. So, we have to use the initial configuration with already moving walls.

The one-bubble solution (8.8) is the approximate solution to exact Eq. (8.6). It also satisfies the correct boundary conditions at infinity up to the terms of order $(\epsilon/\lambda)^2$ and hence can be chosen as the new initial condition at specific moment t or at definite radius of the expanding bubble $R = R(t)$. The only thing that remains to be done is to connect the radius R and the velocity v. To find the velocity v in the one-bubble solution (8.8) at an arbitrary moment t or at definite bubble radius $R(t)$ we note that the energy

$$E = \int \left(\frac{1}{2}(\frac{d\phi}{dt})^2 + \frac{1}{2}(\nabla\phi)^2 + V(\phi) \right) d^3x$$

is conserved if the field ϕ is governed by Eq. (8.6). The substitution of the field ϕ in form (8.8) leads after simple calculations to the expression

$$E \simeq \frac{8\pi}{3}\frac{1}{\lambda}R(t)^2[\gamma - R(t)\epsilon/\lambda] = Const. \qquad (8.9)$$

The *Const* can be determined at $t = 0$, because we know the values of the parameters at this moment: $\gamma(t = 0) = 1$ and $R(t = 0) = 2\lambda/3\epsilon$ [299]. Substituting it into expression (8.9), we find the connection between the bubble radius $R = R(t)$ and γ-factor (or, equivalently, the velocity v):

$$\gamma = R\frac{\epsilon}{\lambda} + \frac{4}{27}\frac{\lambda^2}{\epsilon^2 R^2}. \qquad (8.10)$$

Thus, the initial conditions for one bubble of radius R is represented by formula (8.8) with the γ-factor (8.10).

2. Wall motion through thermal background

Two factors effect on the kinetic energy of the colliding walls: the distance between the bubble centers and the interaction of the walls with thermal plasma. As was mentioned above, the horizon heats the Universe by the evaporation of all possible sorts of particles. It is the interaction with these particles that leads to a friction which slows down the wall motion. For example the walls move with the velocity ~ 0.5 in the case of late electroweak phase transitions [262].

In Chapter 4 we have investigated friction of a wall which was born after the end of inflation. It was shown for specific form of potential (4.44) that the friction is negligible in a wide range of parameters. As is shown below, this conclusion takes place also during inflation with different sorts of interaction. The only restriction is that particles must not change their mass when crossing the wall.

Let us investigate separately the scattering of the inflaton particles and other 'light' particles with the masses $m << T$ by the wall at rest. It is well-known that there are two regimes in dependence on whether a mean free path of the particles is much more or much less, than the wall width L [262]. The wall width can be estimated easily: $L \approx 1/m_\varphi = 1$ in the chosen units. The main uncertainty is contained in the mean free path that depends on cross-sections. To estimate them suppose the interaction in the form:

$$L_{int} = g_1 \varphi \psi^+ \psi + g_2 \varphi^2 \phi^2 + g_3 \phi \psi^+ \psi, \qquad (8.11)$$

where ψ is an operator of spinor particles, ϕ is an operator for scalar particles, other than that is responsible for the inflation. The interaction of the inflaton (scalar) particles can be obtained from Eq. (8.1) and equals $\lambda \varphi^4$. No symmetries are assumed to be broken yet and all the constants are supposed of the same order $g_1 \sim g_2 \sim g_3 \sim \lambda << 1$. The temperature $T \sim \epsilon/\lambda$ (see (4.79)) is small compared to the mass of inflaton quanta and large compared to the masses of other particles. The interaction of light particles leads to the largest cross-section that can be estimated as $\sigma \sim g_3^4/T^2$. Here the expression $E_p \approx 3T$ for the energy of relativistic particles at the temperature T is presumed.

The mean free path equals to $l = 1/\sigma n$, where

$$n = \kappa \sigma_s/3 \cdot T^3 \qquad (8.12)$$

is the number density of the light particles at the temperature T [313]. The constant is $\sigma_s = \pi^2/15$ in our units and a total number of sorts of the light particles is $\kappa \sim 50$. Thus one can easily check that the mean free path $l \sim 1/(10g_3^4 T)$ is much more than the wall width $L \approx 1$ (note that we are working in the limits $g_3, \epsilon/\lambda << 1$). Therefore we can consider the interaction of the wall with the particles supposing them to be free, on the contrary to the electroweak phase transition [262]. Another, and maybe more essential difference consists of the equality of the particle masses at both sides of the wall because the inflaton field is not responsible for the mass generation.

To estimate the strength of the interaction at the end of inflation we neglect further the asymmetry of the potential $V(\varphi)$, that is proportional to the small parameter ϵ/λ, and a curvature of the wall.

To describe the scattering of inflaton particles one can consider them as small fluctuations around classical solution $\varphi = \varphi_{cl} + \varphi'$, where

$$\varphi_{cl} = \varphi_0 th(z/2)$$

is the 'flat wall' solution to the classical equation of motion

$$\Box \varphi + \frac{\partial V}{\partial \varphi} = 0. \tag{8.13}$$

After a linearization Eq. (8.13) has the form [261]

$$\left(\frac{d^2}{dz^2} + 4\omega^2 - 4 + \frac{6}{ch^2 z} \right) \varphi' = 0. \tag{8.14}$$

The one-particle scattering by the wall with the particle energy ω and momentum $\mathbf{k}$ should satisfy the boundary conditions

$$\varphi'(t \to -\infty, z) = \exp(-i\omega t + ikz);$$
$$\varphi'(t \to \infty, z) = A \exp(-i\omega t + ikz) + B \exp(-i\omega t - ikz).$$

It can be shown that the solution to Eq. (8.14) is reflectionless [294], i.e.

$$Re A = 1, B = 0$$

and hence the momentum transfer equals zero. Scalar particles do not scatter by a flat wall made of the same scalar field and do not slow down the wall.

The only thing remaining is to estimate the pressure of the other particles:

$$p_p = qnW. \tag{8.15}$$

Here q is the momentum transfer, the particle number density is equal to $n \approx 10T^3$ according to (8.12) and W is the probability of the scattering of the incoming particle. The wall at rest can be considered as an external field and conservation of energy E_k and of parallel projection of the particle momentum $\mathbf{k}$ determines the momentum transfer: $\mathbf{q} = (0, 0, -2k_z)$.

The probability W can be expressed in the form

$$W = \left| M_{k,k+q} \right|^2 / (8k_z E_k)$$

in the case of a planar wall perpendicular to z axis. The matrix element is obtained using the first term of the expression (8.11), where only classical part φ_{cl} of the field φ is taken into account:

$$M_{k,k+q} = g_1\varphi_0 \lim_{\alpha \to 0} \int_{-\infty}^{\infty} dz e^{iqz - \alpha z^2} th(z/2). \tag{8.16}$$

The integral is easily estimated if one substitutes a linear function for $th(z/2)$ in the region $|z| < 2$, which gives the dominant contribution:

$$M_{k,k+q} = i2g_1\varphi_0 \frac{\cos 2q - \sin 2q/q}{q} \leq i2g_1\varphi_0.$$

The last inequality takes place due to the fact that the q-dependent ratio is less than approximately one. So expression (8.15) becomes

$$P_p \leq n\frac{g_1^2\varphi_0^2}{E_k}.$$

The energy of the incoming particle in the rest frame of the wall is $E_k \sim \gamma T$ and the upper limit of the pressure can be written as

$$P_p \leq 10\frac{g_1^2}{\gamma\lambda}\left(\frac{\epsilon}{\lambda}\right)^2. \tag{8.17}$$

The incoming particles cause the pressure p_p that decelerates the wall. On the other side the wall is accelerated by a pressure difference $p_\varphi = p_V = 2\epsilon/\lambda^2$ in the true and false vacuums that are separated by the wall. Its ratio is given by

$$\frac{p_p}{p_\varphi} \leq \frac{g_1^2}{\gamma}\frac{\epsilon}{\lambda}. \tag{8.18}$$

This ratio is always much less then unity because both g_1 and ϵ/λ are supposed to be small values. Note that expression (8.18) is the upper limit for the pressure. As it follows from the form of the matrix element $M_{k,k+q}$, the pressure tends to zero not only at small momentum transfer but also at large. The same estimation of upper limit of the pressure could be done for the scattering of light scalar particles.

The conclusion on the absence of the friction is in agreement with that for electroweak transition [262] in the limit of the particle mass equality on both sides of the wall. In our case the masses are equal because the absolute value of the fields in the two minimums are approximately equal. If the Lagrangian has the local minimum at $\varphi = 0$ friction would appear. The similar situation was considered in [314].

Thus the friction of the wall that moves through the heated medium at the end of inflation is small enough and the bubble walls collide having large kinetic energy. It could be in its turn the reason for the large density fluctuation of the inflaton field. Now we have all the ingredients to perform numerical calculations, but let us start with qualitative analysis.

3. Bubble collisions – Qualitative analysis

Let us introduce the dimensionless variables $\psi = \phi/\phi_0$, $\lambda^{1/2}\phi_0 t \to t$ and $\lambda^{1/2}\phi_0\mathbf{r} \to \mathbf{r}$. The classical equation of motion for the scalar field of Lagrangian (8.1) has the form

$$\partial_t^2 \psi - \nabla^2 \psi = -\frac{1}{2}\psi(\psi^2 - 1) - \epsilon/\lambda \qquad (8.19)$$

The suitable initial two-bubble configuration has in our dimensionless variables the form

$$\psi = \psi_0\{th[\frac{\gamma}{2}(r_+ - R)] - \epsilon/\lambda\}, z < 0,$$

$$\psi = \psi_0\{th[\frac{\gamma}{2}(r_- - R)] - \epsilon/\lambda\}, z > 0, \qquad (8.20)$$

$$r_\pm = \sqrt{x^2 + y^2 + (z \pm b)^2}, b > R.$$

To start numerical simulations of bubble collisions, we need a set of simple criteria indicating a proper range of parameters favorable to formation of separated false vacuum regions.

Let us first find the condition at which the region of a false vacuum can be formed as a result of a collision of two relativistic bubbles. A field configuration in a bubble wall is just the transition from the true vacuum inside the bubble to the false vacuum outside it. While propagating through the false vacuum before collision, the bubble absorbs the energy of a surrounding false vacuum and transforms it into the kinetic energy of the wall. The kinetic energy is characterized by the Lorentz factor $\gamma = 1/\sqrt{1 - v^2}$. To get a region of a false vacuum between bubbles as a result of a collision, energy absorbed by walls from a false vacuum to the moment of a collision must be sufficient to form a false vacuum state at least at the scale of the wall width. Let us estimate the lower limit for γ at which such minimal region can be formed.

Consider collision of two spherical $O(3)$ bubble walls described by the solution (8.8) with the parameters $R = b$ and γ_{in} to the moment of collision. The leading term in the energy density of a wall, as calculated for the quasiplanar solution (8.8), is

$$\rho_w \simeq \gamma^2/4 \cosh^4(\gamma(R - vt)/2).$$

Before the collision, in the solid angle

$$\Delta\Omega = \frac{\pi r^2}{R^2} \ll 1,$$

each wall has the energy

$$E_{in} = \frac{2}{3}\Delta\Omega R^2 \gamma_{in}.$$

After the collision, the walls reflect with a final kinetic energy E_{fin}. If a false vacuum region of a radius r and width h within the solid angle $\Delta\Omega$ is formed between reflecting walls, we must have

$$E_{in} = E_{fin} = \frac{2}{3}\Delta\Omega R^2 \gamma_{fin} + 2\frac{\epsilon}{\lambda}V_{fvr},$$

where $\rho_{vac} = 2\epsilon/\lambda$ is the false vacuum energy density for the case of the potential (8.2), and V_{fvr} is the volume of a false vacuum region within a cone with a solid angle $\Delta\Omega$, which is given by

$$V_{fvr} = \frac{\pi}{6}h(3r^2 + h^2) \approx \frac{1}{2}\Delta\Omega R^2 h$$

The width of the false vacuum region is of order of the width of the wall, which is equal to $2/\gamma_{fin}$, to the moment of reflection. For $\gamma_{fin} = 1$, the width is $h = 2$. It gives us the constraint for γ_{in}, with which the wall comes to the first collision, in the form

$$\gamma_{in} \geq 1 + \frac{3}{2}\frac{\epsilon}{\lambda}h \geq 1 + 3\frac{\epsilon}{\lambda}. \tag{8.21}$$

Now let us specify the line joining centers of bubbles as z axis. Let us show that the energy conservation puts constraint on the propagation of a false vacuum region in z direction. Consider a slice of a false vacuum region originated from the collision in the element $\Delta\Omega$ of spherical bubble walls, which have radius R in the moment of collision. The acceleration of the considered element of the wall comes from the transformation of the energy of surrounding false vacuum into the kinetic energy of the wall on the way to its first collision, when the true vacuum bubbles grow from the initial radius $R(0)$ to the radius $R \gg R(0)$ in the moment of collision. So, the kinetic energy absorbed by the wall from a false vacuum to the moment of collision, is

$$E_{kin} = \frac{2\epsilon}{\lambda}\Delta\Omega\frac{1}{3}(R^3 - R(0)^3) \simeq \frac{2\epsilon}{\lambda}\Delta\Omega\frac{1}{3}R^3.$$

The walls reflect off each other in the moment of the first collision and move outwards, creating a false vacuum region between them. Each wall stops when all its kinetic energy has been transformed into the energy of a false vacuum region formed between the walls. In this moment the walls' radius is R_{max} and the false vacuum, created by each wall, fills a region between the spherical shells R_{max} and R. The energy balance gives

$$\frac{2\epsilon}{\lambda}\Delta\Omega\frac{1}{3}R^3 \simeq \frac{2\epsilon}{\lambda}\Delta\Omega\frac{1}{3}(R_{max}^3 - R^3),$$

so that

$$R_{max} \simeq 2^{1/3}R. \tag{8.22}$$

Since we are considering overcritical bubbles, the wall surface energy is neglected in this treatment, provided that $\epsilon R_{max}/\lambda \gg 1$. The same result has been obtained for the case of $O(4)$ symmetric bubbles in [308]. One finds from the equation (8.22) that after the collision a false vacuum is formed and occupies a region between the outgoing walls, with a maximal size given by a distance between the planes $z = \pm(2^{1/3} - 1)b$, where $2b$ is the initial separation of the centers of true vacuum bubbles. After the walls stop their outward movement in the region of the walls' intersection at $R_{max} = \pm(2^{1/3} - 1)R$, the parts of walls in this region reflect off one another and next time they collide at $\Delta t \sim 2(2^{1/3} - 1)R$ after the first collision. The shortest interval between the two subsequent collisions is at $r = 0$, when $R = b$. Using the condition (8.21), we find from the Eq. (8.10) the minimal γ, at which the false vacuum region is formed between the walls after the second collision. Before the second collision at $r = 0$, the value of γ for the walls in the second collision is given by

$$\gamma_2 = R\frac{\epsilon}{\lambda} \approx (2^{1/3} - 1)b\frac{\epsilon}{\lambda}. \tag{8.23}$$

Note that before the first collision this factor for the walls at $r = 0$ is given by the Eq. (8.10) as

$$\gamma_1 = b\frac{\epsilon}{\lambda}. \tag{8.24}$$

If we want the false vacuum region to be maintained after the second collision of the reflected parts of the walls, we must satisfy $\gamma_2 > \gamma_{in}$. Then it follows from the Eqs. (8.21), (8.23), (8.24), that γ before the first collision must be

$$\gamma_1 > \gamma_{min} = \frac{1 + 3\epsilon/\lambda}{(2^{1/3} - 1)}. \tag{8.25}$$

It indicates the favorable range of the γ parameter before the first collision needed for numerical simulation, and also, with the use of (8.10) and (8.24), the favorable range for the parameters R and b. In the case $\gamma_1 \gg \gamma_{min}$, a false vacuum region undergoes a succession of oscillations – expansions and contractions – along the z axis in the region confined by

$$-b(2^{1/3} + 1) < z < b(2^{1/3} - 1). \tag{8.26}$$

Repeating the above reasoning for the subsequent collisions we find easily that in the limit of large γ the period of the n-th oscillation decreases as $(\sqrt{2n})^{-1}$. This agrees with the result [308] for the $O(4)$ symmetry case. The reason for such a coincidence can be easily understood.

The main difference between the $O(3)$ and $O(4)$ cases is in the form of the initial wall configurations, taken in our case as a quasiplanar $O(3)$ solution. However, in all the above reasoning the internal structure of the walls was not involved, which just resulted in the similar estimation for the decreasing of the period of oscillations. For large γ we can treat the oscillations of a false vacuum region along the axis z as the continuous propagation of a spherical wave moving with speed of light (in our units $c = 1$). In the frame with the origin in, say, $z = -b$, the element $\Delta\Omega$ of the wall with the angle α with respect to the z axis, follows the trajectory $r = z \tan \alpha$. Assume that to the moment of reflection the considered element of the wall has the coordinate $z = ct$. At the same moment its radial coordinate is $r = z \tan \alpha$. The region of causal contact along the axis r satisfies the condition $r \leq ct$. It follows then that only for the angles $\alpha < \pi/4$, the region of intersection of walls is in the causal contact. Therefore the boundary of the region of causal contact within a false vacuum region is the cone

$$\alpha = \pi/4. \tag{8.27}$$

It means that further evolution of the false vacuum region confined within the boundary (8.27) does occur independently on dynamics of field outside this boundary. So, the considered region of the false vacuum is separated from the bulk false vacuum space in its further causal and hence dynamical evolution. Now we can easily estimate the total energy of the separated false vacuum region. The energy density of a false vacuum is given by $\rho_{vac} = 2\epsilon/\lambda$. The volume of a CE is the volume of two cones whose height is equal b and base area πb^2. So, the mass, confined within this region, is equal to

$$M = \frac{4\epsilon}{\lambda} \frac{\pi b^3}{3}. \tag{8.28}$$

It is evident that separation occurs at the time of order of $t_{sep} = (\sqrt{2} - 1)b$.

4.   Bubble collisions − Numerical results

The qualitative analysis given above, has revealed the possibility of formation of high energy density regions during first-order phase transitions. Numerical calculations represented below confirm this guess, reveal new features and give the range of parameters for which this process could take place.

In the cylindric coordinates the equation of motion for the scalar field (8.19) equation takes the form

$$\partial_t^2 \psi - \partial_r^2 \psi - \partial_z^2 \psi = -\frac{1}{2}\psi(\psi^2 - 1) - \epsilon/\lambda$$

The solution to this equation has the axial symmetry and reflection symmetry with respect to $z = 0$ plane. The initial configuration described by the solution is chosen in the form suitable for numerical calculation (compare with (8.20))

$$\psi = th[\gamma/2(r_+ - R - vt)] + th[\gamma/2(r_- - R - vt)] - 1 - \epsilon/\lambda. \tag{8.29}$$

The profile is shown in Figure 8.3. The walls already have kinetic energy which is indicated by the γ factor. Time evolution of the scalar field in the center of the region of collision $\psi(t, r = z = 0)$ shown in Figure 8.4, was calculated for the parameters $\gamma = 5; b = 52; R = 50$. The qualitative behavior of the field with time has been discussed in the previous section. From the

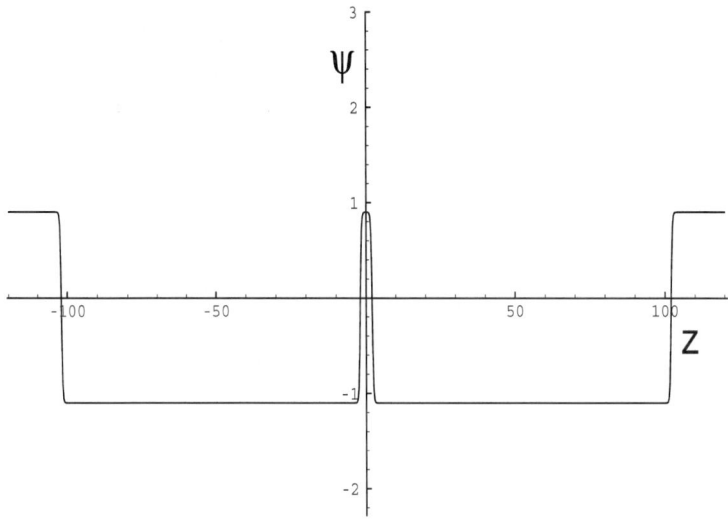

Figure 8.3. The initial two-bubble configuration just before collision.

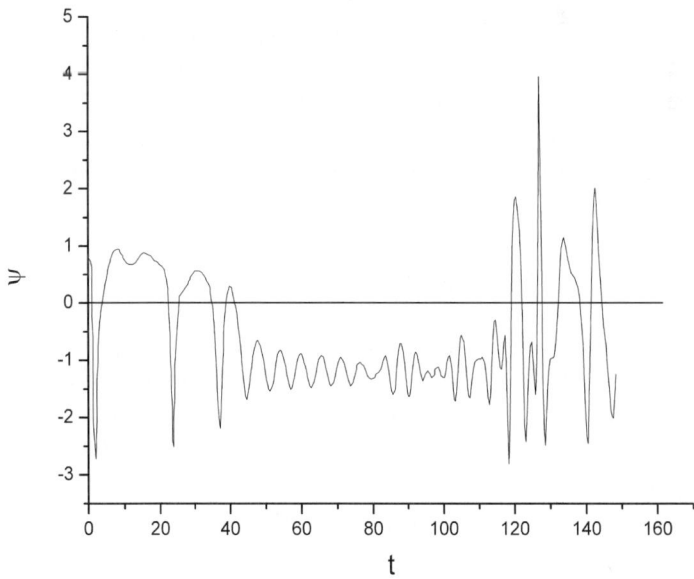

Figure 8.4. Time dependence of the field at the center of collision.

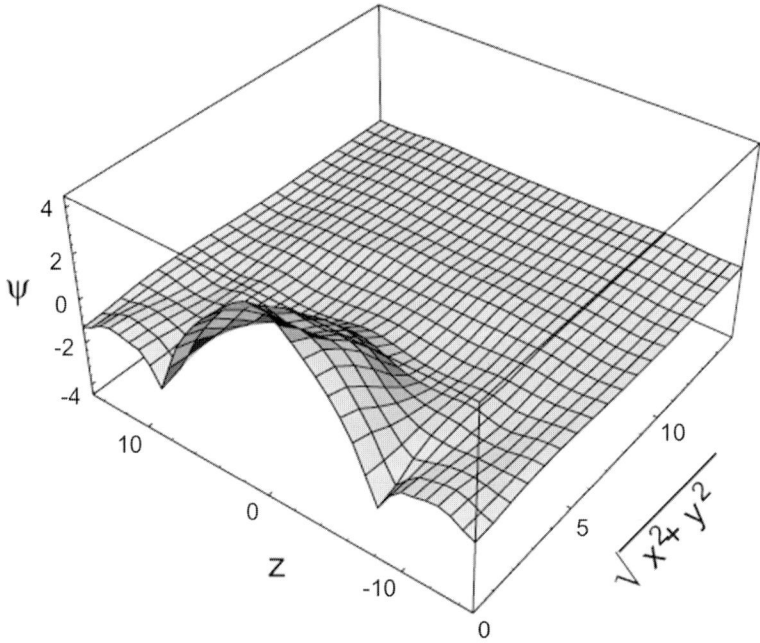

Figure 8.5. The field configuration at large time. It is seen that energy of this configuration is concentrated in the center of the collision with radius ∼ 5.

beginning the field changes in the manner discussed in [307], [308], then, as it is clear from Figure 8.4, it oscillates around the true vacuum for a long time and finally, large secondary fluctuations appear again.

As we shall see below, the energy of oscillations is perfectly localized. This behavior does not change with changing the step in numerical calculations. Figures 8.5 and 8.6 display field configurations at different moments of time.

The energy density profile calculated from scalar field potential is shown in the next series of figures. They demonstrate concentration of field energy in the center of the region of collision. In Figure 8.7 one can see time dependence of energy density in the center of the region of collision. A large secondary peak is created due to the coherent field oscillations which are coming from outside. The density profile at this time is represented in Figure 8.8.

The localized configuration described above oscillates for some time and finally is converted into outgoing radiation. Only gravity could prevent this process.

Until now we did not consider gravitational effects, but they will be estimated below. Consider the evolution of energy contained in the sphere of certain radius as shown in Figures 8.9, 8.10

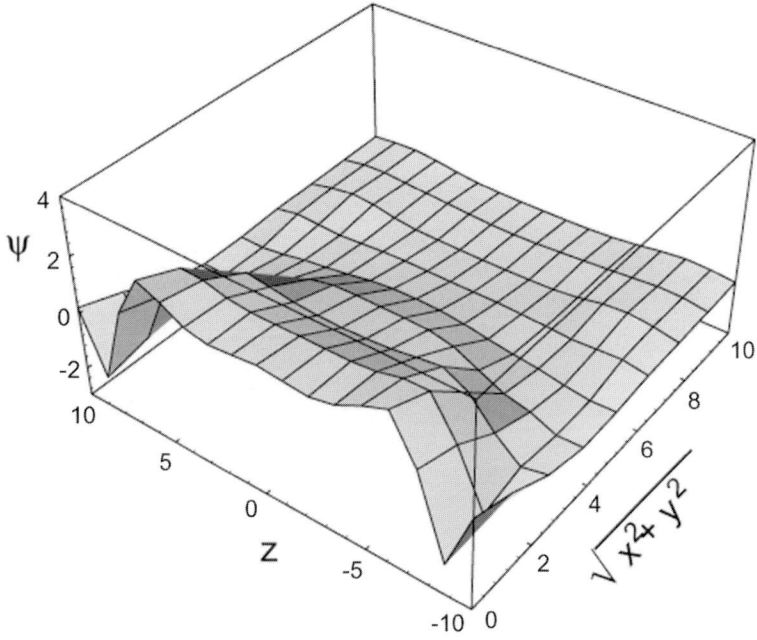

Figure 8.6. The same as in Figure 8.5 at a slightly different moment.

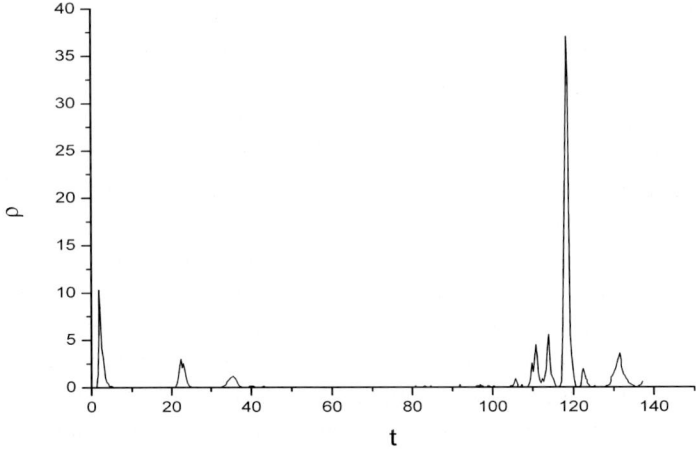

Figure 8.7. Time evolution of density energy in the center of a two-bubble collision, $z = r = 0$. The secondary peak is formed at $t \approx 129$

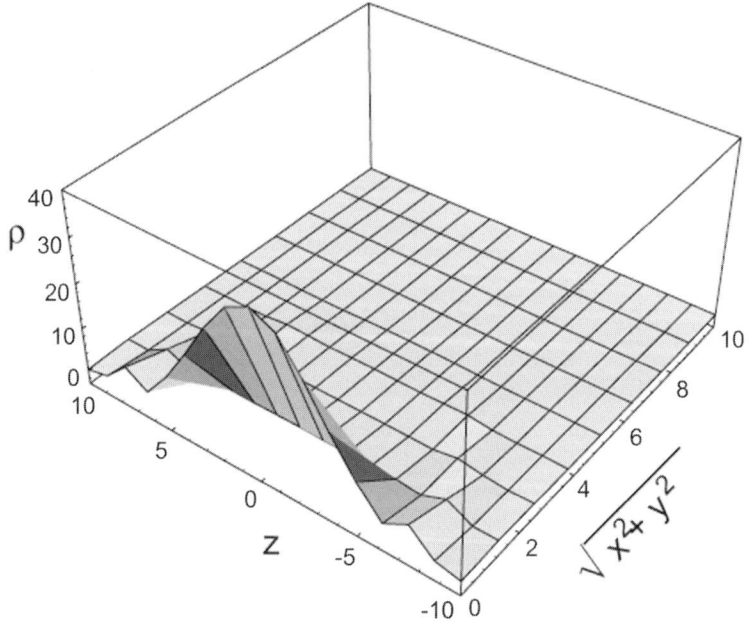

Figure 8.8.   Energy density at the time of secondary peak formation.

These pictures indicate two peaks of energy – the first is due to the energy in the moment of collision, the second is the energy of the clot which is formed as a result of the collision. It becomes evident by comparing these figures that the energy is strongly localized for the second time. Indeed, the energy contained in the internal sphere of radius $r_0 = 5$ of the second peak is only 1.5 times smaller than that contained in the external sphere of radius equal to $r_0 = 10$, while the ratio of their volumes is 8. One could conclude that a substantial part of the energy is contained within the sphere with radius $r_0 = 5$. It has to be compared with the gravitational radius of the clot. The latter is equal to

$$r_g = \left(\frac{m}{M_P} \right)^2 E/\lambda \qquad (8.30)$$

in our units. If the first-order phase transition happens at the end of inflation, the mass m of the inflaton field is rather large and $m_\varphi/m_{pl} \sim 10^{-5}$. Substituting this value into (8.30) and the value of energy $E \approx 1000$ obtained from Figure 8.9, one can easily find the condition when gravitational radius is comparable with the size of the clot, which can be taken as $r_0 \approx 5$ in our dimensionless units. Evidently, this condition is satisfied if coupling constant $\lambda \sim 10^{-8}$. For $\lambda < 10^{-8}$ gravitational forces become essential and the probability of black hole formation grows up to unity, when λ tends to 10^{-8}. If the bubble collision takes place at GUT energies and we deal with a scalar (not

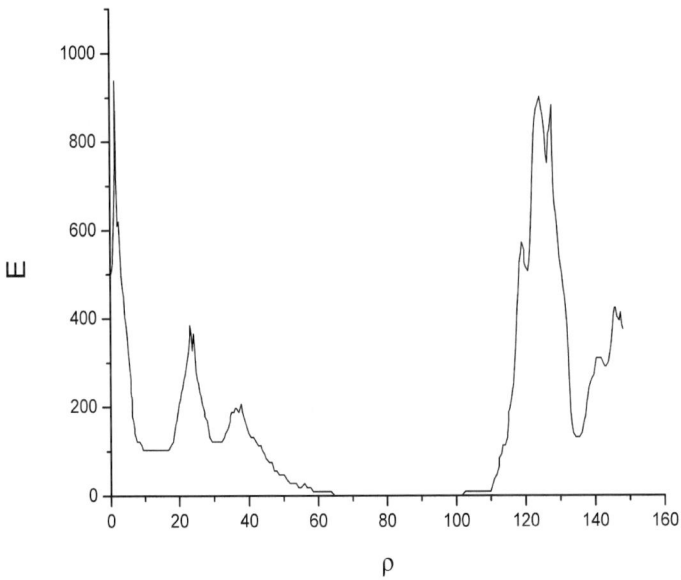

Figure 8.9. Field energy inside a sphere of radius $r_0 = 5$.

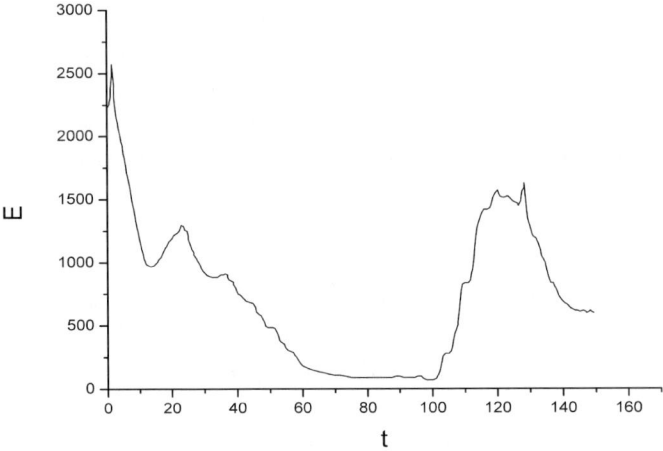

Figure 8.10. Field energy inside a sphere of radius $r_0 = 10$.

inflaton) field with a mass of its quanta of the order of 10^{16} GeV, black holes could be formed at $\lambda < 10^{-4}$.

Thus, in bubble collisions the interaction of bubble walls leads to formation of a nontrivial vacuum configuration. The subsequent collapse of this vacuum configuration induces black hole formation with high probability.

The primordial black holes that have been created in this way in the first-order phase transitions at the end of inflation could give an essential contribution into the total density of the early Universe. The possibilities of establishing some nontrivial restrictions on the inflation models with first-order phase transition are discussed below (Section 7).

5.   Mass distribution of BHs in the early Universe

Previous sections were devoted to capability of black holes formation at bubble collisions. Below we investigate a mass spectrum of such black holes. Consider a theory predicting the probability of false vacuum decay to be equal to Γ and the difference of energy densities between the false and true vacua equal to ρ_V. Initially bubbles are produced at rest, however, the bubble walls quickly increase their velocity up to the speed of light $v = c = 1$ because the conversion of the false vacuum energy of the bubble into the kinetic one is energetically favorable.

Let us discuss following [305] the dynamics of a collision of two true vacuum bubbles which have been nucleated at the points $(\mathbf{r}_1, t_1)$ and $(\mathbf{r}_2, t_2)$ and which are expanding into the false vacuum. Following the papers [308, 307], one could assume for simplicity that the horizon size is much greater than the distance between the bubbles. Just after the collision, mutual penetration of the walls up to distances comparable with their widths is accompanied by a significant potential energy increase [315]. Then the walls are reflected and accelerate backwards. The space between them is filled with the field in the false vacuum state converting the kinetic energy of the wall back to the energy of the false vacuum state and slowing down the velocity of the walls.

Meanwhile, the outer area of the false vacuum is absorbed by the outer wall, which expands and accelerates outwards. Evidently, there is an instant when the central region of the false vacuum is separated. One can note that this CE does not possess spherical symmetry. But, as it was shown above, gravitational forces are very strong when the clot is forming and are able to convert it into a black hole and further evolution of the CE consists of several stages:

1) The CE grows up to a certain size D_M with its energy stored both in kinetic and potential parts;

2) Secondary oscillation of the CE occurs;

3) The waves caused by outer interacted walls are concentrated in the center of bubble collision supplying a new peak of energy.

The process of periodical expansions and contractions leads to CE energy losses in the form of scalar field quanta. It has been confirmed in [308, 307] that several oscillations take place. On the other hand, it is important to note that secondary oscillations might occur only if the minimal size of the CE is greater than its gravitational radius, $D^* > r_g$. The opposite case ($D^* < r_g$) leads to BH creation with a mass close to the mass of the CE. As we will show later, the probability of BH formation is almost unity in a wide range of parameters of theories with first-order phase transitions.

Consider in more detail the conditions of converting a CE into a BH. The mass M of the CE can be calculated in the framework of a specific theory and can be estimated in a coordinate frame K', where the colliding bubbles are nucleated simultaneously. The radius of each bubble b' in this frame equals to half of their initial coordinate distance at the first instant of collision. Apparently, the maximum size D_M of the CE is of the same order as the size of the bubble, since this is the only parameter of necessary dimension on such a scale: $D_M = 2b'C$. The value of the parameter $C \leq 1$ has to be obtained by numerical calculations in the framework of specific theory, but its exact numerical value does not affect the conclusions significantly. One can express the mass of CE that arises at the collision of two bubbles of radius b' in the form:

$$M = \frac{4\pi}{3} \left(Cb'\right)^3 \rho_V.$$
(8.31)

This mass is contained in the shrinking area of the false vacuum. Suppose for estimations that the minimal size of the CE is of the order of the wall width Δ. The BH is created if the CE minimal size is smaller than its gravitational radius. It means that at least under the condition

$$\Delta < r_g = 2GM$$
(8.32)

the CE can be converted into a BH (where G is the gravitational constant).

As an example consider a simple model with the Lagrangian (8.1)

$$L = \frac{1}{2} \left(\partial_\mu \Phi\right)^2 - \frac{\lambda}{8} \left(\Phi^2 - \Phi_0^2\right)^2 - \epsilon \Phi_0^3 \left(\Phi + \Phi_0\right).$$
(8.33)

In the thin-wall approximation the width of the bubble wall can be expressed as $\Delta = 2 \left(\sqrt{\lambda}\Phi_0\right)^{-1}$. Using (8.32), one can easily derive that at least a CE with the mass

$$M > \frac{1}{\sqrt{\lambda}\Phi_0 G}$$
(8.34)

should be converted into a BH of mass M. The last condition is valid only in case the CE is completely contained in the cosmological horizon, namely, $M_H > 1/\sqrt{\lambda}\Phi_0 G$, where the mass of the cosmological horizon at the instant of the phase transition is given by $M_H \cong M_P^3/\Phi_0^2$. Thus for the potential (8.33) under the condition $\lambda > (\Phi_0/M_P)^2$ the BH is formed. This condition is valid for any realistic set of parameters of the theory.

The bubbles do not nucleate simultaneously at the same instant of time. It leads to some mass distribution of CE and/or BH. Besides, they have different velocities because colliding bubbles were created at different instants and hence have different kinetic energy of their walls to the moment of collision. Our nearest aim is to find mass and velocity distributions of the BH which satisfy inequality (8.34). Apparently the mass and velocity of a specific BH depends on coordinates and instants of both bubbles whose collision leads to the BH formation: $M = M(|\mathbf{r}_2 - \mathbf{r}_1|, t_2 - t_1); v = v(|\mathbf{r}_2 - \mathbf{r}_1|, t_2 - t_1)$.

The probability dP of collision of two specific bubbles depends on coordinates $\mathbf{r}_2, \mathbf{r}_1$ and instants $t_1 t_2$ of their creation

$$dP = dP_1 \cdot dP_2 \cdot P_-,$$
$$dP_1 = \Gamma dt_1 d^3 r_1, \qquad\qquad (8.35)$$
$$dP_2 = \Gamma dt_2 4\pi |\mathbf{r}_2 - \mathbf{r}_1|^2 d|\mathbf{r}_2 - \mathbf{r}_1|,$$

where dP_1 is the probability of bubble formation with 4-coordinates $(\mathbf{r}_1, t_1)$, dP_2 – the probability of the second bubble formation at the distance $|\mathbf{r}_2 - \mathbf{r}_1| \equiv 2b$ from the first one (space isotropy was taken into account). Factor $P_- = e^{-\Gamma\Omega}$ is the probability of absence of another bubbles inside 4-volume Ω, which could prevent the collision of the two bubbles in question. Below we treat the probability density of a vacuum decay Γ as a free parameter. Integrating out the variable $\mathbf{r}_1$, we obtain

$$dP/V = 32\pi\Gamma^2 e^{-\Gamma\Omega} b^2 dt_1 dt_2 db. \qquad (8.36)$$

Here V is the volume within the cosmological horizon at the moment of the phase transition.

Let us substitute the variables t_1, t_2, b by more suitable ones M, v, t. Here M is the mass of CE (or BH) created due to bubble collision, v is its velocity and $t = b + (t_1 + t_2)/2$ is the instant of a first contact of the two bubbles. In the following it will be suitable to choose reference frame K' where the two bubbles are nucleated simultaneously. Its velocity is

$$v = (t_2 - t_1)/2b, \quad (c = 1).$$

Apparently it is also the velocity of CE or BH in the initial reference frame. In this frame the bubble radii are equal to each other and is described by the formula $b' = b/\gamma, \gamma = (1 - v^2)^{-1/2}$. Using expression (8.36) one can obtain mass and velocity distribution of the BH

$$dP/V dv dM = \frac{64\pi}{3}\Gamma^2 e^{-\Gamma\Omega}\gamma^4 \left(\frac{M}{C\rho_v}\right)^{1/3} \frac{1}{C\rho_v} dt. \qquad (8.37)$$

To estimate the 4-volume Ω, assume that any bubble whose wall has reached the sphere of radius b' with the center in the point O before instant t', prevents the formation of CE by the two bubbles in question. Then one can easily obtain

$$\Omega = \int_0^{t'} d\tau' d^3\mathbf{r}' \theta \left(r' + \tau' - b' - t'\right) = \frac{\pi}{3}\{(b' + t')^4 - b'^4\}$$

The parameter b' is expressed in terms of mass M according to (8.31), the instant t' of the first instant of the bubbles contact is γt in the reference frame K'. Integrating out the variable t one comes to the desired mass and velocity distribution of the CE

$$dP/VdvdM = \frac{64\pi}{3}\Gamma^2 \exp\{\frac{\pi}{3}\Gamma(\frac{M}{C\rho_v})^{4/3}\}\gamma^4 \left(\frac{M}{C\rho_v}\right)^{1/3} \frac{1}{C\rho_v} I,$$

$$I = \int_{t_-}^{\infty} d\tau \exp\{-\frac{\pi}{3}\Gamma \left[\left(\frac{M}{C\rho_v}\right)^{1/3} + \gamma\tau\right]^4\}, \tag{8.38}$$

$$t_- = (1+v)\,\gamma \left(\frac{M}{C\rho_v}\right)^{1/3}.$$

It is interesting to compare the volume V_{bag}, containing one CE inside it, with the volume V_{bubble} of the bubble at the end of the phase transition. After numerical integration of expression (8.38) one obtains

$$V_{bag} \cong 3.9\Gamma^{-3/4}. \tag{8.39}$$

On the other hand average volume of the bubble is

$$V_{bubble} = \frac{4}{3}\pi \left(\frac{3}{\pi}\right)^{3/4} \Gamma^{-3/4} \cong 4.0\Gamma^{-3/4}. \tag{8.40}$$

An approximate equality $V_{bag} \cong V_{bubble}$ points out that one bubble produces one CE, i.e. the probability of CE formation during bubble collision is close to one. Expression (8.38) can be represented in terms of dimensionless mass parameter

$$\mu \equiv \left(\frac{\pi}{3}\Gamma\right)^{1/4} \left(\frac{M}{C\rho_v}\right)^{1/3}$$

in the form

$$\frac{dP}{\Gamma^{-3/4}Vdvd\mu} = 64\pi \left(\frac{\pi}{3}\right)^{1/4} \mu^3 e^{\mu^4} \gamma^3 J(\mu, v), \tag{8.41}$$

$$J(\mu, v) = \int_{\tau_-}^{\infty} d\tau e^{-\tau^4}, \tau_- = \mu \left[1 + \gamma^2 \left(1 + v\right)\right].$$

Velocity distribution of BHs gives little information and this variable can be integrated out. The distribution in dimensionless mass is represented in Figure 8.11.

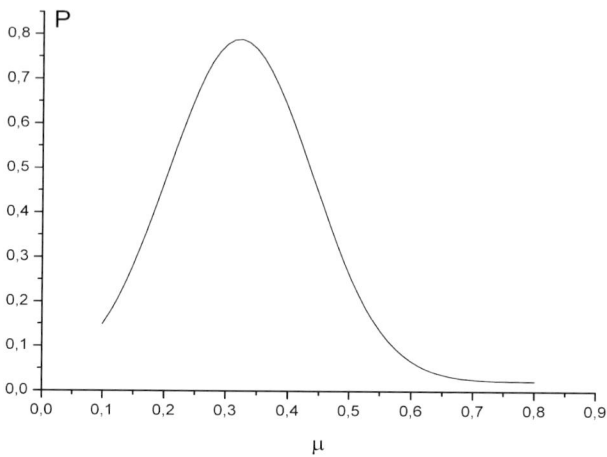

Figure 8.11.   The probability of BH nucleation in dependence on dimensionless mass μ.

Almost any theory incorporating a possibility of first-order phase transition contains at least two parameters. They are the difference of energy density of the two vacua ρ_v and the density probability of false vacuum decay in unit time Γ. The last can be connected with BH concentration at the moment of the phase transition

$$n_{BH} \simeq 0.25\Gamma^{3/4},$$

where expression (8.39) was taken into account. As can be seen from Figure 8.11, the average value of dimensionless mass parameter $\mu \simeq 0.32$, which helps to express the average mass of the BH in terms of the main parameters of the phase transition

$$\langle M_{BH} \rangle \simeq 0.03C\rho_v\Gamma^{3/4}. \tag{8.42}$$

Note that the constant C is a model-dependent parameter, being less or of the order of unity. Constraints on concentration and average mass of the BH lead to constraints on both parameters of a specific theory of a cosmological scenario. The mass and velocity distribution of CEs, supposing their masses are large enough to satisfy the inequality (2), has been found in [316], [305]. This distribution can be written in terms of the dimensionless mass parameter μ

$$\frac{dP}{\Gamma^{-3/4}V\,dv\,d\mu} = 64\pi \left(\frac{\pi}{3}\right)^{1/4} \mu^3 e^{\mu^4} \gamma^3 J(\mu, v), \tag{8.43}$$

$$J(\mu, v) = \int_\tau^\infty dt\,e^{-\tau^4}, \quad \tau_- = \mu\left[1 + \gamma^2\left(1 + v\right)\right].$$

Numerical integration of (8.43) revealed that the distribution is rather narrow. For example, the number of BHs with mass 30 times greater than the average is suppressed by a factor of 10^5. The average value of the dimensionless mass is equal to $\mu = 0.32$. It allows one to relate the average mass of BHs $\langle M_{BH} \rangle$ and the volume containing one BH $\langle V_{BH} \rangle$ at phase transition:

$$\langle M_{BH} \rangle = \frac{C}{4} \mu^3 \rho_v \langle V_{BH} \rangle \simeq 0.012 \rho_v \langle V_{BH} \rangle . \qquad (8.44)$$

6. First-order phase transitions in the early Universe

Inflationary models ending with a first-order phase transition, to which we further refer to as the first-order inflation models, occupy a significant place in modern cosmology of the early Universe (see, e.g., [317, 297, 305]). The interest in these models is due to the fact that such models are able to generate the observed large-scale voids as remnants of the primordial bubbles for which the characteristic wavelengths are several tens of Mpc. A detailed analysis of a first-order phase transition in the context of extended inflation can be found in [297]. Hereafter we will be interested only in the final stage of inflation, when the phase transition has been completed. Recall that a first-order phase transition is considered to be completed immediately after establishing the true vacuum percolation regime. Such a regime is established approximately when at least one bubble per unit Hubble volume has been nucleated. An accurate computation [297] shows that a first-order phase transition is successful if the following condition is valid:

$$Q \equiv \frac{4\pi}{9} \left(\frac{\Gamma}{H^4} \right)_{t_{\text{end}}} = 1. \qquad (8.45)$$

Here Γ is the bubble nucleation rate. In the framework of first-order inflation models the filling of the whole space with the true vacuum takes place due to collisions of bubbles, nucleated at the final moment of exponential expansion. The collisions between such bubbles occur when they have a comoving spatial dimension smaller or equal to the effective Hubble horizon H_{end}^{-1} in the transition epoch. If we take $H_0 = 100h$ Km/s/Mpc in an $\Omega = 1$ Universe, the comoving size of these bubbles is approximately $10^{-21} h^{-1}$ Mpc. In the standard approach one believes that such bubbles are rapidly thermalized without leaving a trace in the distribution of matter and radiation. However, in the previous section it has been shown that for any realistic parameters of theory, a collision between only two bubbles leads to BH creation with a probability close to 100% . The mass of this BH is given by (see (8.42))

$$M_{BH} = \gamma_1 M_{\text{bub}}, \qquad (8.46)$$

where $\gamma_1 \simeq 10^{-2}$ and M_{bub} is the mass that could be contained in the bubble volume in the epoch of collision under the condition of full thermalization of bubbles. The discovered mechanism leads to a new direct possibility of PBH creation in the reheating epoch in first-order inflation models. In the standard picture PBHs are formed in the early Universe if density perturbations are sufficiently large, and the probability of PBH formation from small post-inflation initial perturbations is suppressed exponentially. A completely different situation takes place in the final epoch of first-order inflation: namely, collisions between bubbles of Hubble size in the percolation regime leads to PBH formation with the masses

$$M_0 = \gamma_1 M_{\text{end}}^{\text{hor}} = \frac{\gamma_1}{2} \frac{M_P^2}{H_{\text{end}}}, \tag{8.47}$$

where $M_{\text{end}}^{\text{hor}}$ is the mass within the Hubble horizon at the end of inflation. According to (8.42), the initial mass fraction of these PBHs is given by $\beta_0 \approx \gamma_1/e \approx 6 \cdot 10^{-3}$. For example, for the typical value of $H_{\text{end}} \approx 4 \cdot 10^{-6} M_P$ the initial mass fraction β is contained in PBHs with the mass $M_0 \approx 1$ g. In general the Hawking evaporation of mini-BHs could give rise to a variety of possible final states. It is generally assumed that evaporation proceeds until the PBH vanishes completely [72], but there are various arguments against this proposal (see, e.g., [318, 319]). If one supposes that BH evaporation leaves a stable relic, then it is natural to assume that it has a mass of the order $m_{\text{rel}} = k M_P$, where $k \simeq 1 \div 10^2$. We can investigate the consequences of PBH formation in the percolation epoch after first-order inflation, supposing that a stable relic is a result of BH evaporation. As follows from our previous consideration, the PBHs are preferentially formed with a typical mass M_0 at a single time t_1. Hence the total density ρ at this time is

$$\rho(t_1) = \rho_\gamma(t_1) + \rho_{\text{PBH}}(t_1) = \frac{3(1-\beta_0)}{32\pi t_1^2} M_P^2 + \frac{3\beta_0}{32\pi t_1^2} M_P^2. \tag{8.48}$$

The evaporation time scale can be written in the following form:

$$\tau_{BH} = \frac{M_0^3}{g_* M_P^4}, \tag{8.49}$$

where g_* is the number of effective massless degrees of freedom. Let us derive the density of PBH relics. There are two distinct possibilities to consider. The Universe is still radiation-dominated at τ_{BH}. This situation will hold if the following condition is valid $\rho_{\text{BH}}(\tau_{\text{BH}}) < \rho_\gamma(\tau_{\text{BH}})$. It is possible to rewrite this condition in terms of the Hubble constant at the end of inflation

$$\frac{H_{\text{end}}}{M_P} > \beta_0^{5/2} g_*^{-1/2} \simeq 10^{-6}. \tag{8.50}$$

Taking the present radiation density fraction of the Universe to be $\Omega_{\gamma_0} = 2.5 \cdot 10^{-5} h^{-2}$ (h being the Hubble constant in the units of 100 km·s^{-1}Mpc^{-1}), and using the standard values for the present time and the time when the density of matter and radiation became equal, we find the contemporary density fraction of relics:

$$\Omega_{\text{rel}} \approx 10^{26} h^{-2} k \left(\frac{H_{\text{end}}}{M_P} \right)^{3/2}. \tag{8.51}$$

It is easy to see that the relics overclose the Universe ($\Omega_{rel} \gg 1$) for any reasonable k and $H_{\text{end}} > 10^{-6} M_P$. The second case takes place if the Universe becomes PBH-dominated in the period $t_1 < t_2 < \tau_{\text{BH}}$. This situation is realized under the condition $\rho_{\text{BH}}(t_2) < \rho_\gamma(t_2)$, which can be rewritten in the form

$$\frac{H_{\text{end}}}{M_P} < 10^{-6}. \tag{8.52}$$

The present day relics' density fraction takes the form

$$\Omega_{\text{rel}} \approx 10^{28} h^{-2} k \left(\frac{H_{\text{end}}}{M_P} \right)^{3/2}. \tag{8.53}$$

Thus the Universe is not overclosed by relics only if the following condition is valid:

$$\frac{H_{\text{end}}}{M_P} \leq 2 \cdot 10^{-19} h^{4/3} k^{-2/3}. \tag{8.54}$$

This condition implies that the masses of PBHs created at the end of inflation have to be greater than

$$M_0 \geq 10^{11} g \cdot h^{-4/3} \cdot k^{2/3}. \tag{8.55}$$

On the other hand, there are a number of well-known cosmological and astrophysical limits [71] which prohibit the creation of PBHs in the mass range (8.55) with an initial fraction of mass density close to $\beta_0 \approx 10^{-2}$. So one has to conclude that the effect of the false vacuum bag mechanism of PBH formation makes impossible the coexistence of stable remnants of PBH evaporation with first-order phase transitions at the end of inflation.

7. Summary

In this chapter we give qualitative arguments supported by numerical simulation for the existence of long-lived fluctuation that arises as a result of a collision of two expanded bubbles. The two-bubble collision leads, first, to the formation of a short-living false vacuum region in the center of collision. Numerical results indicate separation of a false vacuum region at the time $t \sim b$. Then it evolves into a rather compact object – a clot made up of a scalar field oscillating around its true minimum, with lifetime enough to be captured by its gravitational field. At small coupling constants a black hole can be produced. Unil now the similar object discussed in literature was the oscillon [320, 303].

The main difference between these two objects is as follows:

i) Oscillons represent a subcritical bubble of true vacuum inside a false vacuum, that arises due to temperature fluctuations. Our object is the fluctuation of scalar field in the true vacuum background that arises as a result of dynamical process.

ii) To be long-lived, an oscillon should have a rather large initial radius, though less, than the critical one, and rather flat initial distribution of scalar field. The evolution of the oscillon consists of oscillation of field value with almost constant radius of the field configuration. Our clot of energy is a much more compact object with the amplitude value of scalar field being much larger than that of the field in its potential minimum. For $\lambda \leq 10^{-4} \div 10^{-8}$ gravitational forces are essential and the probability of PBH creation is of order unity.

iii) An oscillon, being produced in spite of small probability [321], is an extremely long-lived object with lifetime $10^3 - 10^4 m^{-1}$, m being the mass of scalar field. The lifetime of our clot of energy is of the same order of magnitude but it could be produced with much bigger probability, because it results from collisions of overcritical bubbles, whose rate of nucleation is much bigger than for subcritical bubbles.

Chapter 9

FINE-TUNING OF MICROPHYSICAL PARAMETERS IN THE UNIVERSE

*One of the advantages of being dis-
orderly is that one is constantly making
exciting discoveries.*
A. A. Milne

The general way of development of physics is supported by a whole number of observational and experimental data. On the other hand, some phenomena which we expected to be discovered for a long time, are still only hypotheses. Other data, being very impressive, are yet not explained. As an example, it is worth mentioning the existence of dark matter and dark energy, of super high energy particles in cosmic rays and 'bursts' – an almost instantaneous energy explosion with energy release of order 10^{53} erg. These astrophysical data specify presence of the new phenomena, which should be comprehended.

One of the cornerstones of modern particle physics is the Weinberg–Salam model of electroweak interaction. Its success became evident after the discovery of $W^{\pm}$ and Z bosons. Meanwhile, another prediction of this model – the existence of Higgs particles – is not yet confirmed. Their detection is the challenge for experimental searches at the modern accelerators. It would specify the properties and the parameters of the Standard Model, being the necessary step for its further development, related with the predictions of new physics beyond it. So, the theoretical arguments make us expect the discovery of SUSY particles at future accelerators.

The high level of precision in the measurements of cosmological data makes observational cosmology more close to proper experimental physics. The coming era of precision cosmology provides an additional source of information about new physics, hardly accessible to accelerator study. In particular, it is just astronomical and cosmic data that is expected to provide the information on neutrallino, a popular SUSY candidate for the dominant form of the modern cosmological dark matter.

One could conclude that cosmology and microphysics stand before quantitative progress that has to reconcile theoretical investigations with essentially new experimental data. The important aspect of this progress, with which the development of cosmoparticle physics is related, is the question of the mutual relationship between the fundamental parameters of cosmology and particle physics. It is worth analyzing some basic postulates, which underlie the modern theory, and to study whether the fundamental physical parameters are eternal and given *ad hoc*, or their choice is specific just for our Universe. As we will see, it is really possible that the physical laws, governing the modern state of our Universe, come from the same process that led to its creation.

The widespread approach consists of choosing from the beginning dynamical variables and the form of a Lagrangian. It is supposed that the smaller the number of parameters the better the quality of the theory. This approach being successful in a field of specific calculations, suffers from some problems. These problems are not discussed and not usually mentioned:

A) Why do we suppose that minimum of a potential is exactly zero?

B) What mechanism is responsible for the specific form of potential with concrete values of parameters in our Universe?

C) To what extent are the quantum corrections important to the form of potential?

D) The question of renormalizability of a theory appears to be not so simple if one takes into account an interaction of particles with gravitons. The latter is usually extremely small and surely should not be considered in real calculations, but its existence is of principal importance. The general relativity, being unrenormalizable theory, leads to the same property of any theory connected with it.

These problems are not very important for low energy physics and scientists usually wave them off. But the modern accelerator physics approaches the energies of the order of 1 TeV and higher. Not only experimenters, but theoreticians as well feel the necessity to deal with new level of energies – theories of the early Universe would operate with Planck densities. Moreover, as we have already discussed in Chapter 7, the low-scale gravity can influence the physical processes even in the range of TeV energies.

The problem A) becomes topical after a discovery of dark energy [73], that could be explained most easily as nonzero vacuum energy density. The last, being ~ 120 orders of magnitude smaller than the Planck scale, allows the formation of the large-scale structure of our Universe.

The problem B), or, more widely a problem of creation of the Universe with observable properties, attracted the attention of a large number of scientists and gave rise to a prolonged discussion [322, 323]. This discussion is continued up to the present time [75, 324, 325, 326].

Neglecting quantum corrections (problem C)) seems doubtful at high energies. Moreover, new terms in a Lagrangian appeared due to quantum corrections used for creation of inflationary models [57] and models of elementary particles [327]. It is evident that quantum corrections have a double meaning. On the one hand, they add significant uncertainty to predictions of any model of the early Universe starting from the inflationary stage. On the other hand, the same corrections give new possibilities for elementary particle models and hence for models of the early Universe tightly connected with them. Quantum corrections to gravitational field can directly lead to inflationary scenario [50, 51]. In addition, the fluctuations of the same gravitational field may renormalize the parameters of a Lagrangian responsible for low energy physics. One result of wormhole physics is a continuum of disconnected sectors, with different values of the parameters [328].

Anthropic principle plays a significant role in the discussion of fine-tuning of parameters of the Universe and, in particular, of the problem of nonzero dark energy [324, 325, 329, 330]. In general, this principle proposes an existence of a set of universes with different properties. Some of them are similar to our Universe. The other ones representing the dominant majority, are not suitable for our existence. "The only thing remaining" is to create a theory which could base the existence of such a set of universes. In this chapter we argue that modern quantum field theory can supply us with the necessary ingredients to solve this problem.

We will show that the issues listed above are connected tightly with each other and with the problem of fine-tuning. So, within many years it was supposed that we live in a space with a Friedmann–Robertson–Walker (FRW) metric. From the astrophysical point of view it means an expanding universe with small negative acceleration. From the point of view of modern field theory it means the vacuum energy density being strictly zero or, equivalently, a vanishing cosmological constant. There were no clear theoretical reasons for this, but there were speculations about a hidden symmetry, implying this strict equality (see, for example, [331] and the review [75]). Several years ago observations [73] indicated some positive value of the cosmological constant $\Lambda \approx 0.7\rho_M$, which is only a little less than the average density of matter ρ_M in the Universe. All quantum effects, which give contribution to the vacuum energy, exceed this value by many orders of magnitude.

The mechanism of almost complete cancellation of different contributions is still not understood. And at the same time, if the cosmological constant were approximately 200 times larger its present value, galaxies would not have been formed [332] and life would have been impossible. The impression that the Universe is specially arranged to create life [332] is rather strong.

The restriction on the cosmological constant described above is not the only case where our existence implies a constraint on parameters in nature. In elementary particle physics there are a number of similar examples. We recall here only one aspect of this fine-tuning – the fine-tuning of proton, neutron and electron mass. The electron mass is about 2000 times smaller than the nucleon mass. One might suppose that it would not matter if it were several times larger than its value $0.511 MeV/c^2$. But in this case neutrons would be stable and the process $p^+ + e^- \rightarrow n + \bar{\nu}$ would result in a sharp decrease of proton abundance in the Universe with adverse consequences for our existence [323].

We see that the Universe is "adjusted to life" by a set of parameters and the cosmological constant is only one of these (for a recent review see, e.g., [333]). It looks like nature has in store a large number of worlds and only a small number of them is suitable for our existence.

A mechanism of realization of a large number of worlds is considered below. A new approach, offered in [371, 330, 326], is proposed in this chapter. It is supposed *a priori* that contribution of quantum corrections in a Lagrangian must not be neglected. This supposition permits to validate the existence of the universes with different properties and look from another side at the problem listed above. The worlds differ from each other by microphysical parameters as well as by the global properties that result from the stochastic realization of these parameters. It is shown that worlds with parameters suitable for creation of the life are necessarily produced as a result of quantum fluctuations. Modern accelerators supply us with ways of experimental test of the considered approach. Some of the possibilities are discussed below.

As we have seen in the preceding chapters inflation, baryosynthesis and nonbaryonic dark matter are determined by particle theory and provide necessary conditions for cosmological expansion, creation of baryonic matter and galaxy formation. The microphysical parameters that determine these phenomena are unknown and model-dependent. In the present chapter we concentrate on the problem of fine-tuning for another set of microphysical parameters that determines the energy density of the modern physical vacuum and the properties of known particles, described by the Standard Model. The notions "creation of life", "conditions suitable for life" or "our existence" are, evidently, used below not in the biological, but in the fundamental physical sense. They undermine the set of the observed physical conditions of the modern Universe, essentially determined by the laws of microphysics.

1.   Basic postulates

Microphysical description, based on Lagrange field theory, postulates, first of all, a concrete Lagrangian for elementary particles. Coupling constants are assumed to be small such that quantum corrections to the original Lagrangian are also considered to be small. Nevertheless, corrections are small only for weak fields, while for strong fields this does not hold.

To be more specific, let us consider the Lagrangian of a scalar field φ

$$L = \frac{1}{2} \left(\partial_\mu \varphi \right)^2 - \frac{m^2}{2} \varphi^2 - \frac{\lambda}{4} \varphi^4 . \tag{9.1}$$

One can compute one-loop quantum corrections to the potential and find [57]

$$\delta V = \frac{\left(3\lambda\varphi^2 + m^2\right)^2}{64\pi^2} \ln \frac{\left(3\lambda\varphi^2 + m^2\right)}{2m^2} - a\varphi^2 - b\varphi^4. \qquad (9.2)$$

The last two terms renormalize the mass and coupling constant of the Lagrangian and depend on the scheme of renormalization. The first term significantly complicates the form of the potential. This is the most important term for the nearest considerations. Multi-loop corrections as well as interaction with other fields may add new terms to the potential. It is important to note that any simple interaction causes an infinite number of additional terms to the original Lagrangian.

One can easily see, comparing expressions (9.1) and (9.2), that new terms are small in comparison with the original terms if $\varphi << m \cdot exp(1/\lambda)$. To get an estimate, one may choose $m = 100$ GeV, $\lambda = 0.1$, then quantum corrections to the potential become large at $\varphi \sim 10^6$ GeV. It is a rather large energy for an accelerator. However, at the early inflationary stage of our Universe the average value is assumed to be much larger, $\varphi > 10^{19}$ GeV. Hence, it is necessary to take into account an infinite number of additional terms in the Lagrangian (9.1). Moreover, the amplitude of a scalar field is restricted even more stringently. The logarithm in expression (9.2) is the result of the summation of an infinite number of terms [334], which converges only when $\varphi < m/\sqrt{3\lambda}$. Besides, one can see directly from Lagrangian (9.1) that the interaction term is of order of the mass term when $\varphi \sim m\sqrt{2/\lambda}$.

The last two estimations are in good agreement with each other and give a much smaller value of the field when quantum corrections are really small. A similar problem was discussed in the framework of hybrid inflation [271]. Thus, when considering phenomena in strong fields, i.e. $\varphi > m/\sqrt{\lambda}$, it is necessary to take into account all additional terms, inevitably arising due to quantum corrections. The potential acquires a much more complex form than the one based on the low energy limit of the theory. This can be visualized by the picture of mountains and valleys. In a mountain area it is possible to have smooth surfaces with small curvature only in valleys, i.e. in minima of the potential energy. After climbing to some height, it becomes obvious that the shape of the terrain is much more rocky.

The potential of a scalar field interaction is usually assumed to be of the most simple form. The property of renormalizability of the theory is not so essential if one supposes that gravitational effects on the Planck scale regularize integrals. Usually, the fields are weak and quantum corrections are reduced to the renormalization of parameters of a Lagrangian under the assumption that the final corrections are small.

As a consequence of the previous discussion, at the moment of formation of our Universe, i.e. at large amplitudes of a field, quantum corrections most likely were comparable with original terms of the Lagrangian, and its form was much more complex than the Lagrangian considered above. If we limit ourselves with one scalar fields φ, naive calculations of quantum corrections to its potential result in a polynomial containing all powers of the scalar field

$$V(\varphi) = \sum_{k}^{\infty} a_k \varphi^k. \tag{9.3}$$

Generally speaking, negative powers are not excluded. Calculation of the coefficients a_k seems impossible and a waste of effort, for two reasons. Firstly, it is hard to believe that this Taylor set is correct at large values of the field φ. Secondly, each term of the polynomial is a result of superposition of interactions with particles of every sort. Their contributions vary unexpectedly with increasing of a degree of a term. Consequently, any information about a shape of the potential in a vicinity of a chosen field value φ_0 is useless at $\varphi \gg \varphi_0$. Thus, any model of elementary particles with postulated from the beginning specific form of a potential with a small number of parameters is doomed to failure at large values of the dynamical variables.

The main conclusion is that the choice of any simple form of Lagrangian with specific parameters leads to difficult problems: one must explain *ab initio* the origin of both the form of Lagrangian and numerical values of parameters and finally manage to prove that quantum corrections are small at high energies.

In addition, the field is only a dynamical variable which has no physical meaning. It enters the expressions for the potential and kinetic energy and it contributes the measurable quantities ONLY in such a form. It reminds us of translational invariance in classical physics. It is not clear why we must single out the value $\varphi = 0$ when postulating the form of a potential. The minimum of potential could happen at any field value with the same probability.

Let us take, following [371], the opposite point of view and limit ourselves to the minimal number of specific assumptions about the form of a potential. As a possible solution of the problem a new postulate is proposed. This postulate is an analogue of the concept of attenuation of correlations known in statistical physics. As in the latter, probabilistic approach used below, this allows us to obtain new results and clarify already known problems.

To proceed, let us introduce first of all the concept of probability density $P(V; \varphi)$ to find specific value V of a potential at given field value φ. Then the only requirement to the form of the potential is expressed in the form of the postulate:

($*$)   *Let a value of the potential V_0 be known at a given field value φ_0. Then there exists such a Φ ($0 < \Phi < \infty$), so that for any V and φ, provided $|\varphi - \varphi_0| > \Phi$: $P(V; \varphi) > 0$ and does not depend on φ_0.*

Correctness of this postulate, as any other postulate, is not directly proved. Nevertheless, it must lead to testable consequences if it pretends on description of the reality. Some of them are discussed below.

As an important example, consider the following Lagrangian of a scalar field

$$L = \frac{1}{2}(\partial_\mu \varphi)^2 - V(\varphi) . \tag{9.4}$$

The field φ is determined in the interval $(-\infty, +\infty)$, which appears to be important in the following consideration. The potential $V(\varphi)$ is assumed to obey the postulate (∗). Then, two general corollaries may be proved on the basis of the postulate.

The first direct corollary is that the potential (9.4) possesses infinite but countable set of zeros. Indeed, if we make the inverse supposition that this set is finite, then, starting from some field value $\overline{\varphi}$, the function $V(\varphi)$ is pure positive or pure negative at $|\varphi| > \overline{\varphi}$. Consequently, one of the statements is surely true in this case: $P(V < 0) = 0$, or $P(V > 0) = 0$, which contradicts the postulate (∗). A countable set of zeros means obviously a countable set of extrema of the potential. Thus, if one wishes to express the potential in terms of the Taylor series, it will be a sum with infinite number of terms. It is interesting that the general postulate (∗) leads inevitably to the form of potential (9.3) that is dictated by the quantum corrections.

The second important corollary looks as follows. Let us know that some minimum of the potential takes place at field value $\varphi = \varphi_m$. Then according to postulate (∗), there is probability $P(V_m)dV_m > 0$ to find the potential value in given interval $(V_m, V_m + dV_m)$. It immediately follows that there exists infinite but countable set of such a minima in the interval in question.

Figure 9.1 gives a representative form of the potential of the scalar field within some interval. If one bears in mind the scalar field as inflaton, a universe formation takes place at the minima with numbers $m - 1, m, m + 1, m + 2....$ Hence, the transparent consequence of the postulate (∗) is the prediction of a nonzero cosmological constant because the probability to find out a local minimum with a preset energy density is equal to zero.

The shape of the potential is unique in the vicinity of each minimum and hence the process of the inflation is unique. The minimum values of the potential (energy density of the vacuum) are materialized with some density probability that differs from zero for any interval $V \div V + dV$. Hence, there is a countable set of the low-lying minima which is necessary, but not sufficient condition for the formation of universes similar to our Universe.

It should also be stated that the Lagrangian (9.3) is a special case of a more general Lagrangian, where quantum corrections to the kinetic term would be taken into account.

Logical extension of the previous discussion is the inclusion of matter and gauge fields. It needs a generalization of postulate (∗) to any parameters of the theory, which are influenced by the quantum corrections.

(∗∗) *All quantities of the theory, which are deformed by the quantum corrections, comply with postulate (∗).*

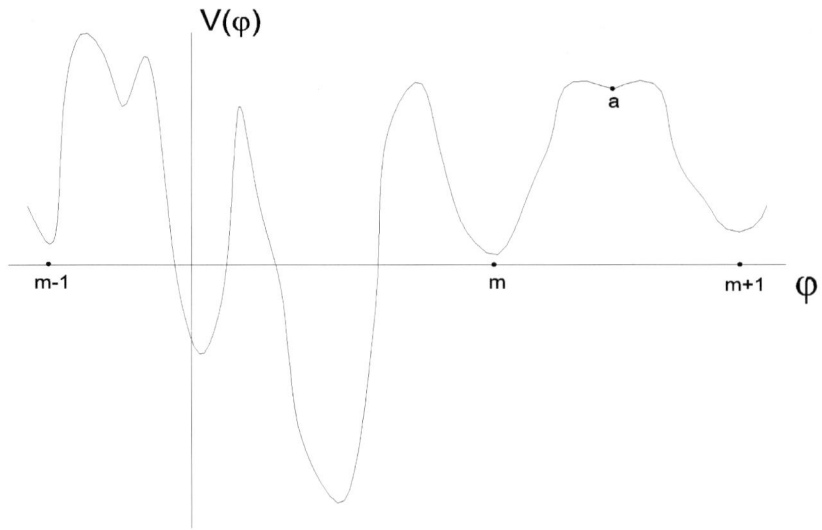

Figure 9.1.   Characteristic form of the potential discussed in the text. Points mark minima where universes of different sorts are formed and evolve. Causal connection between the universes is absent.

In a substantiation of this postulate it is possible to refer to the same reasons, which have resulted in a postulate (∗). Here it is important to note, that, due to the quantum corrections, all parameters of the theory g_n, $n = 1, 2...N$, (N – number of parameters of the theory) turn into (random) functions of a scalar field $g_n(\varphi)$. The values of a scalar field φ_m deliver minima of the potential, but not of the functions $g_n(\varphi)$. The values $g_n(\varphi_m)$ are considered as constants in the ordinary low energy physics.

The Universe is located in one of the minima, where the potential $V(\varphi)$ can be approximated in a simple way:

$$V(\varphi) \approx V(\varphi_m) + a\phi^2 + b\phi^4, \quad \phi = \varphi - \varphi_m . \tag{9.5}$$

Usually, a similar potential is postulated from the beginning with specific constants a and b. The constant a is connected with mass of a quanta of the field φ, $a = m_\varphi^2/2$, if $a > 0$. Other universes occupy other minima which are characterized by a potential with different parameters a and b.

The most unpleasant thing is the occurrence of unstable areas with $V < 0$, a couple of which are shown in Figure 9.1. The similar situation was already revealed and discussed. For example, the quantum corrections from an interaction with fermions can result to potentials of scalar fields, unlimited from below (see for example [335]). Accurate renormalization of a Higgs-like potential reveal new minima [336], some of these being unstable. However, spatial areas with $V < 0$ are not causally connected to the visible part of our Universe. It is as the result of the initial, inflationary period of evolution of the

Universe when its size has increased up to $\sim exp(10^{12})$ cm., that many orders of magnitude greater than the size of the observable Universe $\sim 10^{28}$ cm.

The transition to probabilistic description is the cornerstone of this approach. It greatly enriches the possibilities of our description, as it has happened, when the transition from the classical description to the quantum mechanical was performed. It is worth underlining that a form of the potential is not postulated from the beginning. Instead, only one, rather general property of the potential is proposed. Nevertheless, this property leads to multiple consequences that could be experimentally tested. As we will show below, some of them could be done in the nearest future which would validate or invalidate the postulate.

2. Selection of universes

All (quasi-) stationary states are located in the minima of the potential and our Universe, not being an exception, is also located in such a minimum. There is an enumerable set of minima (note that the potential in question is the polynomial with infinite number of terms), each of which is characterized by some specific energy density. To form a universe similar to our Universe, one has to find first of all a minimum with a very small energy density. For an estimate of this probability, let us assume the uniform distribution of $\rho_V^{(m)}$ in an interval $(0, M_{pl}^4)$. In this case, the probability to find a minimum of the potential with energy density $\rho_V^{(m)} = V(\varphi_m)$ in an interval $d\rho_V^{(m)}$ is given by

$$dP\left(\rho_V^{(m)}\right) = d\rho_V^{(m)}/M_P^4 \qquad (9.6)$$

The estimated value of dark energy density in our Universe is $\rho_V \sim 10^{-123} M_P^4$. Thus, we come to the conclusion that the fraction of universes with vacuum energy density similar to ours is $\approx 10^{-123}$. It is hard to believe that an event with so small probability has happened in Nature. Nevertheless, given an infinite number of universes, we conclude that a set $\mathfrak{R}_0$ of such universes (i.e. those with vacuum energy density $\rho_V \sim 10^{-123} M_P^4$) is still infinite.

As was discussed above, not only small vacuum energy density is necessary to create a universe similar to ours, i.e. with conditions suitable for life of our type. For example, appropriate range of fermion masses is one of the conditions. The generation of the fermion masses gives a nice lesson of how one can overcome in the same framework the difficulties caused by the postulate (∗).

Interaction with fermions

The interaction of a scalar field with fermions is usually considered in the form of Yukawa coupling

$$V_F = g\varphi\bar{\psi}\psi \qquad (9.7)$$

In this case we arrive at a serious problem. The minima of the potential, guaranteeing conditions suitable for life, are very rare and they are most likely to take place far from the value $\varphi = 0$. Hence, the term contributing to the fermion mass $M_F = g\varphi_m$ will be huge compared with the experimentally measured fermion masses.

A hint to the way of solution becomes clear if one notices that the choice (9.7) *a priori* selects the field value $\varphi = 0$ which contradicts the main postulates. The parameter g should also be changed by quantum correction. Hence, according to the postulate (∗∗) it must be substituted with a function of the scalar field φ. Then interaction (9.7) acquires the form

$$V_F = G(\varphi)\bar{\psi}\psi, \tag{9.8}$$

which is a generalization of the expression (9.7). The function $G(\varphi)$ is chosen to be a polynomial with random factors in analogy with the scalar potential $V(\varphi)$. In this case the fermion mass M_F and the constant g of interaction with the field $\phi = \varphi - \varphi_m$ depend on the number m of the universe

$$M_F = G(\varphi_m); \quad g = G'_\varphi(\varphi_m). \tag{9.9}$$

These expressions are obtained by expansion of Eq. 9.8 in a power series around the minimum φ_m.

As we have an infinite number of universes, it is obvious that for any given interval of fermion mass $(\mu_F; \mu_F + \delta)$ and function $G(\varphi)$, one can find an appropriate universe such that the value of the potential at the minimum $V(\varphi_m)$ satisfies the equality $\mu_F \cong G(\varphi_m)$ with the desired accuracy.

Retrieval of the universes

It now becomes possible to use this mechanism for fine-tuning of other parameters, specifying the universe, but not only its vacuum energy density. For example, the existence of life is possible if the fermion mass lies in an interval $(\mu_{life}, \mu_{life} + \delta m)$. Then from an infinite set $\mathfrak{R}_0$ of universes with energy density suitable for life, one can always extract a subset of universes $\mathfrak{R}_1$ with suitable values $G(\varphi_m)$, such that the fermion mass appears in the given interval. Moreover, this new restricted set of universes still contains an infinite number of universes and we can choose a subset of universes with other parameters suitable for life.

Let us introduce physical parameters ℓ_k of a universe. It includes various coupling constants and masses of particles, for example. The creation of life is possible only if the values of these parameters are in some, rather tight, intervals. Total number N_{life} of such a parameter is supposed to be finite. The process of fine-tuning now looks as follows. Fix an interval of values for a first parameter (for example, the vacuum energy density). It gives us an enumerable subset of universes $\mathfrak{R}(\{\ell\}_1)$. Here $\{\ell\}_n$ is a set of n parameters $\ell_1, \ell_2, ..., \ell_n$. The next step consists of fixing the interval for a second parameter (for example, a mass of an electron). It gives us a more weak, but still infinite subset of the universes $\mathfrak{R}(\{\ell\}_2) \in \mathfrak{R}(\{\ell\}_1)$. Thus, the process of finding a suitable universe looks like a consequent choice of more and more weak but enumerable set

$$\mathfrak{R}(\{\ell\}_0) \Rightarrow \mathfrak{R}(\{\ell\}_1) \Rightarrow \mathfrak{R}(\{\ell\}_2) \Rightarrow ... \Rightarrow \mathfrak{R}(\{\ell\}_{N_{life}}).$$

The last subset of the universes satisfies the conditions for all necessary parameters, the number of which is N_{life}. This subset is very weak compared with the initial one, but still contains infinite number of terms. We are not able to estimate exactly what part of the universes are suitable for life. This defect seems not very meaningful because we have no possibility to even visit a neighboring universe.

Neighboring universes

Suppose that a process of inflation takes place at high values of a potential, as is discussed in Chapter 2. Strong quantum fluctuations supply us with field values inside causally connected areas in a wide range of the values. The further destiny of the area strongly depends on field configurations within this area. Configurations being important for our considerations are those where spatial derivatives are small, i.e. $(\partial \varphi_\mu)^2 << V(\varphi)$. In this case we can use well-developed methods of inflation theory – see Chapter 2 – and in particular chaotic inflation [57]. If a causally connected domain starts its evolution at the potential value $V(\varphi)$ the size of this domain in modern epoch could be easily estimated (see (2.32))

$$a(t) = H(t=0)^{-1} \exp\left[\int H(t)dt\right] = H(\varphi_{in})^{-1} \exp\left[\int_{\varphi_{in}}^{\varphi_f} H \frac{d\varphi}{\dot{\varphi}}\right]$$

$$= H(\varphi_{in})^{-1} \exp\left[-\int_{\varphi_{in}}^{\varphi_f} \frac{3H(\varphi)^2 d\varphi}{V'(\varphi)}\right] = H(\varphi_{in})^{-1} \exp\left[\frac{-8\pi}{M_P^2} \int_{\varphi_{in}}^{\varphi_f} \frac{V d\varphi}{V'(\varphi)}\right]$$

The estimation for the simplest form of the potential $V(\varphi) = m^2 \varphi^2 / 2$ is

$$a(t) = H(\varphi_{in})^{-1} \exp\left[2\pi \frac{\varphi_{in}^2}{M_P^2}\right]. \tag{9.10}$$

Here we have take into account that final field value φ_f is much smaller than initial one φ_{in}.

The highest energy density which can be treated theoretically is the Planck density where $V(\varphi) \sim M_P^4$. It is an upper limit where the concept of time can be used. Consequently, the initial field value $\varphi_{in} \sim M_P^2 / m$. The spatial size of fluctuations is then equal to the Planck scale $H^{-1} \sim M_P^{-1} (\sim 10^{19} GeV \approx$

$10^{-33} cm$) as it follows from the relationship between the Hubble parameter H and the energy density $\rho \approx V(\varphi)$, $H = \sqrt{8\pi V(\varphi)/3M_P^2}$. Thus, using formula (9.10) one can obtain the size of a metauniverse

$$a(t) = M_P^{-1} \exp \left[2\pi \frac{M_P^2}{m^2} \right] \approx 10^{-33} e^{10^{13}} cm. \qquad (9.11)$$

Numerical estimation was done for the mass of inflaton $m = 10^{13} GeV$.

Let us consider the evolution of two neighboring causally disconnected domains with slightly different field values marked by letters C and C' in Figure 9.2. Their destiny is rather different because their initial field values are separated by the maximum of potential. The domain which was nucleated in point C will reach a minimum marked as A, while its neighbor nucleated in point C' will reach point B. Two metauniverses will be produced from the two initially neighboring domains. The size of at least one of them is huge compared with the size of our Universe ($10^{28} cm$), the last being incorporated by one of the metauniverses. It means that we never reach neighboring universes.

Another important question arises if one looks at Figure 9.2 more thoroughly. Indeed, a probability of a strict equality $V(\varphi_A) = V(\varphi_B)$ is zero. It means that one minimum is able to decay in the manner discussed in Chapter 10. It does not sound very optimistic for our civilization and it is worth estimating the probability of such a decay. Suppose that the maximum in Figure 9.2 is of order $V_{\max} \sim M_P^4$. Function that approximates correctly the local form of the potential presented in Figure 9.2 may be chosen in the form

$$U = \frac{\lambda}{8} \left(\varphi^2 - a^2 \right)^2 + \frac{\varepsilon}{2a} (\varphi - a). \qquad (9.12)$$

Density probability of the vacuum decay in one-loop approximation was calculated in papers [298, 337] (see also Chapter 10, Sections 1, 3) and has the form

$$\Gamma \approx A e^{-S_E}.$$

Here S_E is the Euclidean action on a classical trajectory connected to the two minima and multiplier A depends on quantum corrections. Exact value of the multiplier is not very important and we put $A = M_P^4$ for estimation. That seems reasonable in the considered scale of Planck energies. The Euclidean action is expressed in terms of Lagrangian parameters [298, 337]

$$S_E \approx \frac{\pi^2}{6} \frac{a^{12} \lambda^2}{\varepsilon^3}.$$

As we have supposed that $V_{\max} \sim M_P^4$, the parameter a can be expressed in the form $a = (8/\lambda)^{1/4} M_P$. The action acquires final form

$$S_E \approx \frac{2^8 \pi^2}{3} \frac{M_P^{12}}{\lambda \varepsilon^3}.$$

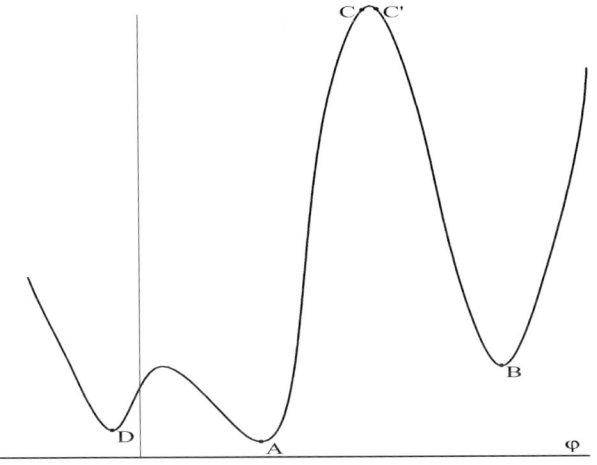

Figure 9.2. A part of the potential.

This result was obtained in the thin-wall approximation limit, that looks like $\varepsilon << M_P^4$ in our case. Coupling constant λ is usually chosen less than unity, $\lambda < 1$. The most unpleasant situation, i.e. the largest value of decay probability, is realized at the parameters values $\varepsilon \sim M_P^4$, $\lambda \sim 1$. Then $S_E \approx 2^8 \pi^2 / 3 \approx$ 800.

Straightforward utilization of the formulae written above leads to the estimation of the decay probability of our Universe within the scale $L_U \sim 10^{28} cm$

$$\Gamma_U \sim \Gamma L_U^3 \sim 10^{-200} s^{-1}.$$

Thus, lifetime of our Universe is about $10^{200} s$ and we may not worry about this problem at least during those time intervals.

Nearly disposed minima

As it follows from the preceding discussion, the inflaton potential may have rather complicated form. It may happen accidentally that two minima are situated closely. The minima in question are marked as 'A' and 'D' in Figure 9.2. The potential can be approximated by the same function (9.12) with another value of parameters λ, a and ε. Inflation takes place when Hubble parameter $H(\varphi) >> m$ which is supposed to hold in this case, $m^2 = U''(\varphi_A)$. In the vicinity of the local minimum, point 'D' the equation of motion becomes simpler,

$$\ddot{\varphi} + 3H(\varphi_D)\dot{\varphi} + M^2 (\varphi - \varphi_D) \approx 0. \qquad (9.13)$$

If $H(\varphi_D) >> M$, dissipation of energy is large and the field could be located in the local minimum for a long time. We encounter serious problems, which were discussed in connection with the old inflationary models [297], where our universe was formed from the domain in a local potential minimum. Nevertheless, if a height of the local maximum happens to be small, first-order phase transition at the very end of inflation could take place. It could lead to effective transformation of the inflaton into particles. The last problem is some shortage of many inflationary scenarios because weak self-coupling of the inflaton takes place only if couplings with fermions are also small. Otherwise, quantum corrections lead inevitably to strong self-interaction of inflaton. It means that the fermion production is suppressed.

In the case $H(\varphi_D) \leq M$ the situation differs from the previous one. The field slowly decreases, according to equation (9.13) until it appears in the vicinity of the local minimum $\varphi = \varphi_D$, where the equation of motion can be reduced to

$$\ddot{\varphi} + M^2 (\varphi - \varphi_D) \approx 0 , \qquad (9.14)$$

and the total energy of the field is approximately conserved. In this case the classical field could overcome the local maximum between the two minima 'A' and 'D' and approach the nearest deeper minimum of the potential. It gives rise to fractal structures in the future. The process of the fractal structure formation performs in the same manner that was discussed in Chapter 4 and we shortly repeat these arguments, applying them to the case in question.

Let classical motion of the field be governed by the equation

$$\ddot{\varphi} + 3H\dot{\varphi} = -dV/d\varphi . \qquad (9.15)$$

The destiny of spatial areas where the field just overcomes potential maxima is rather interesting. The fact is that classical motion is accompanied by quantum fluctuations. Consider the fluctuations of the field in a nearest vicinity of such a maximum (right slope). The initial spatial size of this fluctuation is $\sim 1/H$. After some time of the order of $\sim 1/H$ has passed this spatial area will be separated into e^3 causally disconnected domains with different field values. The average value of the field φ inside some of these domains could fluctuate into the other side of the maximum (left slope). In its turn, each of these domains will be divided in e^3 subdomains of the size of $\approx 1/H$ in time $1/H$ and some of them will pass back through the maximum of the potential. This process continuously reproduces itself and already after several steps a picture of a fractal structure will be observed.

Meantime, classical field moves away from the maximum that could prevent the formation of the fractal structure. Hence, the development of rich fractal structure in a final stage can take place only if the fluctuations are large. More specifically, let us assume that the classical field changes its value by $\Delta\varphi_{cl}$ in the time $1/H$. Then a fractal structure arises if the condition of the fluctuation dominance $\Delta\varphi_{fluct} >> \Delta\varphi_{cl}$ is satisfied. It gives enough time for formation of a fractal structure due to the fluctuations around a maximum.

An average value of fluctuations is well-known, given by $\Delta\varphi_{fluct} \simeq H/2\pi$. The classical motion can be computed explicitly if one approximates the potential around a maximum by the function

$$V(\varphi) \simeq V_0 - (\varphi - \varphi_{Max})^2 a^2/2 .$$

An approximate solution of Eq. (9.15) has the form

$$\varphi(t) \simeq \varphi_{Max} + [\varphi(t=0) - \varphi_{Max}] \, exp \left(\frac{a^2 M_P}{\sqrt{24\pi V_0}} t \right),$$

where the second time derivative is neglected as is usually done at the inflation stage. The initial field value $\varphi(t=0) \approx \varphi_{Max} + \Delta\varphi_{fluct}/2$ and the condition of quantum fluctuation dominance is easily found to be

$$\eta \equiv \frac{\Delta\varphi_{fluct}}{\Delta\varphi_{cl}} \approx \frac{16\pi V_0}{a^2 M_P^2} > 1. \tag{9.16}$$

The larger the parameter η the richer fractal structure will be formed. These fractal structures being small in comparison with the size of our Universe could result in observable consequences.

It is well-known that two domains with field values separated by a potential maximum, are separated by a wall [261]. Classically, fields in such domains tend to various (neighboring) minima 'A' and 'D' and hence the energy density of the wall grows relative to the rest of the space.

The picture, represented above, includes the picture of the eternal inflation [130] and provides realization of the set of universes with intrinsically different properties.

3. Weinberg–Salam model

Let us consider modifications of the standard $SU(2) \otimes U(1)$ Weinberg–Salam model (SM), which are the results of postulates $(*)$, $(**)$.

3.1 Higgs field

Potential of the Higgs field $\chi = \begin{pmatrix} \chi_1 \\ \chi_2 \end{pmatrix}$ is supposed usually in the form

$$V_{Higgs}(\chi) = \frac{\lambda}{4} \left(|\chi|^2 - v^2/2 \right)^2. \tag{9.17}$$

i.e. nonzero vacuum average is postulated from the beginning. The consequence of our postulates is that it is not now necessary – quantum corrections lead to multiple minima for any potential, including the Higgs one. The problem is, however, more complicated because the Higgs field should interact with inflaton (at least owing to multiloop corrections) and we get inevitably, instead of (9.17), an expansion with infinite number of terms,

$$V(\varphi, \chi) = \sum_{k,n=0}^{\infty} a_{kn} \varphi^k |\chi|^{2n}. \tag{9.18}$$

The application of a postulate $(*)$ simultaneously to both fields results in a potential of a complex form with a lot of extrema and valleys in various directions. Minima of the potential are situated in points with field value (φ_m, v_m), $v_m \equiv \sqrt{|\chi_m|^2}$. One can see that there is no need to include nonzero vacuum average artificially. Of course, the probability that two neighboring vacua appears to be symmetrical, as it is in the SM is very small. It should not disturb us for this symmetry is not an important feature of the model. In this connection, we would like to mention an article [338] devoted to the Weinberg–Salam model with two asymmetrical vacua. One of them is placed in Planck scale.

3.2   Modernization of the Standard Model

The Weinberg–Salam model is described by the Lagrangian

$$L = L_{gauge} + L_{lept} + L_{scalar} + L_{int};$$

$$L_{gauge} = -\frac{1}{4} F^i_{\mu\nu} F^{i\mu\nu} - \frac{1}{4} B_{\mu\nu} B^{\mu\nu};$$

$$L_{lept} = \bar{R} i \left(\hat{\partial} + i g' \hat{B} \right) R + \bar{L} i [\hat{\partial} + i \frac{g'}{2} \hat{B} - i \frac{g}{2} \tau^i \hat{A^i}] L;$$

$$L_{scalar} = \left| [\hat{\partial} - i \frac{g'}{2} \hat{B} - i \frac{g}{2} \tau^i \hat{A^i}] \chi \right|^2 - V(|\chi|^2);$$

$$L_{int} = -G \left[\bar{R} \chi^+ L + \bar{L} \chi R \right], \tag{9.19}$$

with the notations

$$F^i_{\mu\nu} = \partial_\mu A^i_\nu - \partial_\nu A^i_\mu + g \varepsilon^{ijk} A^j_\mu A^k_\nu,$$
$$B_{\mu\nu} = \partial_\mu B_\nu - \partial_\nu B_\mu.$$
$$V(|\chi|^2) = \lambda (|\chi|^2 - v^2/2). \tag{9.20}$$

Here A and B are gauge fields, L and R are left and right fermions. The Higgs field χ has nonzero vacuum expectation value $\left(\begin{smallmatrix} 0 \\ v/\sqrt{2} \end{smallmatrix} \right)$. The parameters g, g', v, G of the model are expressed in terms of observable values – a mass of a charged W-boson M_W, a mass of Z-boson M_Z, electric charge e and Fermi constant G_F.

$$M_W = gv/2, \quad M_Z = \frac{v}{2} \sqrt{g^2 + g'^2}, \quad e = \frac{gg'}{\sqrt{g^2 + g'^2}}, \quad G_F = \frac{1}{\sqrt{2} v^2}. \tag{9.21}$$

As was shown in the paper [339], a procedure of renormalization could be chosen in such a manner to preserve gauge invariance of the Lagrangian (9.19). That is why, following [371] we limit ourselves by an investigation of results of renormalization of parameters g, g', G. According to postulates (∗), (∗∗), these parameters are transformed into polynomials of the type (9.18)

$$g, g', G \Rightarrow g(\varphi, |\chi|^2), g'(\varphi, |\chi|^2), G(\varphi, |\chi|^2). \tag{9.22}$$

At the same time, the potential of the Higgs field $V(|\chi|^2)$ acquires form (9.18) due to the same postulates $(*), (**)$.

Let us discuss now the last term in expression (9.19) describing fermion interaction with the Higgs particles. It will be shown that it gives a nontrivial result that differs from the prediction of SM and is accessible to experimental test. The substitution of the constants g, g', v, G with functions (9.22) leaves a trace in an effective Lagrangian.

The expression L_{int} in the Standard Model permits us to obtain a fermion mass m_f one-to-one related to the vacuum expectation value $v/\sqrt{2}$ embedded by hand,

$$m_f = Gv/\sqrt{2}. \tag{9.23}$$

If one takes into account that the parameter G is now a polynomial of the form (9.18), the result appears to be rather different. Indeed, let us parameterize the Higgs doublet as usual: $\chi = e^{i\Theta(x)} \binom{0}{\chi^0}$, where $\Theta(x)$ is $SU(2)$ matrix that is removed from the final Lagrangian by a gauge transformation. As a result, the term in question is

$$L_{int} = -G(\varphi, (\chi^0)^2)\chi^0 \overline{f_R} f_L. \tag{9.24}$$

Potential (9.18) is disposed in some minimum number m at the field values φ_m, χ_m^0 ($\chi_m^0 = v/\sqrt{2}$ in usual notations). First terms in the Taylor expansion of the expression (9.24) in the powers of $h = \chi^0 - \chi_m^0$ have the form

$$L_{int} \simeq -\chi_m^0 \cdot G(\varphi_m, (\chi_m^0)^2)\overline{f_R} f_L - G_f h \overline{f_R} f_L,$$

$$G_f = \frac{\partial G(\varphi_m, (\chi_m^0)^2)}{\partial \chi_m^0} \chi_m^0 + G(\varphi_m, (\chi_m^0)^2). \tag{9.25}$$

Fermion mass

$$m_f = \chi_m^0 \cdot G(\varphi_m, (\chi_m^0)^2) \tag{9.26}$$

appears to be dependent on number m of specific minimum (see also [340]), in which a universe evolves. The field values φ_m, χ_m^0 represent the minimum of the potential (9.18), but not the minima of the other functions, such as $G(\varphi_m, (\chi_m^0)^2)$. That is why the usual proportionality (9.23) of the fermion mass m_f to the constant of its coupling to Higgs particle is absent in this case.

Another deviation from the SM could be found in couplings of a selfinteraction of the Higgs particles. For example, the constant of three-linear interaction in the framework of the Weinberg–Salam model is equal to $\lambda_{hhh} = 3\sqrt{2}\lambda v$ in 0th order of a perturbation theory and hence is proportional to the known value of the vacuum expectation value v. Another prediction follows from the potential (9.18). This constant has the form $\lambda_{hhh} = (1/6)\partial^3 V(\varphi_m, \chi_m)/\partial \chi_m^3$ and it does not have any connection with the other parameters.

It may be shown [371] that this approach is able to restore the Weinberg–Salam model with the exception of those interactions with Higgs particles h.

It is instructive to consider the part of the Lagrangian L_{lept} responsible for an interaction of leptons (electrons) with gauge fields. After standard substitution of the fields B_μ, A^i_μ by physical fields $W^\pm_\mu$, Z_μ, A_μ one obtains [341]

$$
\begin{aligned}
L_{lept} \;=\; & C_1 \left[\bar{v}_e \gamma^\mu (1 - \gamma_5) e W^+_\mu + h.c. \right] \\
& - \; C_2 \left(2\bar{e}_R \gamma^\mu e_R + \bar{v}_e \gamma^\mu v_e + \bar{e}_L \gamma^\mu e_L \right) Z_\mu \\
& + \; C_3 \left(\bar{e}_L \gamma^\mu e_L - \bar{v}_e \gamma^\mu v_e \right) Z_\mu - e A_\mu \bar{e} \gamma^\mu e.
\end{aligned}
\tag{9.27}
$$

The quantities C_i, e are expressed in an ordinary way in terms of the initial parameters

$$
C_1 = \frac{g}{2\sqrt{2}}; \quad C_2 = \frac{e}{2} \tan \theta_W; \quad C_3 = \frac{e}{2} \cot \theta_W,
$$

$$
e = \frac{gg'}{\sqrt{g^2 + g'^2}}; \quad \tan \theta_W = g'/g
\tag{9.28}
$$

and, according to (9.22), are functions of the fields φ and χ. Both fields are placed in the vicinity of some minimum of potential (9.18) and we can limit ourselves with first terms in Taylor expansion

$$
C_i \;\simeq\; C_i(\varphi_m, \chi^0_m) + \frac{\partial C_i(\varphi_m, \chi^0_m)}{\partial \chi^0_m} h;
$$

$$
e \;\simeq\; e(\varphi_m, \chi^0_m) + \frac{\partial e(\varphi_m, \chi^0_m)}{\partial \chi^0_m} h.
\tag{9.29}
$$

The inflaton field is supposed to be sufficiently massive so that we can neglect any interaction with its quanta. The first terms in expansion (9.28) were determined experimentally as far as they are connected with known parameters – electron charge e and Weinberg angle θ_W according to expressions (9.28). Second terms are responsible for new vertices of interaction of the leptons with the quanta of Higgs field h. More definitely, these vertices are: $\bar{v}_e W e h, \bar{e} e Z h, \bar{v}_e v_e Z h, \bar{e} e A h$, in the Lagrangian (9.27)

$$
\begin{aligned}
L'_{lept} \;=\; & \Gamma_1 \left[h \bar{v}_e \gamma^\mu (1 - \gamma_5) e W^+_\mu + h.c. \right] - \\[6pt]
& - \; \Gamma_2 \left(2h\bar{e}_R \gamma^\mu e_R + h\bar{v}_e \gamma^\mu v_e + h\bar{e}_L \gamma^\mu e_L \right) Z_\mu + \\[6pt]
& + \; \Gamma_3 \left(h\bar{e}_L \gamma^\mu e_L - h\bar{v}_e \gamma^\mu v_e \right) Z_\mu - \Gamma_e h A_\mu \bar{e} \gamma^\mu e.
\end{aligned}
\tag{9.30}
$$

As far as the four vertices are expressed in terms of two unknown parameters, $B \equiv \partial g / \partial \chi^0_m$ and $B' \equiv \partial g' / \partial \chi^0_m$, they are connected with each other as follows

$$\Gamma_2 = \Gamma_e \frac{1}{2} \left(\frac{g}{e}\right)^2 \left(\frac{e^2 g'}{g^3} + 1\right)$$

$$+ \Gamma_1 \sqrt{2} \left[\frac{eg'}{g^2} \left(\frac{e^2 g}{g'^3} - 1\right) - \frac{e}{g} \left(\frac{g}{g'}\right)^3 \left(\frac{e^2 g'}{g^3} + 1\right)\right],$$

$$\Gamma_3 = \Gamma_e \frac{1}{2} \left(\frac{g}{e}\right)^3 \frac{eg}{g'^2} \left(\frac{e^2 g'}{g^3} - 1\right) \qquad (9.31)$$

$$+ \Gamma_1 \sqrt{2} \left[\frac{e}{g'} \left(\frac{e^2 g}{g'^3} + 1\right) - \frac{eg}{g'^2} \left(\frac{g}{g'}\right)^3 \left(\frac{e^2 g'}{g^3} - 1\right)\right].$$

The existence of new vertices (9.29) obeying the relations (9.31) is a direct consequence of the initial postulates (∗), (∗∗). On the other hand, they could be checked experimentally in the nearest future. For example, properties of the Higgs bosons will be investigated in modern accelerators. Such a possibility is discussed in the paper [342] for $e^+ e^-$ annihilation at the energies 500 GeV in the center of mass system. In case of discovering the Higgs bosons, it allows to validate or invalidate connections (9.31) and hence the postulates proposed above.

4. Discussion

In this chapter were considered some consequences of the postulates (∗), (∗∗), proposed in [371]. These postulates are used instead of strict fixing of a Lagrangian from the very beginning. It gives rise to nontrivial consequences both in cosmology and in physics of elementary particles. First of all, it allows one to prove the existence of a countable set of universes disposed in potential minima with different values of microscopic parameters. It serves as a necessary ingredient of the anthropic principle which permits us to explain the origin of a universe similar to ours.

The problem listed in the beginning looks quite solvable if the solution is based on probability language used for formulation of the postulates. In this framework, the answer to the problem A) is: "Minimal value of a potential is not equal to zero. Instead, there exists infinite set of universes with sufficiently small values of the potential minimum. It is a necessary condition for nucleation and formation of universes similar to ours". One could extract from this set of universes a subset with parameters close to those presented in our Universe. The last statement is the answer to the problem B) mentioned above. An additional pleasant property of the potential (9.3) is its absolute renormalizability (problem D)). Indeed, it contains terms with any power of the scalar field from the beginning and any quantum correction could be included in the Lagrangian parameter. The answer to the question C) is evident from the above discussion.

Postulates (∗), (∗∗) lead with necessity to scalar–tensor theories of gravity. Different sorts of such theories are discussed in, e.g., [343, 345, 346]. If we wish to be consistent, then we have to admit, according to postulates (∗), (∗∗), that the quantum corrections convert *all* parameters of a Lagrangian, including the gravitational constant G_N into polynomials. In this case, general form of the Lagrangian can be readily written

$$L = -\frac{F(\varphi)}{16\pi G_N} R + \frac{K(\varphi)}{2} (\partial\varphi)^2 - V_{ren}(\varphi). \qquad (9.32)$$

Immediate conclusion from expression (9.32) is that Newton gravitational constant is an effective one and it is different in different universes enumerated by number m: $G_N(m) = G_{N,our} F(\varphi_m)/F(\varphi_{our})$ (index 'our' relates to our Universe).

The important thing is that in spite of generality and brevity of the postulates they not only give new sight to the abstract problems listed in the beginning of this chapter, they also possess predictive power and some of their predictions could be tested in the nearest future. More definitely

a) new vertices of interaction of leptons, gauge fields and Higgs bosons take place in this framework. One-to-one connections (9.31) between them could be checked experimentally in the nearest future;

b) Strict connection between a mass of fermions and coupling constant of their interaction with scalar particles is absent. It could be very important for axion models because it strongly enriches their ability to satisfy the stringent limits following from the observational data [3];

c) The constant of the three-linear interaction appears to be a free parameter on the contrary to the prediction of the standard Weinberg–Salam model.

The serious argument against the proposed postulates would be strict equality to zero of the vacuum energy density (density of the dark energy) because a measure of such universes is zero. The evidences that the cosmological density of the dark energy is not zero are considered rather firm these days, but we still could not exclude the opposite possibility. Formally, if future observations indicate that vacuum energy density is strictly equal to zero, it will make the postulates very doubtful. On the other hand, this hypothetical possibility can never be experimentally proven, since the observational data can only put an upper limit on this quantity and cannot rule out its existence below the level of experimental sensitivity.

In this chapter the mechanism of creation of universes with given set of microscopic parameters is developed. The process of formations of each universe is unique, because the form of potential is unique in the vicinity of each minimum. The formation of universes is described by different types of inflationary models in different minima. Presently, a large number of models with a wide range of different potentials are considered as potentially realistic. Apparently, each of them describes some subset of the universes of our type. It is shown, that at an early stage of formation of our Universe primordial fractal structures could be created in a natural way.

Chapter 10

INFLATION: ADDITIONAL RESOURCES

In the presence of eternity, the mountains are as transient as the clouds.

Robert Green Ingersoll

Up to now our consideration was performed in the framework of chaotic inflation. We revealed that scalar field(s) connected with gravity gives rise to a very interesting and important period of the evolution of our Universe. A lot of observational data can find new nontrivial explanation in this framework. In addition, new phenomena are predicted, some of them having been discussed in this book.

Nevertheless, not everything is as good as it seems. For example, coupling constant must be very small to fit the observed temperature fluctuations of the relic radiation. Another problem is energy transition from inflaton to fermions and photons during the end of inflation. Indeed, if coupling constants are small, the decay of inflaton is suppressed. The problem could be cured if one takes into account an effect of a parametric resonance [347], [348]. Thus, this problem is not unresolved in principle, but it would be instructive to discuss other mechanisms of the inflation.

In this chapter we consider a mechanism of a first-order phase transition at the end of inflation, an effect of auxiliary massive fields on inflationary processes, a decay of cosmological constant as the Bose–Einstein condensate evaporation and a quite general case of inflation based on so-called scalar–tensor theories.

1. First-order phase transition as a terminator of inflation

Let us treat the case when inflation is finished due to vacuum decay with probability density Γ per unit time. The false vacuum is percolated exactly in

219

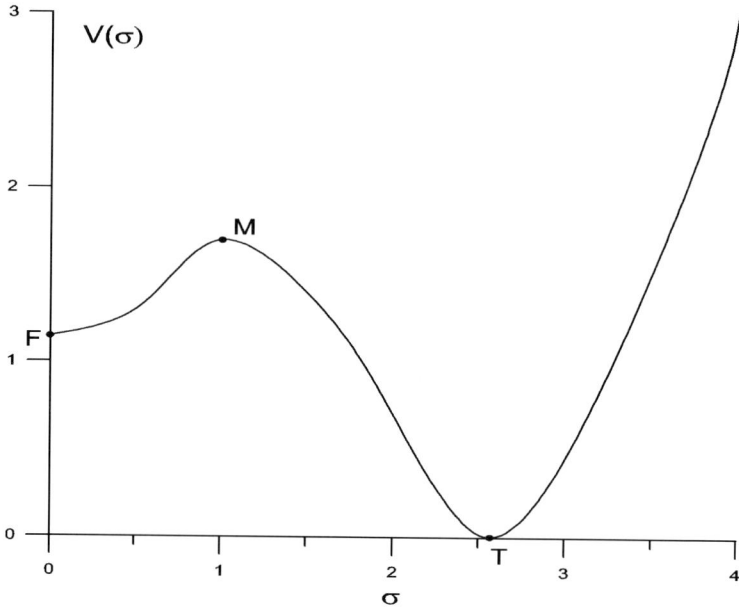

Figure 10.1.   Typical form of a potential which leads to false vacuum decay. False vacuum
state is denoted as 'F' while true vacuum state is marked as 'T' .

in the arrow direction. When the Universe arrives at point 'B' the barrier appears to be small enough to make the process of tunneling considerable. The
vacuum decay from the false vacuum state (point 'B') to the true vacuum state
(point 'C') signifies the end of the inflation. An example of such a model
could be found in [349].

   We see that if we involve new fields in the consideration, it gives new interesting effects and possibilities to solve topical problems. The next section is
devoted to an interplay of inflaton and an auxiliary massive field.

2.   Massive fields and superslow motion during inflation

   Early inflation mechanisms [52, 51] were based on the consistent equations
of scalar and gravitational fields. Nevertheless, the simplest inflation models
could not explain the totality of the observed data. In particular, the predictions
of the chaotic inflation model [55] about temperature fluctuations in cosmic
background radiation do not contradict observations only for a rather unnatural
form of the inflaton field potential.

   At the same time, the interaction of a large number of various fields existing
in nature should give rise to new phenomena in inflation scenario. Further development of the theory has led to the emergence of inflation models involving
additional fields, among which are the models of hybrid inflation [269] and

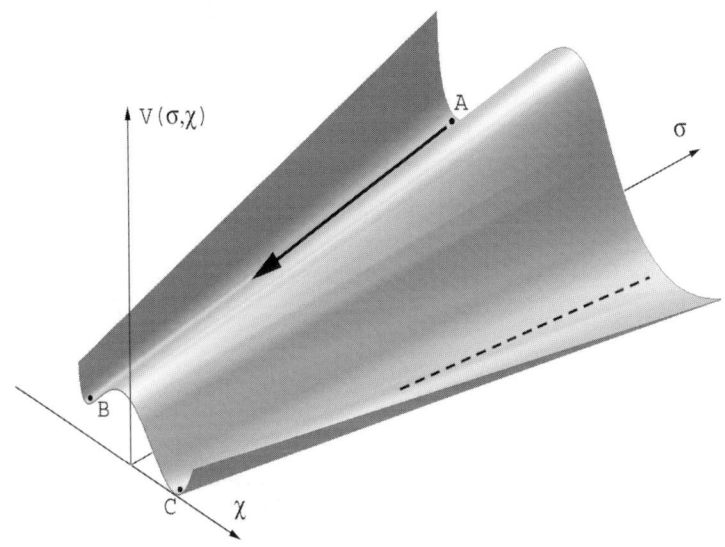

Figure 10.2. One possible form of a two-field potential for solution of the graceful exit problem.

inflation on the pseudo Nambu–Goldstone field [350], for example. The interaction of the classical inflaton field with other particles is one of the basic elements of inflationary models. Back reaction of the produced particles on the dynamics of inflaton field was considered in [311, 351] and in Section 1 of this chapter.

Below we study an effect of an additional field on the classical motion of the basic inflaton field [352]. It is assumed that the additional field is massive enough for it to be at the minimum of its effective potential during final stage of the inflation. Nevertheless, it is shown below that its influence could noticeably decelerate the system motion.

In what follows, the simplest form of interaction is considered allowing the analytical results to be obtained. Namely, we introduce, apart from the inflaton field φ, an additional scalar field χ and write the action in the form [352]

$$S = \int d^4x \sqrt{-g} \left[\frac{1}{2}\varphi_{,\mu}\varphi^{,\mu} - V(\varphi) + \frac{1}{2}\chi_{,\mu}\chi^{,\mu} - \frac{1}{2}m^2\chi^2 - \kappa\chi u(\varphi) \right],$$

$$(10.4)$$

where $u(\varphi)$ is a polynomial of degree no higher than three for the renormalizable theories. Below, particular case $u(\varphi) = \varphi^2$ is taken for definiteness. The first power of the field χ in the interaction is necessary in order to obtain compact analytical results valid for an arbitrary coupling constant κ, rather than the expansion in powers of this constant. The interaction of this type arises in supersymmetric theories and is considered in hybrid inflation scenarios [353]. Dolgov and Hansen [351] used this type of interaction in studying the back reaction of produced particles on the motion of classical field.

The set of the classical equations for both fields is written as

$$\frac{1}{\sqrt{-g}}\partial_\mu\left(\sqrt{-g}\partial^\mu\chi\right) + m_\chi^2\chi + \kappa\varphi^2 = 0, \tag{10.5}$$

$$\frac{1}{\sqrt{-g}}\partial_\mu\left(\sqrt{-g}\partial^\mu\varphi\right) + V'(\varphi) + 2\kappa\varphi\chi = 0. \tag{10.6}$$

Let us consider the case of heavy χ particles. In the inflationary era, this means that

$$m_\chi \gg H(\varphi), \tag{10.7}$$

and the Hubble constant $H(\varphi)$ is determined by the slowly varying classical field φ. The latter plays the role of inflaton. The Eq. (10.5) can be brought to the form

$$\chi(x) = -\kappa\int G(x,x')\varphi^2(x')dx'. \tag{10.8}$$

The right-hand side of Eq. (10.8) can be simplified using the equation for the Green function $G(x,x')$ [354] written as

$$G(x,x') = \frac{1}{m^2}\delta(x-x') - \frac{1}{m^2}\frac{1}{\sqrt{-g}}\partial_\mu\sqrt{-g}\partial^\mu G(x,x'). \tag{10.9}$$

After two iterations, the field χ takes the explicit form

$$\chi(x) \simeq -\frac{\kappa}{m_\chi^2}\varphi^2(x) + \frac{\kappa}{m_\chi^4}\partial^\mu\sqrt{-g}\partial_\mu\left(\frac{1}{\sqrt{-g}}\varphi^2(x)\right), \tag{10.10}$$

which is valid if the derivatives of the inflaton field φ are small. Substituting this expression into the Eq. (10.6), one arrives at the following classical equation for the inflaton field:

$$\partial_\mu\sqrt{-g}\partial^\mu\varphi + \sqrt{-g}V'_{ren}(\varphi) + \frac{2\alpha^2}{m_\chi^2}\varphi\partial_\mu\sqrt{-g}\partial^\mu\varphi^2 = 0, \tag{10.11}$$

where $\alpha \equiv \frac{\kappa}{m_\chi}$ a dimensionless parameter and

$$V_{ren}(\varphi) = V(\varphi) - \frac{\alpha^2}{2}\varphi^4 \tag{10.12}$$

is the potential of inflaton field renormalized due to interaction with the field χ. The last term on the left-hand side of Eq. (10.11) is usually treated as a back reaction of radiation [351]. Equation (10.11) corresponds to the effective action for inflaton field

$$S_{eff} = \int d^4x\sqrt{-g}\left[\frac{1}{2}\varphi_{,\mu}\varphi^{,\mu} - V_{ren}(\varphi) - \frac{1}{\sqrt{-g}}\frac{\alpha^2}{2m^2}\varphi^2\partial_\mu\sqrt{-g}\partial^\mu\varphi^2\right]. \tag{10.13}$$

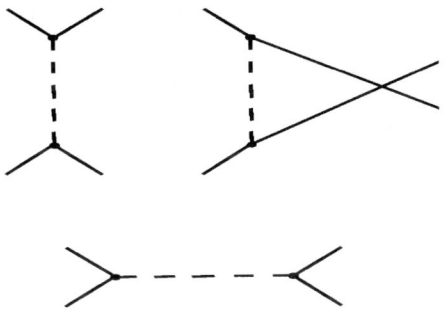

Figure 10.3. Feynman diagrams for first-order corrections to coupling constant λ. Solid lines represent field φ, dotted lines represent field χ.

Note that the correction $\delta V = -\frac{\alpha^2}{2}\varphi^4$ to the potential follows from the analysis of classical Eqs. (10.5)–(10.6). At the same time, the same expression can be obtained by calculating the first quantum correction to the potential of the field φ due to interaction with the field χ. Necessary diagrams for the first-order corrections are shown in Figure 10.3. External lines correspond to the quanta of φ-field at zero 4-momenta. The internal lines correspond to the χ-field propagator in the s and t channels. The calculation of these diagrams leads to a renormalized coupling constant

$$\lambda = \lambda_0 - \frac{\kappa^2}{2m^2},$$

which gives renormalized potential (10.12).

The last term in Eq. (10.11) is important for further consideration. A non-minimal kinetic term arises in equations for density fluctuations in the early Universe [355]. Morris [356] showed that a change in the form of the kinetic term in the scalar–tensor theory leads to the inflation at a lower than ordinary energy scale, in agreement with the conclusions of the work [352]. Similar results can be obtained by introducing a nonminimal interaction between an inflaton and a gravitational field [357], [358].

In general, the renormalized potential contains the sum of contributions from the corrections due to interaction with all the existing fields. In the first model of chaotic inflation with the $\lambda\varphi^4$ potential, the observational data implied a value of $\lambda \sim 10^{-13}$. This means that the corrections contributed to the expression for the potential $V = \lambda\varphi^4$ by all fields, including the correction $\delta V = -(\alpha^2/2)\varphi^4$ considered above, must cancel each other with high accuracy.

We demonstrate below that the renormalization of the kinetic term allows one, in particular, to weaken significantly the conditions imposed by the observations on the parameters of the theory. In weak fields, the contribution from the last term in Eqs. (10.11), (10.13) is negligible. As to the inflation stage, it can be substantial at large field magnitudes.

During inflation, the field is assumed to be uniform; i.e. $\varphi = \varphi(t)$, and Eq. (10.11) is greatly simplified. Taking into account that the scale factor a is expressed in terms of the Hubble constant H in the ordinary way, $a = H^{-1}exp(\int Hdt)$, Eq. (10.11) can be rewritten as

$$\frac{d^2\varphi}{dt^2} + 3H\frac{d\varphi}{dt} + V'_{ren}(\varphi) + \tag{10.14}$$

$$\frac{4\alpha^2}{m_\chi^2}\left[3H\varphi^2\frac{d\varphi}{dt} + \varphi^2\frac{d^2\varphi}{dt^2} + \varphi\left(\frac{d\varphi}{dt}\right)^2\right] = 0.$$

Slow time variation of the field φ implies that the terms proportional to $d^2\varphi/dt^2$ and $(d\varphi/dt)^2$ are small. Neglecting them, one obtains the easily integrable equation

$$\left(3H + \frac{12H\alpha^2}{m_\chi^2}\varphi^2\right)\dot{\varphi} + V'_{ren}(\varphi) = 0. \tag{10.15}$$

In what follows, the nonrenormalized potential is taken in the form $V(\varphi) = \lambda_0\varphi^4$, and, therefore, $V_{ren} = \lambda\varphi^4$ where, $\lambda = \lambda_0 - \alpha^2/2$. Taking into account the usual relation $H = \sqrt{(8\pi V_{ren}(\varphi)/3)}/M_P$ between the Hubble constant and the potential, one can easily obtain the field variable φ as an implicit function of time:

$$t = \frac{\sqrt{3\pi/2}}{M_P\sqrt{\lambda}}\left[\ln(\varphi_0/\varphi) + \frac{2\alpha^2}{m_\chi^2}(\varphi_0^2 - \varphi^2)\right]. \tag{10.16}$$

Here, the first term reproduces the result of the standard inflation model. The second term results from the interaction of the inflaton field and the field χ. It follows from Eq. (10.15) that the second term dominates at

$$\varphi \geq \varphi_c \equiv \frac{m_\chi}{2\alpha}. \tag{10.17}$$

Therefore, there are two inflationary stages: the ordinary stage at $\varphi \leq \varphi_c$ and the ultraslow stage at $\varphi \geq \varphi_c$. Indeed, the field motion velocity obtained from (10.15) with allowance made for Eq. (10.17) is much smaller than its ordinary value $\dot{\varphi} = V'/3H$. The first inflation stage is completed when the condition (10.17)) ceases to be true. Then the ordinary inflation stage $\ddot{\varphi} \ll 3H\dot{\varphi}$ begins and continues as long as the condition is satisfied.

Since the second stage has been widely studied, we will analyze the first stage, for which the second term in Eqs. (10.15), (10.16) dominates, i.e. for $\varphi > \varphi_c$. In this case, the field depends on time as

$$\varphi(t) = \sqrt{\varphi_0^2 - t\frac{M_Pm_\chi^2}{\alpha^2\sqrt{6\pi}}}. \tag{10.18}$$

This expression is derived under the 'ultraslow roll-down' condition, which, according to Eq. (10.14), has a rather unusual form

$$\ddot{\varphi} << 12H\varphi^2\dot{\varphi}\alpha^2/m_\chi^2.$$

Let us determine the amplitude of quantum fluctuations arising at the first inflation stage for the potential $\lambda\varphi^4$. This can most easily be done by taking into account that the first term in Eqs. (10.11) and (10.15) is much smaller than the third and introducing a new effective field $\tilde{\varphi}$, after which the substitution

$$\tilde{\varphi} = (\alpha/m_\chi)\varphi^2$$

brings action to the form

$$S = \int dx\sqrt{-g}\left[\frac{1}{2}\partial_\mu\tilde{\varphi}\partial^\mu\tilde{\varphi} - \frac{1}{2}\tilde{m}^2\tilde{\varphi}^2\right], \qquad (10.19)$$

corresponding to a free massive field with mass $\tilde{m} \equiv m_\chi\sqrt{2\lambda}/\alpha$. The new field $\tilde{\varphi}$ does not represent a new physical field, but is only a suitable dynamical variable. Indeed, this variable is always positive, which is not true for real fields. An interaction with fermions, having the ordinary form in terms of inflaton, $\varphi\bar{\psi}\psi$, looks rather strange in terms of the field $\tilde{\varphi}$ - $\sqrt{\tilde{\varphi}}\bar{\psi}\psi$.

Nevertheless, this substitution is useful and valid at the inflation stage under consideration, when the field value is positive. The fluctuation amplitude for the massive noninteracting field is known to be $\Delta\tilde{\varphi} = \sqrt{3/(8\pi^2)}H^2/\tilde{m}$ [360], see also Chapter 3 of this book. On the scale of modern horizon, the constraint on the mass of quanta of this field is also known: $\tilde{m} \sim 10^{-6}M_P$, as is obtained from the comparison with the COBE measurements of the energy density fluctuations, $\delta\rho/\rho \approx 6 \cdot 10^{-5}$ [254]. Expressing $\tilde{m}$ in terms of the initial parameters, one obtains the following relation between them:

$$\frac{m_\chi}{M_P}\frac{\sqrt{\lambda}}{\alpha} \sim 10^{-6} \qquad (10.20)$$

Let us determine the field φ_U at which a causally connected area was formed, which generated the visible part of the Universe. The number of e-foldings necessary to explain the observed data is $N_U \approx 60$. Then, using the relation $N_U = \int_{\varphi_U}^{\varphi_{end}} Hdt$

$$N_U = \int_{\varphi_U}^{\varphi_c}\frac{H(\varphi)}{\dot{\varphi}}d\varphi + \int_{\varphi_c}^{\varphi_{end}}\frac{H(\varphi)}{\dot{\varphi}}d\varphi = \qquad (10.21)$$

$$= \frac{2\pi\alpha^2}{M_P^2 m_\chi^2}\left(\varphi_U^4 - \varphi_c^4\right) + \frac{\pi}{M_P^2}\left(\varphi_c^2 - \varphi_{end}^2\right).$$

Here we have taken into account that the time dependence of the field φ at the first and the second inflation stages are different. The second stage is completed at $\varphi = \varphi_{end}$. Assuming that the first term containing the initial value of the field φ_U dominates, one obtains the desired expression

$$\varphi_U \simeq \left(\frac{N_U}{2\pi}\right)^{1/4} \sqrt{\frac{M_P m_\chi}{\alpha}}. \tag{10.22}$$

Note that the visible part of the Universe in this case can be formed at $\varphi < M_P$, i.e. rather late. This is explained by the fact that at the first stage the field moves ultraslowly and the Universe had enough time to expand up to the suitable size.

Expression (10.22) differs substantially from the standard result $\varphi_U \sim M_P$, which is obtained for the inflaton field with potential $\lambda\varphi^4$ without regard for the interaction with the massive fields of other sorts $\varphi_U = \sqrt{N_U/\pi}\, M_P$.

The second term in Eq. (10.21) determines the number N_2 of e-foldings at the second inflation stage. Assuming that $\varphi_c^2 \gg \varphi_{end}^2$ and substituting the value φ_c from Eq. (10.17), one has

$$N_2 = \frac{\pi}{4}\left(\frac{m_\chi}{\alpha M_P}\right)^2. \tag{10.23}$$

Evidently, over a wide range of parameters α and m_χ, the second stage may be short or even absent.

The above arguments are valid if the field m_χ is massive enough so that it is placed at the minimum of its effective potential during inflation. As is known, the field rapidly rolls down to the minimum if the Hubble constant becomes smaller that the field mass, i.e. if $H < m_\chi$. The Hubble constant depends on the magnitude of the inflaton field and it is enough to estimate its value at the instant when the visible part of the Universe was originated ($\varphi = \varphi_U$). It corresponds to the largest scale in the modern Universe. Simple mathematics gives

$$m_\chi > H(\varphi_U) \quad \rightarrow \quad \frac{\sqrt{\lambda}}{\alpha} \leq \sqrt{\frac{3}{4N_U}} \sim 0.1. \tag{10.24}$$

This restriction indicates that one cannot fully avoid fine-tuning of the parameters. Indeed, as it was shown above, renormalized coupling constant $\lambda = \lambda_0 - \alpha^2/2$ and, according to constraint (10.24), $\alpha^2 \geq 100\lambda$. It means that two values λ_0 and α^2 of order $\geq 100\lambda$ each, must cancel each other to obtain small value of order λ. Nevertheless, this fitting is much more weak than that requiring the cancellation of all quantum corrections down to a value of $\sim 10^{-13}$ in the early inflationary models with the potential $\lambda\varphi^4$. Using Eqs. (10.20) and (10.24), one can easily obtain a rather weak limitation: $m_\chi \geq 10^{-5}M_P$ on the mass of the additional field χ.

Thus, a particular example was taken here to demonstrate the fact that massive fields, even being at their minimum (which in its turn depends on the magnitude of inflaton field), can materially decelerate the motion of the main inflaton field at the first inflation stage. Due to the first, ultraslow, stage, the visible part of the Universe could form at $\varphi < M_P$. The second stage precedes the completion of inflation and evolves in the ordinary way, but it is rather short. In particular, for the parameters $m_\chi = 10^{-3} M_P$ and $\lambda = 10^{-6}$, one has: the visible part of the Universe formed at $\varphi_U \approx 5 \cdot 10^{-2} M_P$; the first and second stages are separated at $\varphi_c \approx 5 \cdot 10^{-4} M_P$; and the second inflation stage is much shorter than the first.

The inclusion of the interaction between the inflaton field and more massive fields enables one to weaken essentially the constraints imposed on the potential parameters by the smallness of energy density fluctuations, although one fails to fully avoid the fine-tuning of the parameters. The effects considered here are associated with the renormalization of the kinetic term for the inflaton field interacting with an additional massive field. Because similar renormalization takes place for every sort of additional field [341], the inclusion of new fields will enhance the effect of deceleration of classical motion at high energies.

3. Suppression of first-order phase transitions by virtual particles in the early Universe

In the previous section we discussed the role of massive fields on inflation. It was shown that the interaction of inflaton with these fields could influence significantly the classical motion of the inflaton. On the other hand, first-order phase transitions are an important cosmological consequence of high energy physics. In the following we show [359] that the effects of virtual particles decrease significantly the probability of vacuum decay even at zero temperature. It could even lead to a permutation of the order of phase transitions in the early Universe.

Let us start with the double-well potential of the scalar field with nondegenerate vacua. Following the logic, discussed in Section 2 of this chapter, consider an auxiliary field χ with action (10.4). Phase transitions are investigated usually in Euclidean space [337, 298], which means the substitution $t \to i\tau$ in the formulae written above.

The calculations similar to those in the previous section lead to an effective Euclidean action for the scalar field φ

$$S_E = \int d^4x \left[\frac{1}{2}(\partial\varphi)^2 + V_{ren}(\varphi) \right] + \frac{\alpha^2}{2m_\chi^2} \int d^4x \left[\frac{\partial u(\varphi(x))}{\partial x_\mu} \right]^2. \quad (10.25)$$

(Here, and below, gravitational effects are omitted). The last term can be interpreted as an influence of the virtual χ-particles.

Let the field φ be placed initially in a metastable minimum of the potential V_{ren}. In this case the decay of the vacuum goes by nucleation and expanding of bubbles with true vacuum φ_T within. The outer space is filled with a metastable phase φ_F. This process is described by O(4)–invariant solution $\varphi_B(r)$ of the classical equation of motion in Euclidean space with boundary conditions $\varphi_B(0) = \varphi_T; \varphi_B(\infty) = \varphi_F$. The probability of the vacuum decay was obtained in [298] and has the form

$$\Gamma/V = \left(\frac{S_E(\varphi_B)}{2\pi}\right)^2 \left|\frac{Det'\hat{D}(\varphi_B)}{Det\hat{D}(\varphi_F)}\right|^{-1/2} e^{-S_E(\varphi_B)}, \tag{10.26}$$

where the kernel K of the operator $\hat{D}(\varphi)$ is

$$K(x,y) \equiv \frac{\delta^2 S_E(\varphi)}{\delta\varphi(x)\delta\varphi(y)}.$$

The determinant in the denominator is usually reduced to well-known determinant of operator $Det(-\partial^2 + Const)$, where $Const = \varphi_F$. In our case, however, effective action is nonlocal and evaluation of the determinant is not easy. To proceed, let us determine

$$\Omega^2 \equiv V''(\varphi_F) + \alpha^2 \left(\frac{du}{d\varphi_F}\right)^2 \tag{10.27}$$

$$M^4 \equiv \kappa^2 \left(\frac{du}{d\varphi_F}\right)^2.$$

and the problem consists of calculation of the determinant of operator $D(\varphi_F)$ with the kernel

$$K_F(x,y) = \left\{\delta(x-y)\left(-\partial_x^2 + \Omega^2\right) - M^4 G_E(x-y)\right\}. \tag{10.28}$$

This kernel is not a diagonal one, but it can be expressed using diagonal operators (see [361]). It is useful to determine the operator

$$\hat{G}_E^{-1} \equiv (-\partial_x^2 + m_\chi^2).$$

Then, some operator algebra simplify the expression

$$Det\hat{D}(\varphi_F) = DetG_E^{-1}\hat{D}(\varphi_F)/DetG_E^{-1} = \tag{10.29}$$
$$= Det\left[(-\partial_x^2 + \Omega^2)(-\partial_x^2 + m_\chi^2) - M^4\right]/DetG_E^{-1} =$$
$$= Det(-\partial_x^2 + \omega_1^2) \cdot Det(-\partial_x^2 + \omega_2^2)/DetG_E^{-1},$$

where parameters $\omega_{1,2}$ are

$$\omega_{1,2}^2 = \frac{1}{2}\left[\Omega^2 + m_\chi^2 - \sqrt{(\Omega^2 - m_\chi^2)^2 + 4M^4}\right]. \tag{10.30}$$

The problem of calculation of the determinant of the nonlocal operator in the predexponent factor of the expression (10.26) is reduced to the calculation of known determinants of oscillator-like operators.

Nevertheless, the main factor of the expression (10.26) is the effective action in the exponent. On the other side, as it is shown below, the value of effective action increases in orders of magnitude under the influence of virtual particles. It gives rise to a huge suppression of tunneling processes, much more than a 'naive' estimation gives.

Let us choose the potential in the form

$$V_{ren} = \frac{\lambda}{8} \left(\varphi^2 - a^2 \right)^2 + \frac{\varepsilon}{2a} (\varphi - a). \tag{10.31}$$

The instanton solution of the Euclidean equation of motion for the field φ may be parameterized in the following way

$$\varphi(x) = \varphi_B(r) = A \tanh \left(\frac{M}{2}(r - R) \right) - B, \tag{10.32}$$

where $r^2 \equiv \sum_{\alpha=1}^{4} x_{\alpha}^2$. Parameters R and M are to be determined by minimization of the action (10.25), while parameters A and B are chosen to satisfy boundary conditions

$$\varphi_B(r \to \infty) = \varphi_F;$$
$$\varphi_B(r \to 0) = \varphi_T. \tag{10.33}$$

Higher derivatives are neglected in the expression (10.25) of the action. This approximation is correct if $\partial_x \varphi_B / m_\chi \varphi_B \ll 1$ (m_χ is the mass of χ-particles which have formed the virtual cloud). On the other hand, the derivative of instanton trajectory $\partial_x \varphi_B$ is of the order $m_\varphi \varphi_B$, where m_φ is the mass of φ-particles. Numerical calculations indicate that the transition from false vacuum to true becomes noticeably wider due to the influence of virtual particles, i.e. $M \ll m_\varphi$. Thus, our approximation is valid at least if $m_\varphi / m_\chi \ll 1$. Numerical O(4)–symmetrical solution of the equation

$$\partial_r^2 \varphi + \frac{3}{r} \partial_r \varphi - V'(\varphi) + \tag{10.34}$$

$$+ \frac{\alpha^2}{m_\chi^2} u'(\varphi) \left[\frac{3}{r} u'(\varphi) \partial_r \varphi + u''(\varphi)(\partial_r \varphi)^2 + u'(\varphi)\partial_r^2 \varphi \right] = 0$$

is supposed to have the form (10.32). The results of calculations are represented in Figure 10.4. The standard result, when the effect of virtual particles is not taken into account, corresponds to the curve at $\zeta = \kappa = 0$. It is clearly seen that the effective action increases in orders of magnitude and vacuum decay is exponentially suppressed as compared with the well-known result.

As result, we found that virtual particles at high energies could significantly influence the classical motion and vacuum decay. It is shown that in a wide range of parameters the first stage of inflation consists of 'superslow' motion of inflaton field. One of the useful effect of virtual particles is the considerable weakening of conditions on model parameters which is given by observational data. Tunneling processes could be strongly forbidden compared also with textbook results.

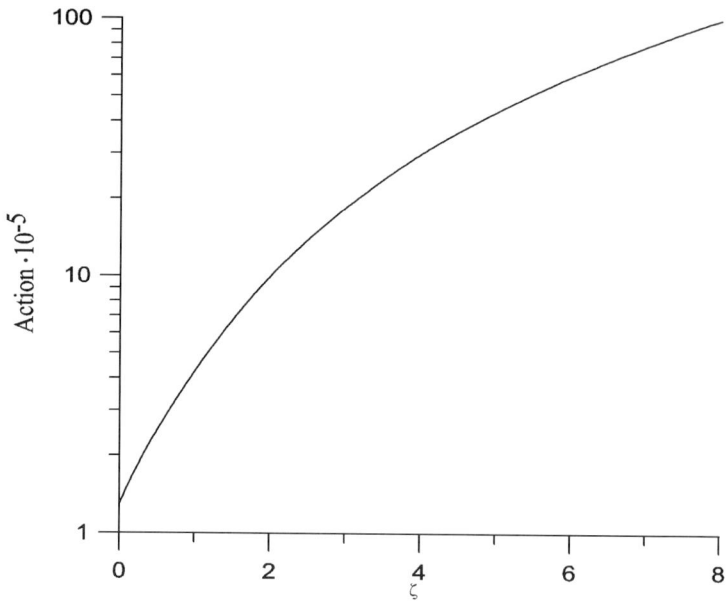

Figure 10.4. Dependence of effective action on parameter $\zeta = 2\kappa^2/m_\chi^4$. The parameters of potential (10.31) have the values: $a = 1, \lambda = 0.1, \varepsilon = 0.01$ in units $m_\chi = 1$.

4. Inflation from slow evaporation of Bose condensate

Here we consider, following [311, 113], the process of decay of symmetric vacuum state as evaporation of a Bose–Einstein condensate of physical Higgs particles, defined over asymmetric vacuum state. Energy density of their self-interaction is identified with cosmological constant Λ in the Einstein equation. Λ decay then provides dynamical realization of spontaneous symmetry breaking. The effective mechanism is found for damping of coherent oscillations of a scalar field, leading to slow evaporation regime as the effective mechanism for Λ decay responsible for inflation without special fine-tuning of the microphysical parameters. This mechanism is able to incorporate reheating, generation of proper primordial fluctuations, and nonzero cosmological constant today.

4.1 Introduction

Anything which contributes to the stress–energy tensor as $T_{\mu\nu} = \rho_{vac}g_{\mu\nu}$, behaves like cosmological term $\Lambda g_{\mu\nu}$ in the Einstein equation. Developments in particle physics and quantum field theory, as well as confrontation of models with observations in cosmology, compellingly indicate that the cosmological constant Λ ought to be treated as a dynamical quantity (for recent review see [123]). At the very early stage of the cosmological evolution a huge value of cosmological constant is needed corresponding to $\rho_{vac} \geq \rho_{GUT}$, to drive inflation as providing the reason for the expansion of the Universe and its iso-tropy and large-scale homogeneity [48, 49, 52, 110, 35, 121]. Now cosmolo-gical constant is estimated from the variety of observational data at the level $\rho_{vac} \sim (0.6 - 0.8)\rho_{today}$ [97, 122, 364].

Several mechanisms have been proposed involving and supporting a negat-ive vacuum energy density growing with time to cancel initial pre-existing positive cosmological constant Λ. All these mechanisms utilize the basic property of de Sitter space–time – its quantum and semiclassical instabilities [51, 120, 115, 108, 109, 119, 117].

In scalar field dynamics the potential of a scalar field $V(\varphi)$ plays the role of an effective cosmological constant in regimes when its derivatives are close to zero. Starting from such a regime as an initial state, field equations typically lead to successive coherent field oscillations [36, 121]. In the context of infla-tionary models with effective Λ related to inflaton field in slow rolling regime, the further decay of coherent oscillations involves inflaton interactions with other fields in models of preheating [348, 240, 118, 127, 105, 129].

On the other hand, we can treat the classical scalar field as the Bose–Einstein condensate of physical quanta of this field defined over its ground state (true vacuum). In the papers [111, 106] it was shown that process of emergence of massive scalar particles in (from) de Sitter vacuum looks like evaporation of a Bose condensate. In the paper [112] the model of self-consistent inflation was proposed in which the same self-interacting scalar field is responsible for both initial value of Λ and its further decay.

In the case of the Higgs field this approach involves both space–time and particle internal symmetries. In gauge theories mechanisms of spontaneous symmetry breaking imply that unbroken symmetry state is false vacuum state. At the same time this is the highly symmetric state of space–time geometry invariant under de Sitter group. In Ref. [112] this state was interpreted as Bose–Einstein condensate of physical Higgs particles whose self-interaction energy density corresponds to a scalar field potential in the state with the un-broken symmetry. The process of decay of inflationary vacuum appears then as slow evaporation of a Bose condensate responsible for inflation and further transition to the standard FRW model without special fine-tuning in initial con-ditions for inflation. Dynamics of the cosmological term is directly related to the hierarchy of particle symmetries breaking, which makes such an approach physically self-consistent [112].

In the present section, following [311, 113], we identify cosmological con-stant Λ with the energy density of self-interaction of scalar bosons bound in the condensate. We show the existence of the slow evaporation mechanism which seems to be generic for dynamics of Bose condensates. This mechan-ism for Λ decay produces an effective kinetic damping involving decoherence of coherent scalar field oscillations due to self-interaction and back reaction of decay products.

4.2   The condensate of Higgs bosons

Let us start with the simplest example of the Higgs field φ

$$V(\varphi) = \frac{\lambda}{4}(\varphi^2 - \varphi_0^2)^2. \tag{10.35}$$

In the context of particle theories, constant part of the Higgs potential is usually omitted [4]. However, as we saw in the preceding chapters, in the cosmological context this term becomes important. In theories with spontaneous symmetry breaking the vacuum expectation value of the Higgs field, which couples to bosons and fermions to give them masses, plays the role of an order parameter. The nonzero vacuum expectation value in the asymmetric vacuum state $<\varphi> = \varphi_0$, is interpreted as the development of a condensate of φ particles, leading to the spontaneous symmetry breaking [36, 4].

On the other hand, φ particles are not physical particles, they are tachionic, with the imaginary mass, that reflects the instability of the symmetric state of the theory. In the asymmetric, physically stable vacuum state (the state with the broken symmetry) physical particles are χ particles, related to φ particles by $\varphi = \chi + \varphi_0$. It is χ particles which acquire the mass by the Higgs mechanism, and whose vacuum state is the true vacuum with zero potential and with zero expectation value, $<\chi> = 0$. Therefore, in terms of particles χ, the true vacuum of theories with spontaneous symmetry breaking, cannot be a condensate, and we would rather have to treat the symmetric vacuum state of a theory as a condensate of χ particles in bound state. Replacing φ condensate by χ condensate sheds some light on the origin of Λ, which in φ condensate picture, where an effective Λ is related to the state of zero field [52], looks mysterious. As we shall see below, in χ condensate picture effective Λ is related to nonzero value of the field χ as its energy density in symmetric state.

The Higgs field χ in this simple model is described by the Lagrangian [112]

$$L = \sqrt{-g}\left[\frac{1}{2}g^{\mu\nu}\chi_{;\mu}\chi_{;\nu} - \frac{\lambda}{4}(2\varphi_0\chi + \chi^2)^2\right], \tag{10.36}$$

where we dropped, for simplicity, the indices of internal variables. The potential takes the form

$$V(\chi) = \lambda\varphi_0^2\chi^2 + \lambda\varphi_0\chi^3 + \frac{\lambda}{4}\chi^4. \tag{10.37}$$

In the state with unbroken symmetry $<\chi> = -\varphi_0$, $<\chi^2> = \varphi_0^2$, $<\chi^3> = -\varphi_0^3$, and the energy density of a condensate of χ particles is given by

$$<V(\chi)> = \frac{\lambda}{4}\varphi_0^4. \tag{10.38}$$

We see that the constant term, playing the role of the cosmological constant Λ in the Einstein equation, is identified with the energy density of condensate of χ particles being precisely the energy density of their self-interaction.

The potential (10.37) describes physical particles χ with masses of Higgs bosons $m = \sqrt{2\lambda}\varphi_0$, self-interacting and interacting with condensate.

The term $\lambda\varphi_0\chi^3$ (if re-written as $\lambda < \chi(t) > \chi^3$ with account of the time variation of $< \chi(t) >$ that we discuss below) corresponds to decay of a condensate $< \chi(t) >$ via the channel $(< \chi(t) > \rightarrow 3\chi$, and describes evaporation of Higgs bosons from condensate of χ particles. Note that in the considered picture the initial state of condensate $< \chi(0) >= -\varphi_0$ corresponds to the state of local maximum with $\varphi = 0$ in the terms of usually considered φ particles. This state possesses the energy density (10.38), which formally makes its decay energetically possible. However, if $< \chi >= -\varphi_0$, i.e. remains constant, no evaporation is possible. Only the existence of fluctuations, causing the time dependence of $< \chi(t) >$, can provide the condensate evaporation (as we will see in the next subsection).

The term $\lambda\chi^4/4$ reproduces the runaway particle production discovered by Myhrvold [120]. The difference from the Myhrvold result is in the origin of Λ. Gravity-mediated decay of Λ in Myhrvold approach is due to particle creation by gravitational field generated by pre-existing Λ not related to created particles. In the approach [311] $\Lambda \sim m^2\varphi_0^2$ is the energy density of self-interaction of the same particles bound within a condensate.

The Hubble parameter H during the Λ dominated stage is $H \sim \sqrt{\Lambda}$. In the context of particle theories with $\varphi_0 \ll M_P$, the case [311] corresponds to creation of bosons with $m \gg H$. Therefore, the mechanism [311] differs from that proposed by Mottola who studied creation of particles by gravitational field via the Hawking quantum evaporation which leads to the exponential suppression of masses $m \gg H$ [119]. In the Mottola mechanism light scalar particles with $m < H$ are evaporated from de Sitter horizon induced by pre-existing Λ. In the mechanism [311] Higgs bosons with $m \gg H$ are evaporated from the bound state within a condensate into the free states.

The approach [311] differs also from the Parker and Zhang theory of relativistic charged condensate as a source of slow-rolling inflation [124, 125]. In the aproach [311] condensate of χ particles is essentially globally neutral, since it corresponds to the state with unbroken symmetry – totally symmetric state in both space–time and internal degrees of freedom [112]. The condensate decays by evaporation, as well as by runaway production of χ particles which corresponds to conversion of energy of initial globally neutral state into thermal energy of χ particles.

4.3 Decaying Λ-term

Roughly, scenario [311] of Λ decay looks as follows. Within a χ condensate χ particles have four-momenta $k \rightarrow 0$. It corresponds to the trivial solution $\chi = -\varphi_0$ of the Klein–Gordon equation that for the potential (10.37) in the condensate regime ($V'(\chi) = 0$) reduces to $\ddot{\chi}+3H\dot{\chi}+\Gamma\dot{\chi} = 0$, where Γ is a decay rate. Its solution reads $\chi = \chi_0+\dot{\chi}_0 e^{-(3H+\Gamma)t}/(3H+\Gamma)$ and gives $\chi = -\varphi_0$

Let us show, following [311, 113], that decoherence of χ particle states and back reaction of their relativistic decay products lead to the effective damping of fluctuations. The kinetic equations can be written in the standard way (see, e.g., [36]). The kinetic equation describing the growth and decay of fluctuations is

$$\frac{dn_\chi}{dt} = mn_\chi - \Gamma n_\chi - n_r n_\chi \sigma - 3Hn_\chi, \tag{10.45}$$

where σ is the cross-section of the interaction of χ particles with relativistic products of their decay. In the units $\hbar = c = 1$ the reaction rate in the kinetic equations coincides with σ. The first term in the right-hand side describes creation of χ particles, the second – their decay, the third – their interaction with products of decay, and the fourth – their redshifting.

The kinetic equation for products of decay, effectively relativistic matter with the equation of state $p = \varepsilon/3$, reads

$$\frac{dn_r}{dt} = -3Hn_r + n_r n_\chi \sigma + \Gamma n_\chi. \tag{10.46}$$

In the equilibrium

$$mn_\chi - \Gamma n_\chi - n_r n_\chi \sigma - 3Hn_\chi = 0,$$

$$-3Hn_r + n_r n_\chi \sigma + \Gamma n_\chi = 0. \tag{10.47}$$

Decay of χ particles into light species implies $m \gg \Gamma$, which corresponds to applicability of perturbation theory for calculations of decay, and is valid in models with coupling less than the unity. Then, taking into account Eq. (10.41), we get

$$\rho_r = \frac{m^2}{\sigma}; \qquad \rho_\chi = \frac{3mH}{\sigma}. \tag{10.48}$$

Equilibrium density of relativistic particles ρ_r is achieved when the density of evaporated and decayed χ particles

$$\rho_{\chi d} \sim \Gamma \frac{1}{m} \rho_{vac} = \Gamma \frac{1}{m} \frac{m^2 \varphi_0^2}{8}$$

satisfies the condition

$$\Gamma \frac{1}{m} \frac{m^2 \varphi_0^2}{8} > \frac{m^2}{\sigma}.$$

This gives the lower limit on the characteristic width of χ particles decay

$$\Gamma > \frac{8m}{\varphi_0^2 \sigma}. \tag{10.49}$$

If this condition is satisfied, the potential evolves not to the value $V(\delta) = 0$, but owing to the effective decay of physical χ particles and their destruction by the decay products, the χ field oscillation is damped and the equilibrium moves the potential to the value $V_{max} - \rho_\chi - \rho_r$, becoming successively more flat. Slow evaporation of χ condensate acts in such a way to flatten the potential near its symmetric state.

We see that back reaction of evaporated particles and products of their decay produces an effective damping of scalar field oscillations which leads to effective flattening of initially nonflat potential and provides a mechanism responsible for inflation without fine-tuning of the potential parameters. This is qualitatively similar to effective flattening found around spinoidal line in the Hartree-truncated theory of spinoidal inflation [107] (which involves fine-tuning at the slow-rolling stage preceding the spinoidal regime).

4.5 Regime of slow evaporation

The process of Λ decay is governed by the equation (2.44) of Chapter 2 which accounting for the equilibrium condition (10.48) has the form [113]

$$\frac{d\rho_{vac}}{dt} = -3H(\rho_\chi + \rho_r + \frac{1}{3}\rho_r) = -4H\frac{m^2}{\sigma}. \tag{10.50}$$

We can estimate the characteristic time of decay for two limiting cases of minimal and maximal cross-section σ. The lower limit on cross-section σ is $\sigma = 4\pi/m^2$ which is the hard ball approximation cross-section for scattering of particles of masses $m/2$. In this case [113]

$$\frac{d\rho_{vac}}{dt} = -\frac{H}{\pi}m^4. \tag{10.51}$$

Taking into account that at the stage of Λ dominance $H \propto \sqrt{\rho_{vac}}$ the law for Λ decay follows from the solution of this equation

$$\rho_{vac} = \rho_0\left(1 - \frac{t}{\tau}\right)^2; \quad \tau = \sqrt{\frac{3\pi}{2}\frac{M_P\sqrt{\rho_0}}{m^4}} = \sqrt{\frac{3\pi}{32\lambda}\frac{M_P}{m^2}}. \tag{10.52}$$

The solution (10.52) (as well as the Eq. (10.51) itself) is obtained under the condition of Λ dominance and thus is valid only at $t \leq \tau$. It gives the e-folding number

$$H\tau = \frac{\pi}{8}\frac{1}{\lambda} \tag{10.53}$$

and sufficient inflation for reasonable values of coupling $\lambda < 6 \cdot 10^{-3}$. The characteristic time for reheating is $\tau \sim \lambda H^{-1}$ and the reheating temperature $T_{RH} \sim \lambda^{1/4}m$.

The upper bound for σ is given by $\sigma = \pi/H^2$. In this case

$$\frac{d\rho_{vac}}{dt} = -\frac{4}{\pi}H^3m^2; \quad \rho_{vac} = \frac{\rho_0}{(1 + t/\tau)^2}, \tag{10.54}$$

where

$$\tau = \left(\frac{3^3}{2^{11}\pi}\right)^{1/2}\frac{1}{\lambda^{3/2}}\left(\frac{M_P}{\varphi_0}\right)^4\tau_{Pl}. \tag{10.55}$$

The *e*-folding number is then

$$H\tau = \frac{3}{32}\frac{1}{\lambda}\left(\frac{M_P}{\varphi_0}\right)^2 \tag{10.56}$$

and, for the considered case $\varphi_0 \ll M_P$, inflation is sufficient for any λ. Reheating temperature is $T_{RH} \sim \lambda^{1/4}H$.

More detailed investigation of dynamics of a vacuum decay needs a particular model for calculating σ, but the results will be within the range between the cases of minimal and maximal σ. For example, the picture of evaporation investigated in Ref. [111] corresponds to the case of evaporation of Higgs bosons and their reheating to the Hawking temperature $T_{RH} \sim H$ [116]. The rate of Λ decay is given in this case by

$$\frac{d\rho_{vac}}{dt} = -3Hm(mH)^{3/2}. \tag{10.57}$$

4.6   Discussion

Here we considered Λ decay in the case of vacuum dominance. When radiation density starts to exceed vacuum density at the last stage of evaporation we would have to change the equation (10.50), taking into account the evolution of Hubble parameter as well as of matter and radiation density, which in the standard FRW cosmology evolves as $\propto t^{-2}$. The Eq. (10.54) reproduces this behavior starting at the stage of vacuum dominance for the case of maximum possible cross-section σ. Provided that this behavior remains dominant at successive stages, this corresponds to the existence of remnant evaporating condensate today with density comparable to the total density in the Universe, which seems to agree with results of recent analysis of observational data [97, 364].

The generalization of this approach to the case of arbitrary scalar field potential is straightforward. Any cosmologically reasonable potential must satisfy the condition $V(\varphi^2 - \varphi_0^2) > 0$. True vacuum state $< \varphi >= \varphi_0$ is determined as the minimum of the potential $V = 0$. The physical particles $\chi = \varphi - \varphi_0$ are defined over the true vacuum state. Their mass is given, as usual, by $\partial^2 V/\partial\varphi^2$. Any state with $V(\chi) > 0$ we can treat as bound state of χ particles trapped inside a Bose condensate. The equilibrium fluctuations' density corresponds to deviation of the potential from its initial value at the given point by the quantity $\rho_\chi + \rho_r$ (see Eq. (10.48)). It means that in a characteristic time m^{-1} the field is not completely moved to its ground state, but, instead, is stabilized near its initial value having slightly changed by the magnitude $\rho_\varphi + \rho_r \ll V(\chi)$.

We see that the decoherence of χ particles and the back reaction of their decay products leads to effective freezing of the field near its initial value. Near this value the potential becomes locally flat, and the energy density of condensate of χ particles starts to play a role of an effective cosmological constant. It realizes the case of chaotic inflation for initially nonflat potential in the case $m \gg H$ and $\varphi_0 \ll M_P$.

At the first sight, the appearance of slow evaporation regime in the approach [311, 113] seems to lead to the same spectrum of initial density fluctuations as in slow-rolling models. However, the origin of fluctuations is different. In the considered case fluctuations are generated by statistical distribution of evaporated particles, while in the typical slow-rolling picture they originate from nonsimultaneous transitions to the ground state.

Formally the mechanism presented in this section is based on rather trivial solution ($\varphi \simeq const$) of scalar field dynamics, which however appears to have nontrivial consequences leading to kinetic equilibrium regime for slow evaporation of Bose–Einstein condensate. This kind of solution has analogies in experimentally studied Bose–Einstein condensation in atomic physics [126].

The case of Higgs field considered here is the simple illustration of the proposed mechanism of Λ decay, which seems to be more generic and to work also in non-Abelian gauge models without Higgs mechanism, in which symmetry breaking is induced by nonlinearity of gauge interactions as in technicolor models.

Let us summarize. The kinetics of the Bose condensate evaporation can effectively damp the coherent field oscillations leading to slow evaporation regime for a wide range of possible particle interaction parameters. In the cosmological context this provides the effective mechanism for Λ decay responsible for dynamics of symmetry breaking, which can incorporate inflation, reheating, as well as nonzero cosmological constant today.

If we take $\varphi_0 \sim 10^{15} GeV$ and $\lambda \sim 10^{-2}$ the predicted e-folding ranges from 40 to 10^9, depend on the possible particle interaction parameters. For the same parameter range the condition of equilibrium (10.49), being necessary for the realization of the considered mechanism [311, 113], is valid, if the χ particle width Γ exceeds the lower limit, ranging from $\sim \lambda m$ to $\sim \lambda(\varphi_0/M_P)^2 m$, which can be naturally satisfied in realistic particle models.

The reheating temperature, ranging from $\sim \lambda^{3/4}(\varphi_0/M_P)\varphi_0$ to $\sim \lambda^{3/4}\varphi_0$ is determined in the considered model by the parameters φ_0 and λ. For the values of these parameters, taken above, it is rather high ($3 \cdot 10^9$–$3 \cdot 10^{13} GeV$), exceeding by few orders of magnitude the strict upper limit, following from the analysis [286] of ^6Li overproduction by decaying primordial gravitino. This problem may be resolved, taking into account the model dependence of the above estimation as well as the dependence of the constraint [286] on the model of gravitino.

The more general property of the considered model is that the nearly flat spectrum of density fluctuations, generated at large e-foldings, transforms into ultraviolet spectrum when generated in the end of inflation [311, 113]. Such ultraviolet behavior can escape the constraints of WMAP [74], being predicted at small scales, and appeals to the sensitive probes of the small-scale primordial inhomogeneity, such as the analysis of mini-PBH formation (see review in [71, 3]) for its test.

5.   Non-canonical kinetic term

In this section we investigate dynamics of scalar field in the framework of a non-canonical kinetic term and the simplest form of potential. It was revealed in [384] that behavior of the field in the vicinity of singular points of the kinetic term possesses unusual properties. In particular, the singular points could serve as attractor for classical solutions so that a stationary value of the field may occur distant from a minimum of the potential. We also discuss and estimate the probability for formation of some specific types of effective potentials.

As we have seen, scalar fields play an essential role in the modern cosmology. The modern scenario of the origin of our universe is based on the inflationary paradigm and a vast majority of inflationary models use the dynamics of scalar field(s). New observational data on the nonzero density of dark energy could also be explained by the existence of scalar field. Another possible explanation, models of quintessence, are based on the potentials which tends to zero when the scalar field tends to infinity. The defect of all these applications of scalar fields consists in the unnatural form of their potentials. For example, they have to be extremely flat to be applicable to the standard inflationary scenario [57]. Usually, such scalar field interacts very weakly with other fields. The exception could be its interaction with electromagnetic field that was proposed recently as the explanation for observational indications [365] on possible time variation of the fine structure constant (see, e.g., [366] and refs. therein).

So, one of the main problems is to validate those forms of potential that can provide a satisfactory fit to the observational data. It is usually achieved by a reference to more general theories, such as M-theory, supergravity, multidimensional gravity, and so on. Here we show the natural way to produce a bunch of effective potentials of scalar field. It is achieved by an interference of the simplest form of original potential and non-canonical kinetic terms. In this connection it is worthy mention that a non-canonical form of the kinetic term is applied more and more widely in theoretical research. As an example, a small value of the cosmological constant, consistent with the recent experimental data [364] can be explained using a non-canonical form of the kinetic term in the scalar field Lagrangian (like in the model of quintessence [367, 368]). A non-trivial kinetic term could be responsible for a new coupling between adiabatic and entropy perturbations [369] as well as for the existence of the dark matter caused by phantom fields [370].

Here, we consider the action with the non-trivial kinetic term of the following form

$$S = \int d^4x \sqrt{-g} \left[\frac{R}{16\pi G} + \frac{1}{2} K(\varphi) \partial_\mu \varphi \partial^\mu \varphi - V(\varphi) \right] \qquad (10.58)$$

and concentrate on the influence of singular points of the kinetic term $K(\varphi)$ in the form

$$K(\varphi) = M^n/(\varphi - \varphi_s)^n. \tag{10.59}$$

on the scalar field dynamics. The value of a parameter M will be discussed below. This form is correct at least in the vicinity of the singularity if $n > 0$, or zero if $n < 0$. The singular points are not an uncommon exclusion. The well-known Brans–Dicke model [372] does contain a singularity at zero value of the φ. If a multiplier of the Ricci scalar equals to zero at some point in the Jordan frame, the kinetic term will be singular in the Einstein frame (see, e.g., [356]) and vice versa. Models of quintessence with negative power law [373, 374, 375, 376, 377] are another well-known example for the potential of such a kind.

An important remark should be made now, though this point will be intensively discussed below. A well-known hint is the suitable change of variable reducing the function K to $K(\varphi) = \pm 1$. It could be done during some inflationary period [359], when the scalar field has definite sign, but not at recent epoch when the field fluctuates around a singular point. This problem is discussed also in Refs. [385], [386], [387].

To proceed, the equation of motion for a uniform field distribution has the form

$$K(\varphi)[\ddot{\varphi} + 3H\dot{\varphi}] + \frac{1}{2}K(\varphi)'\dot{\varphi}^2 + V(\varphi)' = 0 \tag{10.60}$$

in the Friedmann–Robertson Walker Universe where H denotes the Hubble parameter.

Keeping in mind the expression (10.59) we have

$$\ddot{\varphi} + 3H\dot{\varphi} - \frac{n}{2(\varphi - \varphi_s)}\dot{\varphi}^2 + V(\varphi_s)'(\varphi - \varphi_s)^n/M^n = 0. \tag{10.61}$$

Evidently the field value φ_s is a stationary solution for any smooth potential V and $n > 0$ provided that $\dot{\varphi} = o(\varphi - \varphi_s)$. The last condition is not very restrictive.

The cosmological vacuum energy density is connected usually with one of potential minima. In the case considered here it is not like this – the vacuum state is connected not with a minimum of potential but with a singular point of the kinetic term $K(\varphi)$. To prove this statement, we limit ourselves by the simplest form of potential

$$V(\varphi) = V_0 + m^2\varphi^2/2. \tag{10.62}$$

In the following we will consider only this class of model characterized by the set of parameters m, V_0, M. The stationary state φ_s is chosen in such a manner to fit the cosmological Λ-term, which is widely considered as an important ingredient of modern cosmology (see review [75])

$$V_0 + m^2\varphi_s^2/2 = V(\varphi_s) = \Lambda \tag{10.63}$$

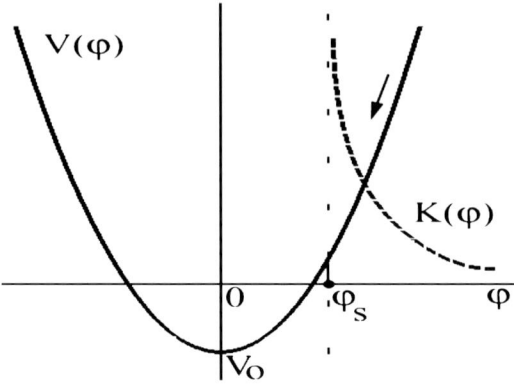

Figure 10.5. Disposition of the potential minima and the singular point of kinetic term for $n = 1$.

The modern vacuum energy density Λ is small compared with any scale at the inflationary stage, what permits us to use the connection

$$\varphi_s \cong \sqrt{2\,|V_0|}/m. \qquad (10.64)$$

To facilitate more detailed analysis, new auxiliary variable χ is usually introduced. We suggest the substitution of variables $\varphi \rightarrow \chi$ in the following manner

$$d\chi = \pm\sqrt{K(\varphi)}d\varphi, \quad K(\varphi) > 0. \qquad (10.65)$$

The action in terms of the auxiliary field χ has the form

$$S = \int d^4x \sqrt{-g}\{\frac{R}{16\pi G} + sgn(\chi)\frac{1}{2}\partial_\mu\chi\partial^\mu\chi - U(\chi)\},$$

where the potential $U(\chi) \equiv V(\varphi(\chi))$ is a 'partly smooth' function. Its form depends on the form of initial potential $V(\varphi)$, the form of kinetic term and a position of the singularities at $\varphi = \varphi_s$. Now let us consider some particular cases of $K(\varphi)$.

5.1   Production of effective potentials

The case $n = 1$

The kinetic term (10.59) has the form $K(\varphi) = M/(\varphi - \varphi_s)$ in this case. The initial situation is presented in Figure 10.5. It will be shown that the singular point should be disposed in the close vicinity of a zero point of potential to supply small energy density $V(\varphi_s)$ of the vacuum state.

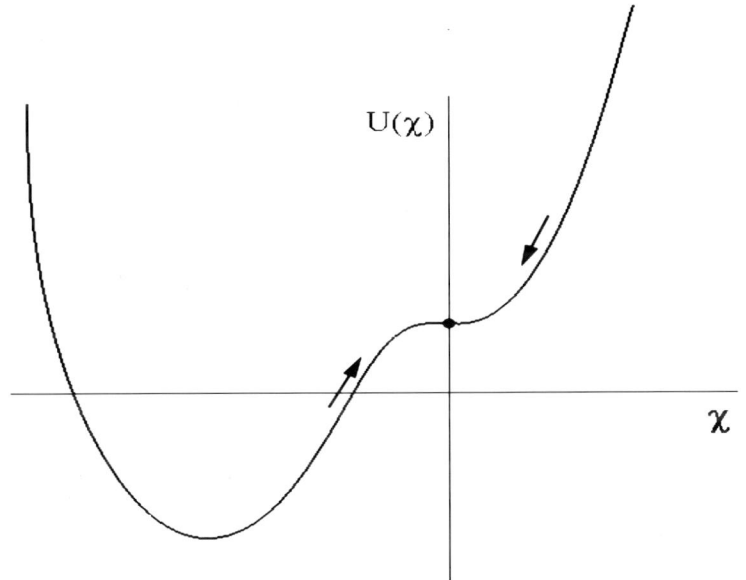

Figure 10.6. The form of the potential for the case $n = 1$. Chosen branches are: $\varphi > \varphi_s \leftrightarrow \chi > 0; \varphi < \varphi_s \leftrightarrow \chi < 0$. In the latter case the auxiliary field χ behaves like the phantom field moving classically to the local extremum at the point $\chi = 0$.

Eq.(10.65) connects the physical field φ with the auxiliary field χ

$$\varphi = \varphi_s + sgn(\chi)\chi^2/4M.$$

Throughout this section we assume that the singularity is placed in the point $\chi = 0$ and choose the one-to-one correspondence between the physical variable φ and auxiliary variable χ in the intervals:

$$\varphi < \varphi_s \rightarrow \chi < 0; \quad \varphi > \varphi_s \rightarrow \chi > 0 \qquad (10.66)$$

using some freedom in the definition of field χ.

The potential of the field χ acquires the form

$$U(\chi) \equiv V(\varphi(\chi)) = V_0 + \frac{1}{2}m^2(\varphi_s + sgn(\chi)\frac{\chi^2}{4M})^2 \quad for \quad \varphi_s > 0; |\chi| < \infty.$$
$$(10.67)$$

If the auxiliary field $\chi > 0$, it finally approaches to the singular point $\chi = 0$ (see Figure 10.6). If the field $\chi > 0$, then the auxiliary field behaves like the phantom field [379, 380], which climbs to the top of a potential and hence also tends to the singular point. For the physical field φ, the motion looks like oscillations around the point φ_s. Finally, it is captured in the vicinity of the singular point $\chi = 0(\varphi = \varphi_s)$, One can conclude that this point is the stationary one and the vacuum energy density equals to $V(\varphi_s)$, see Eq. (10.63) rather than to V_0 as could be expected from the expression (10.62).

The values of parameters can be estimated if we interpret the auxiliary field as the inflaton which is in addition responsible for the dark energy. To provide inflation, slow-roll condition [57] should take place. This condition is satisfied at the values of parameters

$$M \sim M_P; \quad |V_0| \sim M_P^4; \quad m \sim 10^{-12} M_P. \tag{10.68}$$

The small value of the parameter $'m'$ is necessary to fit the data on the temperature fluctuations. Note that it is a rather general case for inflationary models, in which the consistency with the observed smallness of temperature fluctuations implies some small parameters.

After the end of inflation the field moves classically to zero, increasing its velocity. If the latter is large enough, the field could overcome the potential minimum and move infinitely to the left as any phantom field. To prevent that, the potential minimum must be deep enough,

$$|V_0| \gtrsim U(\chi_{end}) \tag{10.69}$$

which gives additional constraint on the parameter V_0. Namely, equations (10.67), (10.69) lead to the inequality $|V_0| \gtrsim \left(\frac{m}{M}\right)^2 M_P^4$ at the field values $\chi \geq \chi_{end} \sim M_P$ which is ordinary for the inflationary stage. This estimation does not contradict the value (10.68) supposed above.

The problem of smallness of the vacuum energy density, $\Lambda = 10^{-123} M_P^4$ remains topical though the situation is changed. As was mentioned above, the smallness of the vacuum energy density is usually connected with the smallness of a potential minima that leads to intensive searches of physical reasons for such a smallness. In the considered case the modern energy density is determined by the singular point φ_s of non-canonical kinetic term, see Eq. (10.63). The smallness of Λ could take place if the singular point φ_s is placed very close to zero of the potential. The suitable interval

$$0 \leq \varphi_s \leq \Delta\varphi_s \equiv \sqrt{-2V_0/m^2 + 2\Lambda/m^2} - \sqrt{-2V_0/m^2} \simeq \frac{\Lambda}{m\sqrt{2|V_0|}} \tag{10.70}$$

is extremely tiny to be easily explained. It seems almost evident that such coincidence is absolutely occasional, and its probability is very small. The next section is devoted to discussion on this subject. We will show that a probabilistic language helps to obtain an intrinsically consistent picture.

The case $n = 2$

The kinetic term has the form $K(\varphi) \equiv K_s(\varphi) = M^2/(\varphi - \varphi_s)^2$, see Figure 10.7. It was considered in [378] in connection with the model of quintessence. We will show, however, that the latter arises naturally, without *ab*

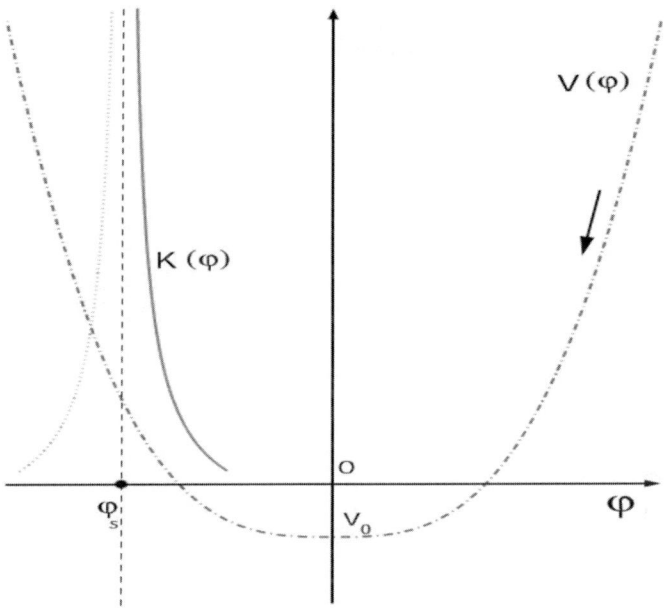

Figure 10.7. Disposition of kinetic term and potential for the case $n = 2$. Arrow indicates the field motion during the inflation.

initio introducing a potential of an exponential form. The new auxiliary field is connected with the physical one by

$$\chi = M \ln \left| \frac{\varphi - \varphi_s}{\varphi_s} \right|,$$

so that the potential has the form

$$U(\chi) = \frac{1}{2} m^2 \varphi_s^2 \left[1 + sgn(\varphi_s) \cdot sgn(\chi) \cdot e^{\chi/M} \right]^2 + V_0. \qquad (10.71)$$

If $\varphi_s > 0$ and $\varphi > \varphi_s$, the potential mimics the model of quintessence with nonzero vacuum energy density $\Lambda = \frac{1}{2} m^2 \varphi_s^2 + V_0$.

The case $\varphi_s < 0$ and $\varphi > \varphi_s$ is much more interesting. The potential (10.71) is highly asymmetric, see Figure 10.8, so that the behavior of inflaton is rather different at $\chi < 0$ and at $\chi > 0$. Let us suppose that the inflation starts with $\chi_{in} > 0$. It corresponds to point 1 on Figure 10.8. The picture is similar to improved quintessence potential [389], without the problem of deviation from radiation-dominated stage during Big Bang nucleosynthesis which could result in modern chemical content. The chosen values of parameters

$$M \sim M_P, m \sim M_P, |V_0| \sim 10^{-14} M_P^4 \qquad (10.72)$$

permit the suitable inflationary stage and are in agreement with observations

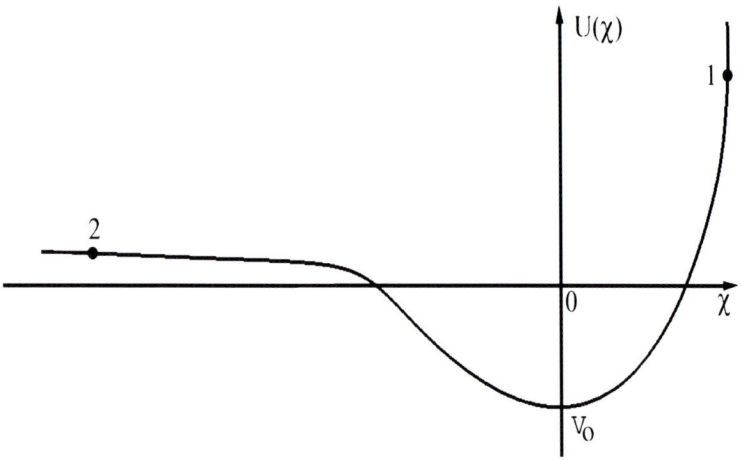

Figure 10.8. Effective potential for the case $n = 2$ mimics the model of quintessence with nonzero vacuum energy density.

A modern epoch is characterized by a large negative value of the field χ. It slowly varies along the flat part of the potential with exponentially slow variation of the vacuum energy density around the value (10.63). The character of future variation will determine the fate of our Universe. There are two possibilities. The field could move all the time to the left slowly decreasing its kinetic energy. The second possibility is that the field turns back at some instant, and after a period of oscillations is rested at the bottom of the potential.

To perform calculations we write the equation of motion of auxiliary field in the form

$$\ddot{\chi} + 3H\dot{\chi} + U'(\chi) = 0.$$

Hubble parameter H depends on the energy density of matter and radiation and varies with time, but for our estimations it will be enough to use the average value that is approximately equal to $H^2 = 10^{-123} M_P^2$. Potential (10.71) has the form

$$U(\chi) \cong \frac{1}{2} m^2 \varphi_s^2 \cdot [(1 - e^{\chi/M})^2 - 1] + \Lambda.$$

Here $\Lambda = \frac{1}{2} m^2 \varphi_s^2 + V_0$ is the modern vacuum energy density.

For numerical calculations, it is suitable to change variables:

$$\tau = tH, \ \zeta = \chi/M - \zeta_*,$$
$$\zeta_* = \chi_*/M \equiv \ln(M^2 H^2/m^2 \varphi_s^2).$$

Equation of motion now acquires the form

$$\frac{d^2\zeta}{d\tau^2} + 3\frac{d\zeta}{d\tau} - e^{\zeta} = 0, \qquad (10.73)$$

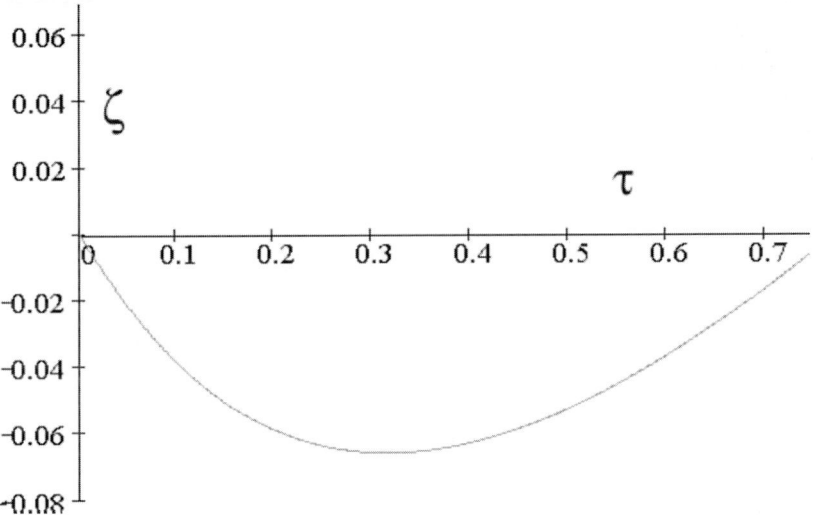

Figure 10.9. Dynamics of dimensionless field ζ in dependence of the cosmological time τ. Initial values are $\zeta(0) = 0, \dot{\zeta}(0) = -0.5$. The field moves away from the potential minima and after some time turns back.

where it was supposed that the field $|\chi| \gg M$ and $\chi < 0$ today.

The initial value of kinetic term in the modern epoch can be estimated using the parameter w of the equation of state

$$w = \frac{\dot{\chi}^2 - U(\chi)}{\dot{\chi}^2 + U(\chi)}$$

and equating the energy density of scalar field to the dark energy density, $\dot{\chi}^2 + U(\chi) = \Lambda$. It leads to the initial value

$$\left|\dot{\zeta}(\tau = 0)\right| \lesssim 1. \tag{10.74}$$

This initial value should be negative to prevent quick falling of the field down to the potential minimum.

Numerical analysis of the equation (10.73) revealed strong sensitivity of its solution to the initial value $\zeta(0)$. The result of calculations with initial values $\zeta(0) = 0, \dot{\zeta}(0) = -0.5$ is shown in Fig.10.9. According to this calculation, the Universe will reach the minimum of potential in the cosmological time. If $\dot{\zeta}(0) \leq 0$ the Universe turns back to the potential minimum during $\sim 1.5 \cdot 10^{10}$ years. This minimum has negative value which leads to the vacuum reconstruction and to the radical change of physical conditions in our Universe. If $\zeta(0) < -2$ the field motion could last exponentially long.

Recent measurements of variation of the fine structure constant α [365] indicate that we could check the field variation. Moreover, in the near future it may be possible to distinguish whether we move towards the minimum or are still moving away from it. Indeed, it could be the consequence of dependence of the fine structure constant on the scalar field, see, e.g., [381, 371]. This idea was developed in the framework of model of quintessence [382] and we may apply it to our case.

Following [366], we can suppose coupling of the scalar field with the electromagnetic one in the form

$$L_{A\chi} = -\frac{1}{4} B(\chi) F_{\mu\nu} F^{\mu\nu}$$

and use first-order expansion

$$B(\chi) \simeq 1 - \varkappa (\chi - \chi_*).$$

So, the effective fine structure constant depends on the value of the field χ as follows

$$\frac{\Delta\alpha}{\alpha} \equiv \frac{\alpha - \alpha_*}{\alpha_*} = \varkappa (\chi - \chi_*)$$

The claimed observational result, giving the value $\frac{\Delta\alpha}{\alpha} \approx -5 \cdot 10^{-5}$ over the redshift range $0.2 < z < 3.7$ [365] can indicate then that the value of the field χ does vary provided the observations will be confirmed. If more accurate measurements supply us with experimental value $|\Delta\dot\alpha(z)/a|$ we could distinguish the direction of motion of field χ. Indeed, the value $|\Delta\dot\alpha(z)/a| \propto |\dot\chi|$ must decrease if we move away to the minimum of potential thus approaching the turning point $\tau \simeq 0.3$, and must increase at times $\tau \gtrsim 0.3$. This remark also indicates the difference between the model in question and models of quintessence and the principal possibility to distinguish the predictions of the two approaches, thus providing their observational test.

The case $n = -1$

A nontrivial situation takes place if the kinetic function has zero value at some point, i.e. $K(\varphi) = (\varphi - \varphi_s)/M$. Eq. (10.65) leads to the classical equation

$$(\varphi_s - \varphi) \cdot (\ddot\varphi + 3H\dot\varphi) - \frac{1}{2}\dot\varphi^2 + M \cdot V(\varphi)' = 0. \qquad (10.75)$$

This equation is quite uncommon. Indeed, if zero point φ_s of the kinetic term does not coincide exactly with the position of the minimum of potential at $\varphi = 0$, the point $\varphi = \varphi_s$ is not a stationary solution of this equation. On the other hand, if the functions $V(\varphi)$ and $K(\varphi)$ behave locally as is shown in Figure 10.10, the point φ_s is some kind of attractor. More precisely, if the field value is larger than φ_s, then the kinetic term $K(\varphi) > 0$ and we have a

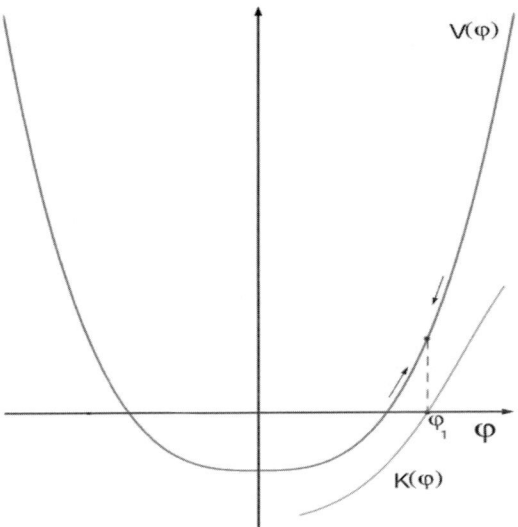

Figure 10.10. Disposition of kinetic term and potential in the case $n = -1$. At the end of inflation the field oscillates around singular point φ_1.

standard rolling of the field down to the singular point φ_s. If the field value is smaller than φ_s, then $K(\varphi) < 0$. The field behaves like the phantom field [388], climbing up to the potential and thus tending toward the point φ_s.

Classically, the situation looks very strange – the singular point attracts the solution, but forbids it to stay there forever. Evidently, the field fluctuates around the singular point which reminds us of stochastic behavior. Additional discussion on the similar problem can be found in the paper [390]. Stationary state could be realized in terms of quantum mechanics like in the case of an electron in the Coulomb field.

Let the initial field value $\varphi = \varphi_{in} > \varphi_s$. An appropriate variable substitution (10.65) is given by

$$\varphi = \varphi_s + sgn(\chi) \cdot \gamma |\chi|^{2/3}, \quad \gamma \equiv (3\sqrt{M}/2)^{2/3} \quad \varphi > \varphi_s.$$

The potential of the auxiliary field χ acquires the form

$$U(\chi) = \frac{1}{2}m^2(\varphi_s + sgn(\chi) \cdot \gamma |\chi|^{2/3})^2 + V_0. \tag{10.76}$$

$U(\chi)$ is finite at $\chi = 0$ but its derivative is singular,

$$U'_{\chi \to +0} = -\frac{2}{3}m^2\varphi_s^2\gamma |\chi|^{-1/3}. \tag{10.77}$$

The potential (10.76) behaves like $\chi^{4/3}$ at large field values. It leads to standard inflation with moderate fine-tuning of the parameters. Namely, to fit the observations the parameters should have the values

$$M \sim M_P, \quad m \sim 10^{-6} M_P, \quad V_0 \sim 10^{-12} M_p^4 \qquad (10.78)$$

If $\varphi_s > 0$, the field φ will oscillate around the critical point with the energy density given by the Eq. (10.63).

5.2   Probabilistic approach to the form of action

We have investigated several specific forms of effective potential. There are many other potentials and kinetic terms discussed in the literature. A substantial amount of them do not contradict observational data, but evidently only one of them can be realized in our Universe. Exact form of scalar field action has not been derived yet from a more fundamental theory. Thus a wide variety of possibilities still remains topical. In this connection two questions can be raised and need to be answered: (i) What shape of potential and kinetic term can be realized in our universe? The answer would require consistency of the theoretical predictions with the whole set of observational data. Suppose one finds the answer. Then the next problem emerges: (ii) Why is this particular shape of potential and kinetic term realized in nature? What are the underlying reasons?

Some theoretical hints on the form of the potential have been given by the supergravity, which predicts an infinite power series expansion in the scalar field potential [10]. Its minima, if they exist, correspond to the stationary states of the field. In the low energy limit it is reasonable to retain only a few terms (lowest powers in the Taylor expansion) of the scalar field [353]. There is still no physical law which limits the potential to have only a finite number of minima. The potential caused, e.g., by the supergravity could correspond to a function with an infinite set of minima. As was shown in Chapter 9 the supposition that the potential possesses infinite number of randomly distributed minima appears to be self-consistent [371]. In the vicinity of each of the minima the potential has an individual form and hence the universe associated with such a minimum may be individually different from other universes. Our own Universe is associated with a particular potential minimum, not necessarily located at $\varphi = 0$. Similar behavior may hold also for the kinetic term.

Actually, any way of introducing scalar field leads to a theory with a non-trivial kinetic term provided that quantum corrections are taken into account. Indeed, such term arises with necessity in ordinary quantum field theory with a scalar field. It is the standard result that the kinetic term of the effective action acquires a multiplier which is a function of the field [341].

Let us discuss in this context the problem of small vacuum energy density claimed to be favored by recent observations. The smallness of the Λ term is usually referred to some more fundamental theory, such as supergravity. Another approach is to relate it to the anthropic principle. Our point of view is that we can merge these approaches. Namely, more fundamental theory supplies us with infinite set of minima of the potential. These minima having an individual shape are responsible for the formation of those universes used in the anthropic consideration.

The problem of small cosmological constant can be solved in the framework of the random potential as it has been done in [371], see also Chapter 9, and with the use of non-canonical kinetic term of the scalar field discussed in this section. Such potential and kinetic term distributed in finite region of the field φ are represented in Figure 10.11. Fluctuations of the scalar field being generated at high energies move classically to stationary points, corresponding to black points in Figure 10.11. Those that reach stationary points with appropriate energy density could form a universe similar to our Universe. This energy density ($\sim 10^{-123} M_P^4$) is the result of small value of a concrete potential minimum or small value of the difference $\varphi_s - \varphi_m$, φ_m is a zero of the potential, $V(\varphi_m) = 0$. The fraction of such universes is extremely small, but nevertheless is infinite because of the infinite number of stationary states.

How could one decide which of them is the most promising? To get the idea we would like to remind ourselves that the main defect of inflationary scenario is the smallness of some intrinsic parameter as compared with unity. It is the small value of self-coupling $\lambda \sim 10^{-13}$ for the potential $V_4 = \lambda \varphi^4$ or smallness of the mass of inflaton field in Planck units, $m/M_P \sim 10^{-6}$ for the potential $V_2 = m^2 \varphi^2/2$. If we have an infinite set of different potentials (which are power sets (9.5) near the infinite sets of minima of random potential), we can use the concept of probability to find a potential with the particular value of a parameter. This concept is discussed in, e.g., [391, 392]. To do this, let us propose that

For any chosen minimum of potential the probability to find the value of parameter g in an interval (g, g + dg) equals to dP = W(g)dg, where W(g) is a uniform distribution. It is suitable but not obligatory to choose the interval (0,1).

Of course it does not relate to those parameters the values of which are determined by observations.

Immediate conclusion is that the probability of potential $\lambda \varphi^4$ is about 10^{-13} while the probability of potential $m^2 \varphi^2/2$ is 10^{-6}. It means that the latter can be realized in 10^6 time more frequently.

Before proceeding we have to mention that in fact the probability is much smaller due to the smallness of the cosmological Λ-term. In this context, the probability to find a universe with such small vacuum energy is $P_\Lambda = 10^{-123}$. This value is very small, but the whole set of minima is infinite. It means that the subset of universes with appropriate vacuum energy density is very weak but still infinite. So the probability to find appropriate potential V_4 is

$$P(V_4) = 10^{-13} P_\Lambda$$

while the same for the potential V_2 is

$$P(V_2) = 10^{-6} P_\Lambda.$$

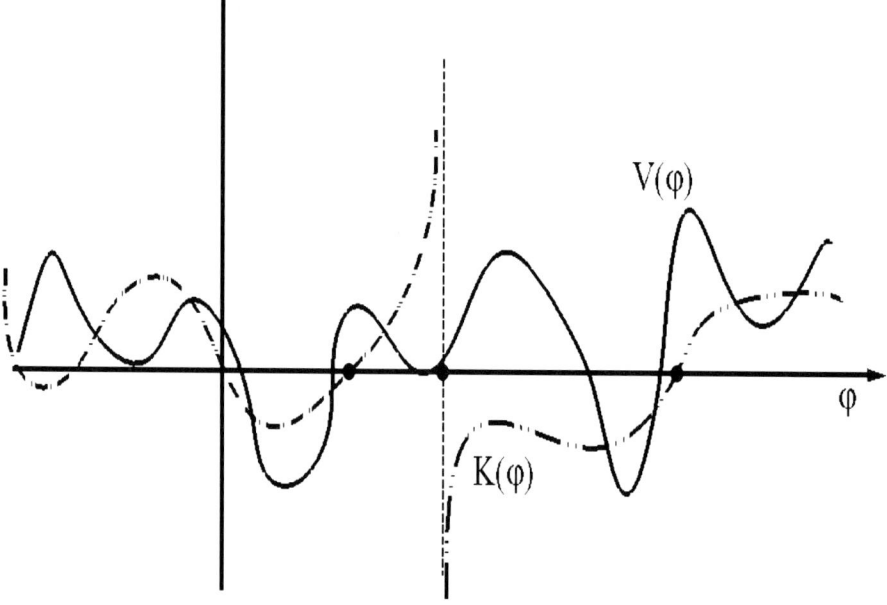

Figure 10.11. Random potential and kinetic term. Black points mark stationary state of the field φ.

It is the ratio of these values but not their absolute value that is important. This ratio gives us the relative number of suitable universes.

It is interesting now to compare these values with probabilities of those universes which are responsible for the singular points of kinetic term, rather than potential minima. In the framework of the previous discussion, we could expect that singular point(s) φ_s may be found near some minimum φ_m of the potential. Now the problem is reformulated as follows: "what part of infinite amount of minima contains singular points disposed closely to them?" It seems evident that this part is very small, but not zero, due to infinite number of the minima. Only this part is important – it represents those vacua where galaxies could be formed [30]

The problem of the cosmological Λ - term acquires another sense. One has to explain the extremal proximity of singular point of kinetic term and a zero point of the potential. Such a fitting seems absolutely accidental. Moreover, a probability of this event is very small.

Following this way we can compare the probability of realization of potentials discussed above. Their common multiplier P_0 is connected with the smallness of the interval for the singular point given by the Eq. (10.70)

$$P_0 = \Delta\varphi_s/M_P \cong \frac{\Lambda}{M_P m \sqrt{2\,|V_0|}} = P_\Lambda \frac{M_P^3}{m\sqrt{2V_0}}. \qquad (10.79)$$

For the case $n = 1$ the only additional smallness is dictated by last expression (10.68) and the probability for such universes is

$$P_1 \sim \frac{m}{M} P_0 = P_\Lambda \frac{M_P^3}{M\sqrt{2V_0}} \approx P_\Lambda. \qquad (10.80)$$

Universes with the properties described in the case $n = 2$ are distributed with the probability

$$P_2 \sim \frac{V_0}{M_P^4} P_0 \simeq P_\Lambda \frac{\sqrt{2V_0}}{mM_P} \sim 10^{-7} P_\Lambda \qquad (10.81)$$

if the inflation starts at the right branch of the potential. Here, we accepted $m \sim M_P, V_0 \sim M_P^4$. The last considered case, $n = -1$, has the probability in order of magnitude greater compared with the previous one

$$P_{-1} \sim \frac{m}{M} \frac{V_0}{M_P^4} P_0 \simeq P_\Lambda \frac{\sqrt{2V_0}}{M_P^2} \sim 10^{-6} P_\Lambda. \qquad (10.82)$$

An important conclusion from this consideration is that the model with kinetic term $\sim (\varphi - \varphi_s)^{-1}$ is much more probable (at least in 10^6 times) compared with other models discussed above, including models with standard kinetic term and the potentials $\sim \varphi^2$ and φ^4. It means that our Universe is likely governed by the model with kinetic term $\sim (\varphi - \varphi_s)^{-1}$, if there are no physical reasons to exclude it.

The probabilistic approach presented above is valid for those models which explain observational data equally as good. For example, if the variation of fine structure constant will be confirmed, only one of the considered models remains topical. The realistic application of the proposed approach implies the self-consistent account for the physics of baryosynthesis and non-baryonic dark matter that follows from the particle model, underlying the considered scenario.

5.3   Conclusion

It is shown that the region where kinetic term changes its sign gives new possibilities for the scalar field dynamics. It takes place even for the simplest form of the potential. Depending on a position of the singular point of the kinetic term, specific forms of the potential of the auxiliary field could be obtained. One of the main results is that the stationary value of scalar field could be situated in the singular points of the kinetic term rather than in minima of the potential. Another interesting result is that if the singular point is a zero of the kinetic term, the final state is intrinsically quantum state. We also compare the probabilities to find universes with specific values of parameters.

6.   Scalar–tensor models

The results of the previous consideration are connected with the renormalization of kinetic term in the expression (10.25). Meanwhile, a much more general form of Lagrangian of scalar field coupled with gravity is known. One of the most general models is Hyperextended scalar–tensor gravity [346]. Following [356], we write it in the form

$$S = \int d^4 x \sqrt{-g} \left\{ \frac{F(\varphi)}{16\pi G} R + \frac{1}{2} K(\varphi) (\partial \varphi)^2 - V(\varphi) \right\}. \qquad (10.83)$$

As we have seen above, the coupling of the inflaton to the additional massive field χ leads to kinetic function K of the simplest form $K(\varphi) = Const\varphi^2 = 2\alpha^2/m_\chi^2 \varphi^2$, provided the value of $Const$ could be arbitrarily large. Thus, the model discussed in the previous two sections represents one of the realization of action (10.83) with the function $F(\varphi) = 1$.

Generalized Brans–Dicke model is another particular case of the model (10.83) with the nonminimal function $F(\varphi) = \varphi$ and redefinition of kinetic function $K(\varphi) = \omega(\varphi)/(8\pi\varphi)$.

The theory of dilaton gravity is obtained if one denotes $F(\varphi) = exp(-\varphi)$ and $K(\varphi) = -exp(-\varphi)/(8\pi G)$.

Classical equations of motion for gravitation and the scalar field are obtained by variation of the action (10.83) with these dynamical variables. The equations are simplified significantly, if we limit ourselves by FRW metric, which is usually good approximation for cosmological problems. In this case the equations have the form [356]

$$H^2 = \frac{8\pi}{3M_P^2 F} \left(\frac{1}{2} K \dot{\varphi}^2 + V \right) - \frac{F'}{F} H \dot{\varphi}, \qquad (10.84)$$

$$\left(\frac{3F'^2}{2F} + \frac{8\pi}{M_P^2} K \right) (\ddot{\varphi} + 3H \dot{\varphi}) +$$

$$\left[\frac{F'}{2F} \left(3F'' + \frac{8\pi}{M_P^2} K \right) + \frac{1}{2} \frac{8\pi}{M_P^2} K' \right] + \frac{8\pi}{M_P^2} F^2 \left(\frac{V}{F^2} \right)' = 0.$$

A prime denotes the differentiation with respect to the scalar field φ. As was discussed in Chapter 2, there are three different FRW metrics corresponding to $k = 0, \pm 1$. Eq. (10.84) is written for the case of flat Universe ($k = 0$) which is the simplest one. In any case, the two others tend exponentially to the chosen case with time.

One can significantly simplify this system applying slow-rolling conditions. As the result one obtains

$$H^2 = \frac{8\pi}{3M_P^2} \frac{V}{F}, \qquad (10.85)$$

$$3HK\dot{\varphi} + F^2 \left(\frac{V}{F^2} \right)' = 0$$

with one of the slow-roll conditions in the form

$$\varepsilon = \frac{M_P^2}{16\pi} \frac{F}{K} \left[\ln \left(\frac{V}{F^2} \right) \right]^{\prime 2} \ll 1 \qquad (10.86)$$

(see [356] for details). The slow-rolling and hence inflationary stage takes place if $\varepsilon \ll 1$.

As we consider a more or less general case with unknown functions F and K, this condition could be fulfilled at energies smaller than the Planck scale. Indeed, a standard result with the functions $F = 1$ and $K = 1$ is that $\varepsilon < 1$ takes place if $\varphi \geq M_P$ for the potential of the form $V = \lambda\varphi^4$ – see Chapter 2. Increasing the function K and/or decreasing function F could make this condition much weaker so that the inflation extends to smaller values of the inflaton φ, as is evident from Eq. (10.86). A concrete example was considered in Section 3 of this chapter.

Additional 'degree of freedom' is a form of the potential. Many interesting possibilities have been discussed in the literature. A bunch of variants of the hybrid inflation is considered in [353]. Quintessence model [374] with monotonically decreasing potential was proposed to explain nonzero value of energy density in the modern epoch. At last, the possibility of random potential is discussed in Chapter 7, see also [326], [330].

Epilogue

As a result, we realize that there are a large number of inflationary models elaborated up to now. Some of them are purely phenomenological, others have some theoretical basis. The main problem now is to choose one that describes observational data, which became much richer and refined during the last decade.

The era of precision cosmology begins. Cosmology is coming to the fascinating task of choosing one theory among others. This task implies with necessity close links between cosmology and particle physics and the development of cosmoparticle physics, studying the fundamental relationship between macro- and micro-worlds.

This theory must explain observations and experimental data which are still poorly connected now. A small list of those we touched upon in the present book is:

- Dark matter.

- Large-scale structure and CMB fluctuations.

- Nonzero vacuum energy.

- Bursts.

- Massive black holes in galaxy centers.

- Intermediate black holes.

- Baryon domination in our Universe.

- High energy component in cosmic rays.

- Properties of Higgs scalar(s).

- 4-th generation in the fermion family.

- Trans-Planckian physics.

The impressive list of these problems, as well as of many others to be solved by cosmoparticle physics, makes the development of this science the exciting challenge for the new Millennium.

Index

References

[1] A.D. Sakharov, "Cosmoparticle physics – the cross-disciplinary science" , Vestnik AN SSSR, vol. 4, pp. 39–40, 1989.

[2] M.Yu. Khlopov, "Fundamental cross-disciplinary studies of microworld and Universe" , Vestnik of Russian Academy of Sciences, vol. 71, pp. 1133–1137, 2001.

[3] M. Khlopov, Cosmoparticle Physics. Singapore-New Jersey-London-Hong Kong: World Scientific, 1999.

[4] C. Quigg, Gauge Theories of the Strong, Weak, and Electromagnetic Interactions, Addison-Wesley Publ. Company, 1983.

[5] V.B. Berestetsky, E.M. Lifshits, L.P. Pitaevski, Quantum Electrodynamics. Moscow: Nauka, 1989

[6] R.D. Peccei, H.R. Quinn, "CP Conservation in the Presence of Pseudoparticles" , Phys.Rev.Lett. vol. 38, no. 25, pp. 1440–1443, 1977.

[7] S. Weinberg, "A New Light Boson?" , Phys.Rev.Lett. vol. 40, no. 4, pp. 223–226, 1978.

[8] F. Wilczek, "Problem of Strong P and T Invariance in the Presence of Instantons" , Phys.Rev.Lett. vol. 40, no. 5, pp. 279–282, 1978.

[9] J.E. Kim, "Light pseudoscalars, particle physics and cosmology," Phys. Rept., vol. 150, no. 1-2, pp. 1–177, 1987.

[10] H.P. Nilles, "Supersymmetry, supergravity and particle physics" , Phys.Rept., vol. 110, no. 1-2, pp. 1–162, 1984.

[11] Ya.B. Zeldovich, M.Yu. Khlopov, "The neutrino mass in elementary particle physics and in big bang cosmology" , Sov. Phys. Uspekhi, vol. 24, pp. 755–774, 1981.

[12] C.K. Jung, C. McGrew, T. Kajita, T. Mann, "Oscillations of atmospheric neutrinos" , Ann.Rev.Nucl.Part.Sci., vol. 51, pp. 451–488, 2001.

[13] K2K Collaboration (M.H. Ahn et al.), "Indications of neutrino oscillation in a 250 km long baseline experiment" , Phys.Rev.Lett., vol. 90, pp. 041801 (5 pages), 2003.

[14] SNO Collaboration (Q.R. Ahmad et al.), "Measurement of day and night neutrino energy spectra at SNO and constraints on neutrino mixing parameters" , Phys.Rev.Lett. vol. 89, pp. 011302 (5 pages),2002.

[15] KamLAND Collaboration (K. Eguchi et al.), "First results from KamLAND: evidence for reactor anti-neutrino disappearance" , Phys.Rev.Lett. vol. 90, pp. 021802 (6 pages), 2003.

[16] R. Bernabei, et al, "Dark Matter search" , Riv.Nuovo Cim., vol. 261, pp. 1–73, 2003.

[17] H. V. Klapdor-Kleingrothaus, "First evidence for neutrinoless double beta decay" , Found. Phys., vol. 33, no. 5, pp. 813–829, 2003.

[18] M.Yu. Khlopov, "Cosmoarcheology. Direct and indirect astrophysical effects of hypothetical particles and fields" , in: Cosmion-94, Eds. M.Yu.Khlopov et al. Editions frontieres, 1996. pp. 67–76.

[19] K.M. Belotsky, M.Yu. Khlopov, A.S. Sakharov, A.L. Sudarikov and A.A. Shklyaev, "Experimental cosmoparticle physics: experimental probes for dark matter physics at particle accelerators" , Grav.Cosmol.Suppl., vol. 4, pp. 70–78, 1998.

[20] M. Green, J. Schwarz, E. Witten, Superstring theory. Cambridge University Press, 1989

[21] Ia.I. Kogan, M.Yu. Khlopov, "Homotopically stable particles in superstring theory" , Sov.J.Nucl.Phys., vol. 46, no. 1, pp. 193–194, 1987.

[22] Ia.I. Kogan, M.Yu. Khlopov, "Cosmological consequences of $E_8 E_8$ superstring models" , Sov.J.Nucl.Phys., vol. 44, no. 5, pp. 873–874, 1987.

[23] E. W. Kolb, D. Seckel and M. S. Turner, "The Shadow World" , Nature, vol. 314, pp. 415–419, 1985.

[24] M.Yu. Khlopov and K.I. Shibaev, "New Physics From Superstring Phenomenology" , Grav.Cosmol.Suppl., vol. 8, pp. 45-52, 2002.

[25] Yu.A. Golubkov, D. Fargion, M.Yu. Khlopov, R.V. Konoplich and R. Mignani, "Possible signatures for the existence of neutrino of fourth generation" JETP Lett., vol. 69, no. 6, pp. 434–440, 1999.

[26] G. Aldazabal, L.E. Ibanez, F. Quevedo and A.M. Uranga, "D-branes at singularities: A bottom-up approach to the string embedding of the Standard Model" , JHEP, vol. 0008, pp. 002, 2000.

[27] L.F. Alday and G. Aldazabal, "In quest of 'just' the Standard Model on D-branes at a singularity" , JHEP, vol. 0205, pp. 022, 2002.

[28] A.S. Sakharov, M.Yu. Khlopov, "Horizontal unification as the phenomenology of the theory of "everything"" , Phys.Atom.Nucl., vol. 57, no. 4, pp. 651–658, 1994.

[29] Ya.B. Zeldovich, I.D. Novikov, Structure and Evolution of the Universe. Moscow: Nauka, 1975.

[30] S. Weinberg, Gravitation and Cosmology. New York - London - Sydney - Toronto: John Wiley and Sons, Inc., 1972.

[31] C. Misner, K. Thorne, and J. Wheeler, Gravitation. San Francisco: W.H. Freeman and Company, ed., 1973. .

[32] A. Lightman, W. Press, R. Price, and S. Tukolsky, Problem book in relativity and gravitation, vol. of . Princeton, New Jersey: Princeton University Press, ed., 1975.

[33] D.A. Kirzhnits, A.D. Linde, "Symmetry Behavior In Gauge Theories" , Annals Phys., vol. 101, pp. 195–238, 1976.

[34] A.D. Linde, "High-density and high-temperature symmetry behavior in gauge theories" , Phys.Rev., vol. D14, no. 12, 3345–3349, 1976.

[35] A.D. Linde, Particle Physics and Inflationary Cosmology, Harvard Acad. Press, Geneva (1990)

[36] E. W. Kolb and M. S. Turner, The Early Universe, Addison-Wesley, 1990.

[37] T.W.B. Kibble, "Topology Of Cosmic Domains And Strings" , J. Phys., vol. A9, no. 8, pp. 1387–1398, 1976.

[38] P. A. M. Dirac, "Quantised Singularities In The Electromagnetic Field" , Proc.Roy.Soc.Lond., vol. A133, pp. 60–72, 1931.

[39] P. A. M. Dirac, "The Theory of Magnetic Poles" , Phys. Rev., vol. 74, no. 7, 817–830, 1948.

[40] G. t'Hooft, "Magnetic monopoles in unified gauge theories" , Nucl. Phys., vol. B79, no. 2, pp. 276–284, 1974.

[41] A.M. Polyakov, "Particle Spectrum In Quantum Field Theory" , JETP Lett., vol. 20, pp. 194–195, 1974.

[42] Ya.B. Zeldovich and M.Yu. Khlopov, "On the concentration of relic magnetic monopoles in the universe" , Phys. Lett., vol. 79, no. 3, pp. 239–241, 1978.

[43] M.Yu. Khlopov, "Primordial magnetic monopoles" , Priroda, no. 12, pp. 99–101, 1979.

[44] J.P. Preskill, "Cosmological Production Of Superheavy Magnetic Monopoles" , Phys. Rev. Lett., vol. 43, no. 19, pp. 1365–1368, 1979.

[45] Ya.B. Zeldovich, "Cosmological fluctuations produced near a singularity" , Mon.Not.Roy.Astron.Soc., vol. 192, pp. 663–667, 1980.

[46] A. Vilenkin, "Cosmic strings" , Phys.Rev., vol. D24, no. 8, pp. 2082–2089, 1981.

[47] I.Y. Kobsarev, L.B. Okun and Y. B. Zeldovich, "Spontaneous Cp-Violation and Cosmology" , Phys.Lett., vol. B50, no. 3, pp. 340–342, 1974.

[48] E.B. Gliner, "The vacuum-like state of a medium and Friedman cosmology" , Sov. Phys. Dokl., vol. 15, no. 6, pp. 559–561, 1970; E.B. Gliner, "Inflationary universe and the vacuumlike state of physical medium " , Phys. Usp., vol. 45, no. 2, pp. 213-220, 2002.

[49] E.B. Gliner, I.G. Dymnikova, " A nonsingular Friedmann cosmology" , Sov. Astron. Lett., vol. 1, no. 3, pp. 93–94, 1975.

[50] V.Ts. Gurovich and A. Starobinsky, "Quantum effects and regular cosmological models" , Sov.Phys.JETP, vol. 50, pp. 844–852, 1979.

[51] A. Starobinsky, "A new type of isotropic cosmological models without singularity" , Phys. Lett., vol. B91, no. 1, pp. 99–102, 1980.

[52] A.H.Guth, "The inflationary universe: A possible solution to the horizon and flatness problems" , Phys. Rev., vol. D23, no. 2, pp. 347–356, 1981.

[224] S.H. Geer, D.C. Kennedy, "The Cosmic Ray Antiproton Spectrum and a Limit on the Antiproton Lifetime" , astro-ph/9809101, 1998.

[225] G. Backenstoss, et al, "Proton – anti-proton annihilations at rest into pi0 Omega, pi0 eta, pi0 gamma, pi0 pi0 and pi0 eta-prime" , Nucl.Phys., vol. B228, no. 3, pp. 424–438, 1983.

[226] T. Sjostrand, "The LUND Monte Carlo for jet fragmentation" , Comput.Phys.Commun., vol. 27, no. 3, pp. 243–284, 1982.

[227] D.C. Ellison, F.C. Jones, M.G. Baring, "Direct Acceleration of Pickup Ions at the Solar Wind Termination Shock: the Production of Anomalous Cosmic Rays" , astro-ph/9809137, 1998.

[228] H. Kohno, et al, "Anti-p p elastic and charge exchange scattering at 230 MeV" , Nucl.Phys., vol. B41, no. 2, pp. 485–492, 1972.

[229] V. Chaloupka, et al, "Measurement of the total and partial anti-p p cross-section between 1901-MeV and 1950-MeV" , Phys.Lett., vol. B61, no. 5, pp. 487–492 , 1976.

[230] N.N. Chugai, S.I. Blinnikov, T.A. Lozinskaya, "Supernovae,"Preprint ITEP–43, 1986.

[231] PS179 Collaboration(F. Balestra, Yu.A. Batusov, G. Bendiscioli, et al,) "Anti-proton He-4 interactions at 200-MeV/c" , Phys. Lett. B, vol. 305, no. 1-2, pp. 18–22, 1993.

[232] F. Balestra, S. Bossolasco, M.P. Bussa, et al, "Anti-proton - helium annihilation around 44-MeV/c" , Phys. Lett., vol. B230, no. 1-2, pp. 36–40, 1989.

[233] F. Balestra, M.P. Bussa, L. Busso, et al, "Inelastic interaction of anti-protons with He-4 nuclei between 200-MeV/c andD 600-MeV/c" , Phys. Lett., vol. B165, no. 4-6, pp. 265–269, 1985.

[234] G.N. Velichko, A.A. Vorobev, A.V. Dobrovolsky, et al, "Elastic scattering of protons on helium nuclei in the energy range of 700-MeV – 1000-MeV" , Sov. J. Nucl. Phys., vol. 42, no. 6, pp. 837–844, 1985.

[235] A. Bujak, P. Devenski, E. Jenkins, et al, "Proton - helium elastic scattering from 45-GeV to 400-GeV" , Preprint JINR–E1–81–289, 1981.

[236] D. Fargion, M. Khlopov, R. Konoplich, et al, "Ultra High Energy Particle Astronomy, Neutrino Masses and Tau Airshowers" , Recent Res. Devel. Astrophys., vol. 1, pp. 1–60, 2003. astro-ph/0303233.

[237] D. Fargion, "Breaking and Splitting asteroids by nuclear explosions to propel and deflect their trajectories" , astro-ph/9803269, 1998.

[238] Z.G. Berezhiani and M.Yu. Khlopov, "Cosmology of spontaneously broken gauge family symmetry with axion solution of strong CP-problem,"Z. Phys. vol. C49, pp. 73–78, 1991.

[239] Z.G. Berezhiani and M.Yu. Khlopov, "The physics of dark matter of the universe in the theory of broken generation symmetry" , Sov. J. Nucl. Phys. vol. 52, no. 1, pp. 60–64, 1990.

[240] L. Kofman, A. Linde and A.A. Starobinsky,"Nonthermal phase transitions after inflation" , Phys. Rev. Lett. vol. 76, no. 7, pp. 1011–1014, 1996. hep-th/9510119.

[241] M.S. Turner, "Windows on the axion" , Phys. Rep. vol. 197, no. 2, pp. 67–97, 1990.

[242] M. Yoshimura, "A mechanism of generating isothermal perturbation by a strong CP violation" , Phys.Rev.Lett., vol. 51, no. 6, pp. 439–442, 1983.

[243] T. Vachaspati and A. Vilenkin, "Formation and evolution of cosmic strings," Phys. Rev., vol. D30, no. 10, pp. 2036–2045, 1984.

[244] R. Davis, "Cosmic axions from cosmic strings," Phys. Lett., vol. B180, no. 3, pp. 225–230, 1986. .

[245] D. Harari and P. Sikivie, "On the evolution of global strings in the early universe," Phys. Lett., vol. B195, no. 3, pp. 361–365, 1987.

[246] R.A. Battye and E.D.S. Shellard, "Axion String Constraints," Phys. Rev. Lett., vol. 73, no. 22, pp. 2954–2957, 1994.

[247] M. Khlopov, A.S. Sakharov, A.L. Sudarikov, "Cosmoparticle physics: Basic principles and prospects for future development" , Grav. Cosmol.Suppl., vol. 4, pp. 1-14, 1998.

[248] D. Sitter, "On Einstein's theory of gravitation and its astronomical consequences" , Month.Notices Roy.Astron Soc., vol. 78, p. 3, 1917.

[249] S.-J. Rey, "Dynamics of Inflationary Phase Transition, "Nucl.Phys., vol. B284, pp. 706–728, 1987.

[250] R. P. Feynman and A. R. Hibbs, Quantum Mechanics and Path Integrals. New York: MacGraw-Hill, 1965.

[251] G. Lazarides, "Introduction to cosmology, "hep-ph/9904502, 1999

[252] J. Yokoyama, "Chaotic new inflation and formation of primordial black holes" , Phys. Rev., vol. D58, no. 8, p. 083510 (9 pages), 1998.

[253] I.A. Strukov, A.A. Brukhanov, D.P. Skulachev, M.V. Sazhin "The Relikt-1 experiment-new results" , Mon. Not. R. Ast. Soc., vol. 258, pp. 37, 1992; "Anisotropy of relic radiation in the Relic-1 experiment and parameters of grand unification" , Phys. Lett., vol. 315B, pp. 198, 1993.

[254] C.L. Bennett, "Four year COBE DMR cosmic microwave background observations: Maps and basic results" , Astrophys.J.Lett., vol. 464, no. 1, pt. 2, pp. L1–L4, 1996.

[255] A.S. Sakharov and M.Yu. Khlopov, "Cosmological signatures of family symmetry breaking in multicomponent inflation models" , Phys.Atom.Nucl., vol. 56, no. 3, pp. 412–417, 1993.

[256] L. Kofman and A. Linde, "Generation of density perturbations in the inflationary cosmology" , Nucl.Phys., vol. B282, p. 555–588, 1987.

[257] D. Rosenberg and J. Rutgers, "Galaxy formation: Was there a big bang shell?" , astro-ph/0012023, 2000.

[258] S. Veilleux, "The starburst - AGN connection" , astro-ph/0012121, 2000.

[259] M. Stiavelli, "Violent relaxation around a massive black hole" astro-ph/9801021, 1998.

[260] A. Starobinsky, "Relict gravitation radiation spectrum and initial state of the Universe" JETP Lett, vol. 30, pp. 682–685, 1979.

[261] R. Rajaraman, Solitons And Instantons. An Introduction To Solitons And Instantons In Quantum Field Theory. Amsterdam – New-York – Oxford: North-Holland Publishing Company, 1982.

[262] B. Liu, L. McLerran, and N. Turok, "Bubble nucleation and growth at a baryon number producing electroweak phase transition" , Phys.Rev., vol. D46, no. 6, pp. 2668–2688, 1992.

[263] A. Linde, "Scalar field fluctuations in expanding universe and the new inflationary universe scenario" , Phys.Lett., vol. B116, no. 5, pp. 335–339, 1982.

[264] A. Starobinsky, "Dynamics of phase transition in the new inflationary universe scenario and generation of perturbations" , Phys.Lett., vol. B117, no. 3-4, pp. 175–178, 1982.

[265] A. Vilenkin and L. Ford, "Gravitational effects upon cosmological phase transitions" , Phys.Rev.D, vol. 26, no. 6, pp. 1231–1241, 1982.

[266] S. Rubin, M. Khlopov, and A. Sakharov, "Primordial black holes from non-equilibrium second order phase transition" , Gravitation & Cosmology, Supplement, vol. 6, pp. 51–58, 2000. hep-ph/0005271.

[267] S. Labini, M. Montouri, and F. Pietronero, "Scale invariance of galaxy clustering" , Phys.Rept., vol. 293, no. 2-4, pp. 61–226, 1998.

[268] B. Elmegreen and D. Elmegreen, "Fractal structure in galactic star field" , Astron.J., vol. 121, no. 3, pp. 1507–1511, 2001. astro-ph/0012184.

[269] A.D. Linde, "Axions in inflationary cosmology" , Phys Lett., vol. B259, no. 1-2, pp. 38–47, 1991,

[270] G. Dvali, Q. Shafi, and R. Schaefer, "Large scale structure and supersymmetric inflation without fine tuning" , Phys.Rev.Lett., vol. 73, no. 14, pp. 1886–1889, 1994.

[271] D.H. Lyth, "The parameter space for tree-level hybrid inflation" , hep-ph/9904371, 1999.

[272] A.G. Cohen, A. Rujula, and S.L. Glashow, "A matter - antimatter universe?" , Astrophys.J., vol. 495, no. 2, pt. 1, pp. 539–549, 1998.

[273] A. Dolgov, "Non-GUT baryogenesis" , Phys.Rept., vol. 222, no. 6, pp. 309–386, 1992.

[274] T.D. Lee, "A theory of spontaneous T violation" , Phys. Rev., vol. D8, no. 4, pp. 1226–1239, 1973.

[275] M. Y. Khlopov, R. V. Konoplich, R. Mignani, S. G. Rubin, and A. S. Sakharov, "Evolution and observational signature of diffused antiworld" , Astroparticle Physics, vol. 12, no. 4, pp. 367–372, 2000. astro-ph/9810228.

[276] V.M. Chechetkin, M.Yu. Khlopov, and M.G. Sapozhnikov, "Antiproton interactions with light elements as a test of GUT cosmologies" , Rivista Nuovo Cimento, vol. 5, no. 10, pp. 1–80, 1982.

[277] L. Kofman, "What initial perturbations may be generated in inflationary cosmological models" , Phys.Lett., vol. B173, no. 4, pp. 400–404, 1986.

[278] M. Sasaki and B.L. Spokoinyi, "Nonscale invariant isocurvature perturbations produced in the power law inflation" , Modern.Phys.Lett., vol. A6, pp. 2935–2946, 1991.

[279] A. Linde and D. Lyth, "Axionic domain wall production during inflation" , Phys.Lett., vol. B246, no. 3-4, pp. 353–358, 1990.

[280] D. Lyth, "Axions and inflation: Sitting in the vacuum" , Phys.Rev., vol. D45, no. 10, pp. 3394–3404, 1992.

[281] D. Lyth and E. Stewart, "Axions and inflation: String formation during inflation" , Phys.Rev., vol. D46, no. 2, pp. 532–538, 1992.

[282] A. Linde, D. Linde, A. Mezhlumian, "From the big bang theory to the theory of a stationary universe" , Phys.Rev., vol. D49, no. 4, pp. 1783–1826, 1994.

[283] A. Liddle and D. Lyth, "The cold dark matter density perturbation" , Phys.Rep., vol. 231, no. 1-2, pp. 1–105, 1993.

[284] M. Y. Khlopov and A. Linde, "Is it easy to save the gravitino?" , Phys.Lett., vol. B138, no. 4, pp. 265–268, 1984.

[285] F. Balestra, et al, "Annihilation of antiprotons with Helium-4 at low energies and its relationship with the problems of the modern cosmology and models of Grand Unification" , Sov.J.Nucl.Phys., vol. 39, no. 4, pp. 626–630, 1984.

[286] M.Yu. Khlopov, Yu.L. Levitan, E.V. Sedelnikov and I.M. Sobol, "Nonequilibrium cosmological nucleosynthesis of light elements: Calculations by Monte Carlo method" , Phys.Atom.Nucl., vol. 57, no. 8, pp. 1393–1397, 1994.

[287] A.G. Cohen and D.B. Kaplan, "Thermodynamic generation of the baryon asymmetry" , Phys.Lett., vol. B199, no. 2, pp. 251–258, 1987.

[288] A.G. Cohen and D.B. Kaplan, "Spontaneous baryogenesis" , Nucl.Phys., vol. B308, no. 4, pp. 913–928, 1988.

[289] F.C. Adams, "Natural inflation: Particle physics models, power law spectra for large scale structure, and constraints from COBE" , Phys.Rev., vol. D47, no. 2, pp. 426–455, 1993.

[290] A.G. Cohen, D.B. Kaplan, and A.E. Nelson, "Progress in electroweak baryogenesis" , Ann.Rev.Nucl.Part.Sci., vol. 43, pp. 27–70, 1993.

[291] D. Morgan and V. Hughes, "Atomic processes involved in matter-antimatter annihilation" , Phys.Rev., vol. D2, no. 8, pp. 1389–1399, 1970.

[292] C.D. Dermer, "Secondary production of neutral pi-mesons and the diffuse galactic gamma radiation" , Astron.Astrophys., vol. 157, no. 2, pp. 223–229, 1986.

[293] S.D. Hunter, et al, " EGRET Observations of the Diffuse Gamma-Ray Emission from the Galactic Plane" , Ap.J, vol. 481, no. 1, pt.1, pp. 205–240, 1997.

[294] L.D. Landau and E.M. Lifshitz, Quantum mechanics. Moscow: Fiz.-Mat.Giz,, 1963.

[295] P. Sikivie, "Of axions, domain walls and the early universe" , Phys.Rev.Lett., vol. 48, no. 17, pp. 1156–1159, 1982.

[296] I. Affleck and M. Dine, "A new mechanism for baryogenesis?" , Nucl.Phys., vol. B249, p. 361, 1985.

[297] E. W. Kolb, "First-order inflation" , Physica Scripta, vol. T36, pp. 199–217, 1991.

[298] S. Coleman, "The fate of the false vacuum. Semiclassical theory" , Phys. Rev., vol. D15, no. 10, pp. 2929–2936, 1977.

[299] A. Linde, "Decay of the false vacuum at finite temperature" , Nucl.Phys., vol. B216, no. 2, pp. 421–445, 1983.

[300] I. Affleck, "Quantum statistical metastability" , Phys.Rev.Lett., vol. 46, no. 6, pp. 388–391, 1981.

[301] I. Bogolyubsky and V. Makhankov, "On the pulsed soliton lifetime in two classical relativistic theory models" , JETP Lett., vol. 24, pp. 12–15, 1976.

[302] I. Bogolyubsky and V. Makhankov, "Dynamics of heavy spherically–symmetric pulsons" , JETP Lett., vol. 25, pp. 120–123, 1977.

[303] M. Gleiser, "Pseudostable bubbles" , Phys. Rev., vol. D49, no. 6, pp. 2978–2981, 1994.

[304] E. J. Copeland, M. Gleiser and H. R. Muller, "Oscillons: Resonant configurations during bubble collapse" , Phys.Rev., vol. D52, no. 4, pp. 1920–1933, 1995. hep-ph/9503217.

[305] R.V. Konoplich, S.G. Rubin, A.S. Sakharov, and M.Yu. Khlopov, "Formation of black holes in first-order phase transitions as a cosmological test of symmetry-breaking mechanisms" , Phys.Atom.Nucl., vol. 62, no. 9, pp. 1593–1600, 1999.

[306] I. Dymnikova, L. Kozel, M. Khlopov, and S. Rubin, "Quasilumps from first order phase transitions" , Gravitation & Cosmology, vol. 6, pp. 311–318, 2000.

[307] R. Watkins and L.M. Widrow, "Aspects of reheating in first order inflation" , Nucl.Phys., vol B374, no. 2, pp 446–468, 1992.

[308] S. Hawking, I. Moss, and J. Stewart, "Bubble collisions in the very early universe" , Phys. Rev., vol. D26, no. 10, p. 2681–2693, 1982.

[309] A. Guth and E. Weinberg, "Could the Universe have recovered from a slow first order phase transition?" Nucl. Phys., vol. B212, no. 2, pp. 321–364, 1983.

[310] I. Dymnikova, "De sitter-schwarzschild black hole: Its particlelike core and thermodynamical properties" , Int.J.Mod.Phys., vol. D5, pp. 529–540, 1996.

[311] I. Dymnikova and M. Khlopov, "Decay of cosmological constant as bose condensate evaporation" , Mod.Phys.Lett., vol. A15, no. 38-39, pp. 2305–2314, 2000.

[312] J. Goldstone and R. Jackiw, "Quantization of nonlinear waves" , Phys. Rev., vol. D11, no. 6, p. 1486–1498, 1975.

[313] M. Khlopov and V. Chechetkin, "Anti-protons in the universe as cosmological test of grand unification" , Sov.J.Part.Nucl., vol. 18, no. 3, pp. 267–288, 1987.

[314] M. Dine, et al, "Towards the Theory of Cosmological Phase Transitions" , Phys.Rev., vol. D46, no. 2, pp. 550-571, 1992. hep-ph/9203203.

[315] R. Konoplich, "The probability of decay of an unstable vacuum" , Sov.J.Part.Nucl., vol. 32, no. 4, pp. 584–587, 1980.

[316] R.V. Konoplich, S.G. Rubin, A.S. Sakharov, and M.Yu. Khlopov, "Formation of black holes in first-order phase transitions in the Universe" , Pis'ma Astron.Zh., vol. 24, no. 7, pp. 1–5, 1998.

[317] D. La and P.J. Steinhardt, "Extended inflationary cosmology" , Phys.Rev.Lett., vol. 62, no. 4, pp. 376–378, 1989.

[318] I. Dymnikova, "Vacuum nonsingular black holes" , Gen.Rel.Grav, vol. 24, pp. 235–242, 1992.

[319] S. Alexeyev, M. Sazhin, and M. Pomazanov, "Black holes of a minimal size in string gravity" , Int.J.Mod.Phys., vol. D10, no. 2, pp. 225–230, 2001. gr-qc/9911036.

[320] M. Gleiser, E.W. Kolb, and R. Watkins, "Phase transitions with subcritical bubbles" , Nucl. Phys., vol. B364, pp. 411–450, 1991.

[321] A. Riotto, "Are oscillons present during a first order electroweak phase transition?" , Phys.Lett., vol. B365, no. 1-4, pp. 64–71, 1996.

[322] B. Carr and M. Rees, "The anthropic principle and the structure of the physical world" , Nature, vol. 278, pp. 605–612, 1979.

[323] I.L. Rozental, "Physical laws and numerical values of fundamental constants" , Sov.Phys.Usp., vol. 23, pp. 296-305, 1980.

[324] T. Banks, M. Dine, and L. Motl, "On anthropic solution to the cosmological constant problem" , JHEP, p. 031, 2001.

[325] S. Weinberg, "The cosmological constant problem" , astro-ph/0005265, 2000.

[326] S. Rubin, "Fine tuning of parameters of the universe" , Chaos, Solitons and Fractals (CHAOS2013), vol. 14, no. 6, pp. 891–899, 2002. astro-ph/0310182.

[327] J. Wudka and B. Grzadkowski, "Triviality and stability in effective theories" , JHEP,PRHEP-hep2001, p. 147, 2001.

[328] S. Coleman, "Black holes as red herrings: Topological fluctuations and the loss of quantum coherence", Nucl.Phys., vol. B307, no. 4, pp. 867–882, 1988.

[329] J. Garriga and A. Vilenkin, "Solution to the cosmological constant problems" , Phys.Rev., vol. D64, no. 2, p. 023507 (8 pages), 2001.

[330] S.G. Rubin, "Quantum fluctuations and life in the universe" , Gravitation & Cosmology, vol. 8, Supplement, pp. 53–56, 2002.

[331] Y. Zeldovich, "Cosmological constant and elementary particles" , Pis'ma ZhETF, vol. 6, no. 9, pp. 883–884, 1967.

[332] S. Weinberg, "Anthropic bound on the cosmological constant" , Phys. Rev. Lett., vol. 59, no. 22, pp. 2607–2610, 1987.

[333] A.H. Guth, "Inflationary models and connections to particle physics" , astro-ph/0002188, 2000.

[334] S. Coleman in Laws of Hadronic Matter, (NY), p. 186, Ed.: A. Zichichi, Academic Press, 1975.

[335] I. Krive and A. Linde, "On the vacuum stability problem in gauge theories" , Nucl. Phys., vol. B117, no. 1, pp. 265–268, 1976.

[336] S.-J. Chang, "Quantum fluctuations in a φ^4 field theory. i. stability of the vacuum" , Phys.Rev., vol. D12, no. 4, pp. 1071–1088, 1975.

[337] C.G. Callan, Jr., S. Coleman, "The fate of the false vacuum. 2. First quantum corrections", Phys.Rev., vol. D16, no. 6, pp. 1762–1768, 1977.

[338] C. Froggatt, H. Nielsen, and Y. Takanishi, "Standard Model Higgs boson mass from borderline metastability of the vacuum", Phys.Rev., vol. D64, no. 11, pp. 113014 (6 pages), 2001.

[339] J.C. Taylor, "Ward identities and charge renormalization of the Yang-Mills field", Nucl. Phys., vol. B33, no. 2, pp. 436–444, 1971.

[340] J.F.Donoghue, "The weight for random quark masses", Phys.Rev., vol. D57, no. 9, pp. 5499–5508, 1998.

[341] C. Itzykson and J.-B. Zuber, Quantum Field Theory. New York: McGraw-Hill, 1984.

[342] C. Castanier, P. Gay, P. Lutz, and J. Orloff, "Higgs self coupling measurement in e^+e^- collisions at center-of-mass energy of 500 GeV", hep-ex/0101028, 2001.

[343] P. Bergmann, "Comments on the scalar tensor theory", Int.J.Theor.Phys., vol. 1, pp. 25–36, 1968.

[344] B. Boisseau, G. Esposito-Farese, D. Polarski and A.A. Starobinsky, "Reconstruction of a scalar tensor theory of gravity in an accelerating universe", Phys. Rev. Lett. vol. 85, no. 11, pp. 2236–2239, 2000.

[345] E. Fradkin and A. Tseytlin, "Effective field theory from quantized strings", Phys.Lett., vol. B158, no. 4, p. 316–322, 1985.

[346] D. Torres and H. Vucetich, "Hyperextended scalar-tensor gravity", Phys.Rev., vol. D54, no. 12, pp. 7313–7377, 1996.

[347] J. Traschen and R. Brandenberger, "Particle production during out-of-equillibrium phase transitions", Phys.Rev., vol. D42, no. 8, pp. 2491–2504, 1990.

[348] L. Kofman, A. Linde, and A. Starobinsky, "Reheating after inflation", Phys.Rev.Lett., vol. 73, no. 24, p. 3195–3198, 1994. hep-th/9405187

[349] L. Amendola, C. Baccigalupi, F. Occionero, F. Konoplich, and S. Rubin, "Reconstruction of the bubble nucleating potential", Phys.Rev., vol. D54, no. 12, p. 7199–7206, 1996. astro-ph/9610038.

[350] K. Freese, "A coupling of pseudo nambu goldstone bosons to other scalars and role in double field inflation", Phys.Rev., vol. D50, no. 12, pp. 7731–7734, 1994.

[351] A.D. Dolgov and S.H. Hansen, "Equation of motion of a classical field with back reaction of produced particles", Nucl.Phys.B, vol. 548, no. 1-3, pp. 408–426, 1999.

[352] S. Rubin, "Effect of massive fields on inflation", JETP Lett., vol. 74, no. 5, pp. 275–279, 2001. hep-ph/0110132.

[353] D.H. Lyth and E. D. Stewart, "More varieties of hybrid inflation", Phys.Rev., vol. D 54, no. 12, pp. 7186-7190, 1996.

[354] N.D. Birrell and P.C.W. Davies, Quantum Fields in Curved Space. Cambridge London New York Sydney: Cambridge Univ. Press, 1982.

[355] V. Lukash, "Production of phonons in an isotropic universe", Sov.Phys.-JETP, vol. 52, no. 5, pp. 807–814, 1980.

[356] J. Morris, "Generalized slow roll conditions and the possibility of intermediate scale inflation in scalar-tensor theory", Class.Quant.Grav., vol. 18, no. 15, pp. 2977–2988, 2001.

[357] T. Futamase and K. Maeda, "Chaotic inflationary scenario of the universe with nonminimally coupled 'inflaton' field", Phys.Rev., vol. D39, no. 2, pp. 399–404, 1989.

[358] V. Faraoni, "Nonminimal coupling with scalar field and inflation", Phys.Rev., vol. D53, no. 12, pp. 6813–6822, 1996.

[359] S. Rubin, H. Kröger, G. Melkonian, "Role of Quantum Fluctuations at High Energies", Gravitation & Cosmology, Supplement, vol. 8, pp. 27–31, 2002; "hep-ph/0207109".

[360] T. Bunch and P. Davies, "Quantum field theory in de sitter space: renormalization by point splitting", Proc.R.Soc., vol. A360, pp. 117–134, 1978.

[361] A. B. Krebs and S. G. Rubin, "Instanton approach to the lattice polaron problem", Phys.Rev., vol. B49, no. 17, pp. 11808 – 11816, 1994.

[362] M. Wang, "The scalar-tensor inflationary cosmology", Phys.Rev., vol. D61, no. 12, p. 123511(7 pages), 2000.

[363] G. Esposito-Farese, "Scalar tensor theories and cosmology and tests of a quintessence gauss-bonnet coupling", gr-qc/0306018, 2003.

[364] R.A. Knop, et al. (The Supernova Cosmology Project), "New constraints on $\omega_m, \omega_{lambda}$ and w from an independent set of eleven high - redshift supernovae observed with hst", astro-ph/0309368, 2003.

[365] J.K. Webb, et al., "Further evidence for cosmological evolution of the fine structure constant", Phys. Rev. Lett, vol. 87, p. 091301, 2001; M.T. Murphy, J.K. Webb and V.V. Flambaum, "Further evidence for a variable fine structure constant from KECK/HIRES QSO absorption spectra ", Mon.Not.Roy.Astron.Soc., vol. 345, p. 609 (31 pages), 2003, astro-ph/0306483.

[366] N.J. Nunes and J.E. Lidsey, "Reconstructing the dark energy equation of state with varying alpha", astro-ph/0310882.

[367] C. Armendariz-Picon, T. Damour, and V. Mukhanov, "K - inflation", Phys.Lett, vol. B458, no. 2-3, pp. 209–218, 1999.

[368] T. Chiba, T. Okabe, and M. Yamaguchi, "Kinetically driven quintessence", Phys.Rev., vol. D62, no. 2, pp. 023511 (8 pages), 2000.

[369] F. D. Marco, F. Finelli, and R. Brandenberger, "Adiabatic and isocurvature perturbations for multifield generalized einstein models", Phys.Rev, vol. D67, no. 6, pp. 063512 (11 pages), 2003.

[370] P. Singh, M. Sami, and N. Dadhich, "Cosmological dynamics of phantom field.", Phys.Rev, vol. D68, no. 2 , pp. 023522 (7 pages), 2003.

[371] S. Rubin, "Origin of universes with different properties", Gravitation & Cosmology, vol. 9, pp. 243–248, 2003.

[372] C. Brans and R. Dicke, "Mach's principle and a relativistic theory of gravitation", Phys. Rev, vol. 124, no. 3, pp. 925–935, 1961.

[373] C. Wetterich "Cosmology and the fate of dilatation symmetry" , Nucl.Phys., vol. B302, no. 4, pp. 668–696, 1988.

[374] P. Peebles and B. Ratra, "Cosmology with a time variable cosmological 'constant'" , Ap. J., Pt. 2 – Letters to Editor, vol. 325, pp. L17–L20, 1988.

[375] E.J. Copeland, A.R. Liddle and D. Wands, "Exponential potentials and cosmological scaling solution, " Phys.Rev. vol. D57, no. 8, pp. 4686–4690, 1998.

[376] P.G. Ferreira and M. Joyce, "Structure formation with a selftuning scalar field" , Phys.Rev.Lett., vol. 79, no. 24, pp. 4740–4743, 1997.

[377] R.R. Caldwell, R. Dave and P.J. Steinhardt, "Cosmological imprint of an energy component with general equation of state" , Phys.Rev.Lett., vol. 80, no. 8, pp. 1582–1585, 1998.

[378] A. Hebecker and C. Wetterich, "Can quintessence be natural?" , *Heidelberg 2000, Dark matter in astro- and particle physics*, pp. 125–134, 2000.

[379] M.R. Garousi, "Tachyon couplings on non-BPS D-branes and Dirac-Born-Infeld action" , Nucl.Phys., vol. B584, no. 1-2, pp. 284–299, 2000.

[380] R.R. Caldwell, "A Phantom Menace? Cosmological consequences of a dark energy component with super-negative equation of state" , Phys.Lett., vol. B545, no. 1-3, pp. 23–29, 2002.

[381] J.D. Bekenstein, "Fine structure constant: is it really a constant? " , Phys.Rev., vol. D25, no. 6, pp. 1527–1539, 1982

[382] E.J. Copeland, N.J. Nunes and M. Pospelov, "Models of quintessence coupled to the electromagnetic field and the cosmological evolution of alpha" , Phys.Rev., vol. D69, pp. 023501 (13 pages), 2004.

[383] F. Fiziev, "A Minimal Model for Dilatonic Gravity" , gr-qc/9911037 (1999); S. Nojiri, S.D. Odintsov, Phys.Lett., vol. B576, no. 1-2, pp. 5–11, 2003

[384] S. Rubin, H. Kröger and G. Melkonian, "Cosmological dynamics of scalar field with non-minimal kinetic term" , astro-ph/0310182, 2003.

[385] A.A. Starobinsky, "Can the effective gravitational constant become negative?" , Sov. Astron. Lett., vol. 7, pp. 36–41, 1981.

[386] K. Bronnikov, "Scalar - tensor gravity and conformal continuations" , J. Math. Phys., vol. 43, no. 12, pp. 6096–6115, 2002. gr-qc/0204001.

[387] G. Esposito-Farese and D. Polarski, "Scalar tensor gravity in an accelerating universe" , Phys.Rev., vol. D63, no. 6, pp. 063504 (20 pages), 2001.

[388] G. Gibbons, "Cosmological evolution of the rolling tachyon" , Phys. Lett., vol. B537, no. 1-2, pp. 1–4, 2002.

[389] A. Albrecht and C. Skordis, "Phenomenology of a realistic accelerating universe using only planck scale physics" , Phys. Rev. Lett., vol. 84, no. 10, pp. 2076–2079, 2000.

[390] A.V. Frolov, L.A. Kofman, A.A. Starobinsky, "Prospects and problems of tachion matter cosmology" , Phys. Lett., vol. B545, no. 1-2, pp. 8–16, 2002.

[391] S. Weinberg, "The Cosmological constant problems" , astro-ph/0005265.

[392] J. Garriga and A. Vilenkin, "On likely values of the cosmological constant" , Phys.Rev., vol. D61, p. 083502 (24 pages), 2000, astro-ph/9908115.

Fundamental Theories of Physics

Series Editor: Alwyn van der Merwe, University of Denver, USA

1. M. Sachs: *General Relativity and Matter*. A Spinor Field Theory from Fermis to Light-Years. With a Foreword by C. Kilmister. 1982   ISBN 90-277-1381-2
2. G.H. Duffey: *A Development of Quantum Mechanics*. Based on Symmetry Considerations. 1985   ISBN 90-277-1587-4
3. S. Diner, D. Fargue, G. Lochak and F. Selleri (eds.): *The Wave-Particle Dualism*. A Tribute to Louis de Broglie on his 90th Birthday. 1984   ISBN 90-277-1664-1
4. E. Prugovečki: *Stochastic Quantum Mechanics and Quantum Spacetime*. A Consistent Unification of Relativity and Quantum Theory based on Stochastic Spaces. 1984; 2nd printing 1986   ISBN 90-277-1617-X
5. D. Hestenes and G. Sobczyk: *Clifford Algebra to Geometric Calculus*. A Unified Language for Mathematics and Physics. 1984   ISBN 90-277-1673-0; Pb (1987) 90-277-2561-6
6. P. Exner: *Open Quantum Systems and Feynman Integrals*. 1985   ISBN 90-277-1678-1
7. L. Mayants: *The Enigma of Probability and Physics*. 1984   ISBN 90-277-1674-9
8. E. Tocaci: *Relativistic Mechanics, Time and Inertia*. Translated from Romanian. Edited and with a Foreword by C.W. Kilmister. 1985   ISBN 90-277-1769-9
9. B. Bertotti, F. de Felice and A. Pascolini (eds.): *General Relativity and Gravitation*. Proceedings of the 10th International Conference (Padova, Italy, 1983). 1984   ISBN 90-277-1819-9
10. G. Tarozzi and A. van der Merwe (eds.): *Open Questions in Quantum Physics*. 1985
ISBN 90-277-1853-9
11. J.V. Narlikar and T. Padmanabhan: *Gravity, Gauge Theories and Quantum Cosmology*. 1986
ISBN 90-277-1948-9
12. G.S. Asanov: *Finsler Geometry, Relativity and Gauge Theories*. 1985   ISBN 90-277-1960-8
13. K. Namsrai: *Nonlocal Quantum Field Theory and Stochastic Quantum Mechanics*. 1986
ISBN 90-277-2001-0
14. C. Ray Smith and W.T. Grandy, Jr. (eds.): *Maximum-Entropy and Bayesian Methods in Inverse Problems*. Proceedings of the 1st and 2nd International Workshop (Laramie, Wyoming, USA). 1985   ISBN 90-277-2074-6
15. D. Hestenes: *New Foundations for Classical Mechanics*. 1986   ISBN 90-277-2090-8;
Pb (1987) 90-277-2526-8
16. S.J. Prokhovnik: *Light in Einstein's Universe*. The Role of Energy in Cosmology and Relativity. 1985   ISBN 90-277-2093-2
17. Y.S. Kim and M.E. Noz: *Theory and Applications of the Poincaré Group*. 1986
ISBN 90-277-2141-6
18. M. Sachs: *Quantum Mechanics from General Relativity*. An Approximation for a Theory of Inertia. 1986   ISBN 90-277-2247-1
19. W.T. Grandy, Jr.: *Foundations of Statistical Mechanics*. Vol. I: *Equilibrium Theory*. 1987
ISBN 90-277-2489-X
20. H.-H von Borzeszkowski and H.-J. Treder: *The Meaning of Quantum Gravity*. 1988
ISBN 90-277-2518-7
21. C. Ray Smith and G.J. Erickson (eds.): *Maximum-Entropy and Bayesian Spectral Analysis and Estimation Problems*. Proceedings of the 3rd International Workshop (Laramie, Wyoming, USA, 1983). 1987   ISBN 90-277-2579-9
22. A.O. Barut and A. van der Merwe (eds.): *Selected Scientific Papers of Alfred Landé*. [*1888-1975*]. 1988   ISBN 90-277-2594-2

Fundamental Theories of Physics

Fundamental Theories of Physics

46. P.P.J.M. Schram: *Kinetic Theory of Gases and Plasmas*. 1991   ISBN 0-7923-1392-5
47. A. Micali, R. Boudet and J. Helmstetter (eds.): *Clifford Algebras and their Applications in Mathematical Physics*. 1992   ISBN 0-7923-1623-1
48. E. Prugovečki: *Quantum Geometry*. A Framework for Quantum General Relativity. 1992
        ISBN 0-7923-1640-1
49. M.H. Mac Gregor: *The Enigmatic Electron*. 1992   ISBN 0-7923-1982-6
50. C.R. Smith, G.J. Erickson and P.O. Neudorfer (eds.): *Maximum Entropy and Bayesian Methods*. Proceedings of the 11th International Workshop (Seattle, 1991). 1993   ISBN 0-7923-2031-X
51. D.J. Hoekzema: *The Quantum Labyrinth*. 1993   ISBN 0-7923-2066-2
52. Z. Oziewicz, B. Jancewicz and A. Borowiec (eds.): *Spinors, Twistors, Clifford Algebras and Quantum Deformations*. Proceedings of the Second Max Born Symposium (Wrocław, Poland, 1992). 1993   ISBN 0-7923-2251-7
53. A. Mohammad-Djafari and G. Demoment (eds.): *Maximum Entropy and Bayesian Methods*. Proceedings of the 12th International Workshop (Paris, France, 1992). 1993
        ISBN 0-7923-2280-0
54. M. Riesz: *Clifford Numbers and Spinors* with Riesz' Private Lectures to E. Folke Bolinder and a Historical Review by Pertti Lounesto. E.F. Bolinder and P. Lounesto (eds.). 1993
        ISBN 0-7923-2299-1
55. F. Brackx, R. Delanghe and H. Serras (eds.): *Clifford Algebras and their Applications in Mathematical Physics*. Proceedings of the Third Conference (Deinze, 1993) 1993
        ISBN 0-7923-2347-5
56. J.R. Fanchi: *Parametrized Relativistic Quantum Theory*. 1993   ISBN 0-7923-2376-9
57. A. Peres: *Quantum Theory: Concepts and Methods*. 1993   ISBN 0-7923-2549-4
58. P.L. Antonelli, R.S. Ingarden and M. Matsumoto: *The Theory of Sprays and Finsler Spaces with Applications in Physics and Biology*. 1993   ISBN 0-7923-2577-X
59. R. Miron and M. Anastasiei: *The Geometry of Lagrange Spaces: Theory and Applications*. 1994   ISBN 0-7923-2591-5
60. G. Adomian: *Solving Frontier Problems of Physics: The Decomposition Method*. 1994
        ISBN 0-7923-2644-X
61. B.S. Kerner and V.V. Osipov: *Autosolitons*. A New Approach to Problems of Self-Organization and Turbulence. 1994   ISBN 0-7923-2816-7
62. G.R. Heidbreder (ed.): *Maximum Entropy and Bayesian Methods*. Proceedings of the 13th International Workshop (Santa Barbara, USA, 1993) 1996   ISBN 0-7923-2851-5
63. J. Peřina, Z. Hradil and B. Jurčo: *Quantum Optics and Fundamentals of Physics*. 1994
        ISBN 0-7923-3000-5
64. M. Evans and J.-P. Vigier: *The Enigmatic Photon*. Volume 1: The Field $B^{(3)}$. 1994
        ISBN 0-7923-3049-8
65. C.K. Raju: *Time: Towards a Constistent Theory*. 1994   ISBN 0-7923-3103-6
66. A.K.T. Assis: *Weber's Electrodynamics*. 1994   ISBN 0-7923-3137-0
67. Yu. L. Klimontovich: *Statistical Theory of Open Systems*. Volume 1: A Unified Approach to Kinetic Description of Processes in Active Systems. 1995   ISBN 0-7923-3199-0;
        Pb: ISBN 0-7923-3242-3
68. M. Evans and J.-P. Vigier: *The Enigmatic Photon*. Volume 2: Non-Abelian Electrodynamics. 1995   ISBN 0-7923-3288-1
69. G. Esposito: *Complex General Relativity*. 1995   ISBN 0-7923-3340-3

Fundamental Theories of Physics

70. J. Skilling and S. Sibisi (eds.): *Maximum Entropy and Bayesian Methods.* Proceedings of the Fourteenth International Workshop on Maximum Entropy and Bayesian Methods. 1996
ISBN 0-7923-3452-3

71. C. Garola and A. Rossi (eds.): *The Foundations of Quantum Mechanics Historical Analysis and Open Questions.* 1995
ISBN 0-7923-3480-9

72. A. Peres: *Quantum Theory: Concepts and Methods.* 1995 (see for hardback edition, Vol. 57)
ISBN Pb 0-7923-3632-1

73. M. Ferrero and A. van der Merwe (eds.): *Fundamental Problems in Quantum Physics.* 1995
ISBN 0-7923-3670-4

74. F.E. Schroeck, Jr.: *Quantum Mechanics on Phase Space.* 1996
ISBN 0-7923-3794-8

75. L. de la Peña and A.M. Cetto: *The Quantum Dice.* An Introduction to Stochastic Electrodynamics. 1996
ISBN 0-7923-3818-9

76. P.L. Antonelli and R. Miron (eds.): *Lagrange and Finsler Geometry.* Applications to Physics and Biology. 1996
ISBN 0-7923-3873-1

77. M.W. Evans, J.-P. Vigier, S. Roy and S. Jeffers: *The Enigmatic Photon.* Volume 3: Theory and Practice of the $B^{(3)}$ Field. 1996
ISBN 0-7923-4044-2

78. W.G.V. Rosser: *Interpretation of Classical Electromagnetism.* 1996   ISBN 0-7923-4187-2

79. K.M. Hanson and R.N. Silver (eds.): *Maximum Entropy and Bayesian Methods.* 1996
ISBN 0-7923-4311-5

80. S. Jeffers, S. Roy, J.-P. Vigier and G. Hunter (eds.): *The Present Status of the Quantum Theory of Light.* Proceedings of a Symposium in Honour of Jean-Pierre Vigier. 1997
ISBN 0-7923-4337-9

81. M. Ferrero and A. van der Merwe (eds.): *New Developments on Fundamental Problems in Quantum Physics.* 1997
ISBN 0-7923-4374-3

82. R. Miron: *The Geometry of Higher-Order Lagrange Spaces.* Applications to Mechanics and Physics. 1997
ISBN 0-7923-4393-X

83. T. Hakioğlu and A.S. Shumovsky (eds.): *Quantum Optics and the Spectroscopy of Solids.* Concepts and Advances. 1997
ISBN 0-7923-4414-6

84. A. Sitenko and V. Tartakovskii: *Theory of Nucleus.* Nuclear Structure and Nuclear Interaction. 1997
ISBN 0-7923-4423-5

85. G. Esposito, A.Yu. Kamenshchik and G. Pollifrone: *Euclidean Quantum Gravity on Manifolds with Boundary.* 1997
ISBN 0-7923-4472-3

86. R.S. Ingarden, A. Kossakowski and M. Ohya: *Information Dynamics and Open Systems.* Classical and Quantum Approach. 1997
ISBN 0-7923-4473-1

87. K. Nakamura: *Quantum versus Chaos.* Questions Emerging from Mesoscopic Cosmos. 1997
ISBN 0-7923-4557-6

88. B.R. Iyer and C.V. Vishveshwara (eds.): *Geometry, Fields and Cosmology.* Techniques and Applications. 1997
ISBN 0-7923-4725-0

89. G.A. Martynov: *Classical Statistical Mechanics.* 1997   ISBN 0-7923-4774-9

90. M.W. Evans, J.-P. Vigier, S. Roy and G. Hunter (eds.): *The Enigmatic Photon.* Volume 4: New Directions. 1998
ISBN 0-7923-4826-5

91. M. Rédei: *Quantum Logic in Algebraic Approach.* 1998   ISBN 0-7923-4903-2

92. S. Roy: *Statistical Geometry and Applications to Microphysics and Cosmology.* 1998
ISBN 0-7923-4907-5

93. B.C. Eu: *Nonequilibrium Statistical Mechanics.* Ensembled Method. 1998
ISBN 0-7923-4980-6

Fundamental Theories of Physics

Fundamental Theories of Physics

119. M. Pavšič: *The Landscape of Theoretical Physics: A Global View*. From Point Particles to the Brane World and Beyond in Search of a Unifying Principle. 2001   ISBN 0-7923-7006-6
120. R.M. Santilli: *Foundations of Hadronic Chemistry*. With Applications to New Clean Energies and Fuels. 2001   ISBN 1-4020-0087-1
121. S. Fujita and S. Godoy: *Theory of High Temperature Superconductivity*. 2001   ISBN 1-4020-0149-5
122. R. Luzzi, A.R. Vasconcellos and J. Galvão Ramos: *Predictive Statitical Mechanics*. A Nonequilibrium Ensemble Formalism. 2002   ISBN 1-4020-0482-6
123. V.V. Kulish: *Hierarchical Methods*. Hierarchy and Hierarchical Asymptotic Methods in Electrodynamics, Volume 1. 2002   ISBN 1-4020-0757-4; Set: 1-4020-0758-2
124. B.C. Eu: *Generalized Thermodynamics*. Thermodynamics of Irreversible Processes and Generalized Hydrodynamics. 2002   ISBN 1-4020-0788-4
125. A. Mourachkine: *High-Temperature Superconductivity in Cuprates*. The Nonlinear Mechanism and Tunneling Measurements. 2002   ISBN 1-4020-0810-4
126. R.L. Amoroso, G. Hunter, M. Kafatos and J.-P. Vigier (eds.): *Gravitation and Cosmology: From the Hubble Radius to the Planck Scale*. Proceedings of a Symposium in Honour of the 80th Birthday of Jean-Pierre Vigier. 2002   ISBN 1-4020-0885-6
127. W.M. de Muynck: *Foundations of Quantum Mechanics, an Empiricist Approach*. 2002   ISBN 1-4020-0932-1
128. V.V. Kulish: *Hierarchical Methods*. Undulative Electrodynamical Systems, Volume 2. 2002   ISBN 1-4020-0968-2; Set: 1-4020-0758-2
129. M. Mugur-Schächter and A. van der Merwe (eds.): *Quantum Mechanics, Mathematics, Cognition and Action*. Proposals for a Formalized Epistemology. 2002   ISBN 1-4020-1120-2
130. P. Bandyopadhyay: *Geometry, Topology and Quantum Field Theory*. 2003   ISBN 1-4020-1414-7
131. V. Garzó and A. Santos. *Kinetic Theory of Gases in Shear Flows*. Nonlinear Transport. 2003   ISBN 1-4020-1436-8
132. R. Miron: *The Geometry of Higher-Order Hamilton Spaces*. Applications to Hamiltonian Mechanics. 2003   ISBN 1-4020-1574-7
133. S. Esposito, E. Majorana Jr., A. van der Merwe and E. Recami (eds.): *Ettore Majorana: Notes on Theoretical Physics*. 2003   ISBN 1-4020-1649-2
134. J. Hamhalter. *Quantum Measure Theory*. 2003   ISBN 1-4020-1714-6
135. G. Rizzi and M.L. Ruggiero: *Relativity in Rotating Frames*. Relativistic Physics in Rotating Reference Frames. 2004   ISBN 1-4020-1805-3
136. L. Kantorovich: *Quantum Theory of the Solid State: an Introduction*. 2004   ISBN 1-4020-1821-5
137. A. Ghatak and S. Lokanathan: *Quantum Mechanics: Theory and Applications*. 2004   ISBN 1-4020-1850-9
138. A. Khrennikov: *Information Dynamics in Cognitive, Psychological, Social, and Anomalous Phenomena*. 2004   ISBN 1-4020-1868-1
139. V. Faraoni: *Cosmology in Scalar-Tensor Gravity*. 2004   ISBN 1-4020-1988-2
140. P.P. Teodorescu and N.-A. P. Nicorovici: *Applications of the Theory of Groups in Mechanics and Physics*. 2004   ISBN 1-4020-2046-5
141. G. Munteanu: *Complex Spaces in Finsler, Lagrange and Hamilton Geometries*. 2004   ISBN 1-4020-2205-0

Fundamental Theories of Physics

KLUWER ACADEMIC PUBLISHERS – DORDRECHT / BOSTON / LONDON